Ashford Books

Six Sense Strategy LTD.

Proceedings of the
XV International Conference on

LASER
SPECTROSCOPY

Proceedings of the
XV International Conference on

LASER
SPECTROSCOPY

Snowbird, Utah USA
10–15 June 2001

editors

Steven Chu, Vladan Vuletić,
Andrew J. Kerman & Cheng Chin

Stanford University, USA

World Scientific
New Jersey • London • Singapore • Hong Kong

Published by

World Scientific Publishing Co. Pte. Ltd.

P O Box 128, Farrer Road, Singapore 912805

USA office: Suite 1B, 1060 Main Street, River Edge, NJ 07661

UK office: 57 Shelton Street, Covent Garden, London WC2H 9HE

British Library Cataloguing-in-Publication Data
A catalogue record for this book is available from the British Library.

LASER SPECTROSCOPY
Proceedings of the XV International Conference

ISBN 981-02-4781-8

Printed in Singapore by Uto-Print

Preface

The 15th International Conference on Laser Spectroscopy, ICOLS01, was held on June 10–15, 2001 at Snowbird, a resort located in the Wasatch Mountain Range of Utah. Following the tradition of the previous conferences held at Vail, Mégève, Jackson Lake, Rottach-Egern, Jasper Park, Interlaken, Maui, Åre, Bretton Woods, Font-Romeu, Capri, Hangzhou, and Innsbruck, this meeting provided a forum where some of the most active scientists in the field of laser spectroscopy could exchange their latest findings and thoughts in an informal atmosphere.

The meeting was attended by 140 scientists from 18 countries. The 39 invited speakers, chosen by an international program committee, and 97 poster contributions reflected the extraordinary vitality of this set of conferences and of the field as a whole. Areas covered by the conference included: studies of Bose-Einstein condensates and ultra-cold Fermi gases, cavity QED and the coherent manipulation of atomic states, laser cooling and trapping, optical frequency measurements, quantum information studies in condensed matter and atomic systems, cold collisions, fundamental measurements, and new spectroscopies in biophysics. Most of the invited presentations are included in this conference proceeding, along with poster contributions selected by the participants.

In addition to the participants, the success of this conference was due to the generous support of our corporate sponsors. Without their support, we would not have been able to attract the set of distinguished scientists that attended or offer reduced registration fees to students and postdoctoral fellows. Finally, the members of the local committee would like to thank members of the Stanford University Physics Department, particularly Jenifer Conan-Tice, Rosenna Yau, Stewart Kramer, and Fides Rojo for their support, advice, and organizational skills.

Cheng Chin, Steven Chu, Andrew Kerman, Vladan Vuletić
and Yoshi Yamamoto
Stanford University

Steering Committee:

F. T. Arecchi, E. Arimondo, N. Bloembergen, C. J. Bordé, R. G. Brewer, S. Chu, W. Demtröder, M. Ducloy, M. S. Feld, J. L. Hall, P. Hannaford, T. W. Hänsch, S. Haroche, S. E. Harris, M. Inguscio, V. S. Letokov, A. Mooradian , Y. R. Shen, F. Shimizu, T. Shimizu, K. Shimoda, B. P. Stoicheff, S. Svanberg, H. Walther, Y. Z. Wang, Z. M. Zhang

Program Committee:

Enio Arimondo, Victor Balykin, Rainer Blatt, Phil Bucksbaum, Steven Chu, Jean Dalibard, Michael Feld, John L. Hall, Ted Hänsch, Wolfgang Ketterle, Peter Knight, Y. R. Shen, Vladan Vuletić, Yoshi Yamamoto, Peter Zoller

List of Sponsors:

Coherent, Inc.
Thorlabs
New Focus
World Scientific
Polytec, PI

CONTENTS

Precision Measurements

Quantum Manipulation

SELECTED PAPERS

Quantum Degenerate Gases

Precision Measurements

Quantum Manipulation

INVITED TALKS

QUANTUM IMPLOSIONS AND EXPLOSIONS IN A ^{85}RB BEC

CARL E. WIEMAN, ELIZABETH A. DONLEY, NEIL R. CLAUSSEN, SARAH T. THOMPSON, SIMON L. CORNISH, JACOB L. ROBERTS

JILA, National Institute of Standards and Technology and the University of Colorado, and the Department of Physics, University of Colorado, Boulder, Colorado 80309-0440

A Feshbach resonance at 155 G in ^{85}Rb allows the self interactions of a ^{85}Rb BEC to be varied simply by adjusting the magnitude of an applied magnetic field. However this low field Feshbach resonance feature also makes it much more difficult to cool to BEC because of the large inelastic scattering rates near the resonance, and the fact that the interactions are large and negative at low fields. We have been able to find an evaporative cooling procedure that successfully produces condensates in this system. These condensates have been used to study the behavior of condensates when the interactions are suddenly made large and attractive so the condensate becomes unstable. We see a variety of interesting behavior including a burst of "energetic" atoms (hundreds of nK), a large fraction of the condensate disappearing into an as yet unknown form, and highly excited condensate remnants. These remnants can be larger than the critical number. All of this behavior has interesting dependencies on the size of the attractive interactions and the number of atoms in the condensate.

1 Introduction

BEC physics is generally described by mean-field theory[1], in which the strength of the interactions depends on the atom density and on one additional parameter called the s-wave scattering length, a, that is specific to each species. For $a > 0$ the interactions are repulsive, and when $a < 0$ the interactions are attractive and a BEC tends to contract to minimize its overall energy. For a strong enough attractive interaction, there is not enough kinetic energy to stabilize the BEC and it is expected to implode in some fashion. A BEC can avoid implosion only as long as the number of atoms N_0 is less than a critical value given by[2]

$$N_{cr} = ka_{ho}/|a| \; , \tag{1}$$

where the dimensionless constant k is called the stability coefficient.

Under most circumstances, a is insensitive to external fields. This is different in the vicinity of a so-called Feshbach resonance where free atom and bound molecular energy levels cross. There a can be tuned over a huge range by adjusting the externally applied magnetic field[3,4]. For ^{85}Rb atoms a is usually negative, but a Feshbach resonance at ~155 G allows us to tune

a by orders of magnitude and even change its sign. This gives us the ability to create stable ^{85}Rb Bose-Einstein condensates[5] and adjust the inter-atomic interactions. We recently used this flexibility to verify the functional form of equation (1) and to measure the stability coefficient to be $k = 0.46(6)$[6]. This was measured by preparing a condensate sample of a known size and then slowly changing the magnetic field to make the interactions more and more attractive. At a particular critical field value (and hence critical value of a) about half of the atoms suddenly disappeared from the condensate. The particular value of a where this collapse occurred was determined to be inversely proportional to the initial number and showed no fluctuations to within our measurement uncertainty of 4 mG (equivalent to an uncertainty in a of 0.13 Bohr).

We also studied the dynamical response ("the collapse") of an initially stable BEC to a sudden shift of the scattering length to a value more negative than the critical value $a_{cr} = -ka_{ho}/N_0$. We have observed many features of the surprisingly complex collapse process, including the energies and energy anisotropies of atoms that burst from the condensate, the time scale for the onset of this burst, the rates for losing atoms, spikes in the wave function that form during collapse, and the size of the remnant BEC that survives the collapse. The unprecedented level of control provided by tuning a has allowed us to investigate how all of these quantities depend on the magnitude of a, the initial number and density of condensate atoms, and the initial spatial size and shape of the BEC before the transition to instability.

A standard double magneto-optical trap (MOT) system[13] was used to collect a cold sample of ^{85}Rb atoms in a low-pressure chamber[5]. The atoms were loaded into a cylindrically symmetric cigar-shaped magnetic trap. Radio-frequency evaporation was then used to cool the sample to ~3 nK to form pure condensates containing >90% of the sample atoms. The final stages of evaporation were performed at 162 G where the scattering length is positive and stable condensates of up to 15,000 atoms could be formed. After evaporative cooling, the magnetic field was ramped adiabatically to 166 G (except where noted), where $a = 0$. This provided a well-defined initial condition with the BEC taking on the size and shape of the harmonic oscillator ground state.

We could then adjust the mean-field interactions within the BEC to a variety of values on time scales as short as 0.1 ms. For the studies here we jump to some value of $a < a_{cr}$ to trigger a collapse. However, the tunability of a does much more than just induce a collapse. It also greatly aided in imaging the sample, and as discussed below, allowed us to freeze the condensate in midcollapse and look at it. Usually the condensate size was below the resolution limit of our imaging system (7μm FWHM). However, we could ramp the scattering length to large positive values and use the repulsive inter-atomic interactions to expand the BEC before imaging, thus obtaining information on the pre-expansion condensate shape and number. A typical $a(t)$ sequence is shown in Fig. 1a. We have used a variety of such sequences to explore many aspects of the collapse and enhance the visibility of particular components of the sample.

2 Condensate contraction and atom loss.

When the scattering length is jumped to a value $a_{collapse} < a_{cr}$, a condensate's kinetic energy no longer provides a sufficient barrier against collapse. As described in Ref. 8, during collapse one might expect a BEC to contract until losses from density-dependent inelastic collisions[14] effectively stop the contraction. This contraction would roughly take place on the time scale of a trap oscillation, and the density would sharply increase after $T_{rad}/4 \simeq 14$ ms, where T_{rad} is the radial trap period. How does this picture compare to what we have actually seen?

A plot of the condensate number N vs τ_{evolve} for $a_{collapse} = -30\ a_0$ and $a_{init} = +7\ a_0$ is presented in Fig. 1b. N was constant for some time after the jump until atom loss suddenly began at $t_{collapse}$. The condensate widths changed very little with time τ_{evolve} before $t_{collapse}$. A simple Gaussian model of the contraction predicts that there is only a 50% increase in the average density to $2.5 \times 10^{13}/\text{cm}^3$ during this time. Using the three body recombination decay constants from Ref. 14, this density gives an atom loss rate, τ_{decay}, that is far smaller than what we observe and does not have the observed sudden onset.

For the data in Fig. 1b and most other data presented below, we jumped to $a_{quench} = 0$ in 0.1 ms after a time τ_{evolve} at $a_{collapse}$. We believe that the loss immediately stopped after the jump. This interpretation is based on the surprising observation that the quantitative details of curves such as that shown in Fig. 1b did not depend on whether the collapse was terminated by a jump to $a_{quench} = 0$ or $a_{quench} = 250\ a_0$.

We have measured loss curves like that in Fig. 1b for many different values of $a_{collapse}$. The collapse time shows a strong dependence on $a_{collapse}$. The atom loss time constant τ_{decay} depended only weakly on $a_{collapse}$ and N_0.

3 Burst atoms.

As indicated by Fig. 1b, atoms leave the BEC during the collapse. There are at least two components to the expelled atoms. One component (the

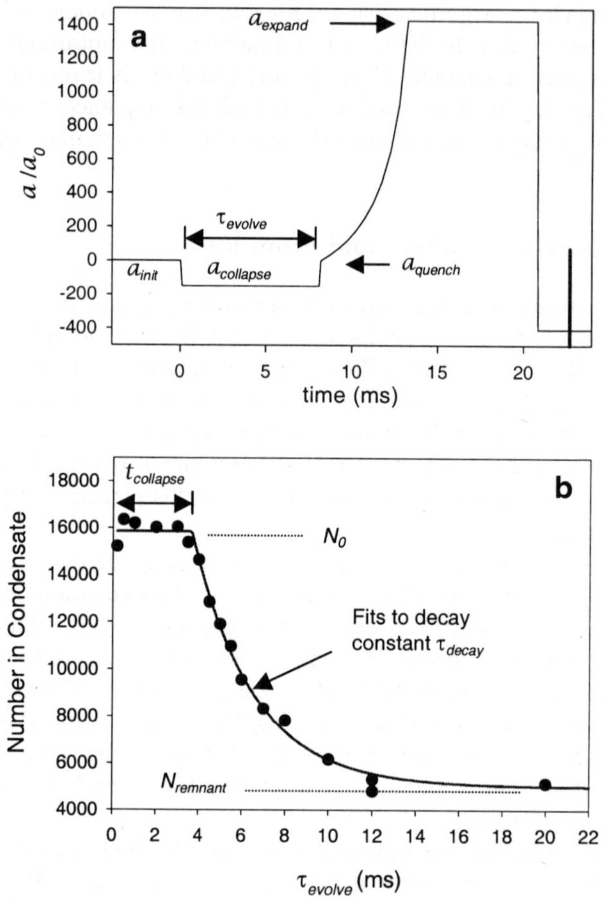

Figure 1. An example of a ramp applied to the scattering length, and a plot of the condensate number versus time after a jump to a negative scattering length. a. A typical $a(t)$ sequence. $a_0 = 0.529$ Å is the Bohr radius. The scattering length is jumped at $t = 0$ in 0.1 ms from a_{init} to $a_{collapse}$, where the BEC evolves for a time τ_{evolve}. The field is carefully controlled so that magnetic-field noise translates into fluctuations in $a_{collapse}$ on the order of $\sim 0.1\ a_0$ in magnitude. The collapse is then interrupted with a jump to a_{quench}, and the field is ramped in 5 ms to a large positive scattering length that makes the BEC expand. After 7.5 ms of additional expansion, the trap is turned off in 0.1 ms and 1.8 ms later the density distribution is probed using destructive absorption imaging with a 40 μs laser pulse (indicated by the vertical bar). The increase in a from $a_{collapse}$ to a_{expand} is far too rapid to allow for the BEC to expand adiabatically. On the contrary, the smaller the BEC before expansion, the larger the cloud at the moment of imaging. Thus we can readily infer the relative size of the bulk of the BEC just prior to the jump to a_{quench}. b. The number of atoms remaining in the BEC versus τ_{evolve} at $a_{collapse} = -30\ a_0$. We observed a delayed and abrupt onset of loss. The solid line is a fit to an exponential with a best-fit value of $t_{collapse} \simeq 3.7(5)$ ms for the delay.

"missing atoms") is not detected. The other component emerges as a burst of detectable spin-polarized atoms with energies much greater than that of the initial condensate but much less than the magnetic trap depth. The burst-energy dependencies on $a_{collapse}$ and N_0 are complex.

The angular kinetic-energy distribution with which the burst atoms are expelled from the condensate can be accurately measured. Figure 2c shows a cross section of the burst after it has been through one half oscillation period in the trap. The burst energy distributions were well fit by Gaussians characterized by a temperature that was usually different for the two trap directions. The burst energy fluctuated from shot to shot by up to a factor of 2 for a given $a_{collapse}$. This variation is far larger than the measurement uncertainty or the variation in initial number (both $\sim 10\%$).

Although the burst energies varied from shot to shot, the average value was well-defined and showed trends far larger than the variation. The axial and radial burst energies versus $a_{collapse}$ are shown in Fig. 3a and 3b for $N_0 = 6,000$ and 15,000, respectively.

When we interrupted the collapse with a jump back to $a = 0$ as discussed above, we also interrupted the growth of the burst. The "interrupted" burst atoms still refocused after sitting at $a = 0$ for the requisite half trap period. The energy of the atoms in the interrupted bursts appeared to be the same, but the number of atoms was smaller. By changing the time at which the collapse was interrupted we could measure the time dependence of the creation of burst atoms. For the conditions of Fig. 1b, the number of burst atoms N_{burst} grew with τ_{evolve} with a time constant of 1.2 ms starting at 3.5 ms and reaching an asymptotic final number of $\sim$2500 for all times $\geq$7 ms. N_{burst} varied randomly by $\sim$20% for the data in Fig. 3, but on average the fraction of atoms going into the burst was about 20% of N_0 and did not depend on $a_{collapse}$ or N_0.

4 Remnant Condensate and Jet Formation.

After a collapse, a "remnant" condensate containing a fraction of the atoms survived with nearly constant number and oscillated in a highly excited collective state with the two lowest breathing modes. The number of atoms in the remnant depended on $a_{collapse}$ and N_0, and in general was not limited by the critical number, N_{cr}. A fixed fraction of N_0 went into the remnant independent of N_0, so that smaller initial condensates often ended up with $N_{remnant} < N_{cr}$, but larger initial condensates did only when a was very close to the critical value. The fraction of atoms that went into the remnant decreased with $|a_{collapse}|$, and was $\sim$40% for $|a_{collapse}| < 10\ a_0$ and $\sim$10% for $|a_{collapse}| > 100\ a_0$. We do not think that the surface-wave excitations[16]

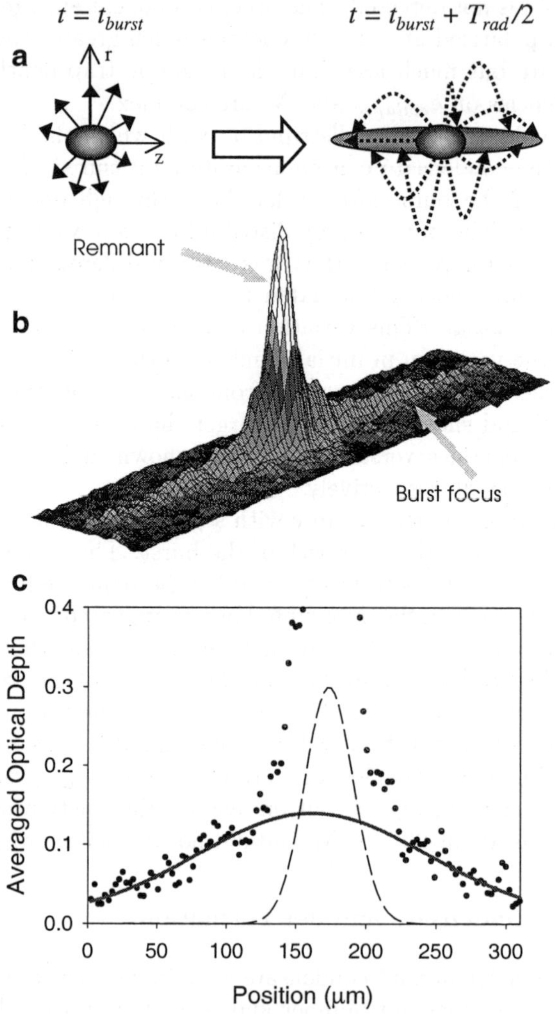

Figure 2. A burst focus. a. Conceptual illustration of a radial burst focus. b. An image of a radial burst focus taken 33.5 ms after a jump from $a_{init} = 0$ to -30 a_0 for $N_0 = 10,000$. $T_{rad}/2 = 28.6$ ms, which indicates that the burst occurred 4.9(5) ms after the jump. The axial energy distribution for this burst corresponded to an effective temperature of 62 nK. The image is 60×310 μm. c. Radially averaged cross section of b with a Gaussian fit to the burst energy distribution. The central 100 μm were excluded from the fit to avoid distortion in the fit due to the condensate remnant ($\sigma = 9\mu$m) and the thermal cloud ($\sigma = 17\mu$m). The latter is present in the pre-collapse sample due to the finite temperature and appears to be unaffected by the collapse. The dashed line indicates the fit to this initial thermal component. Note the offset between the centers of the burst and the remnant. This offset varies from shot to shot by an amount comparable to the offset shown.

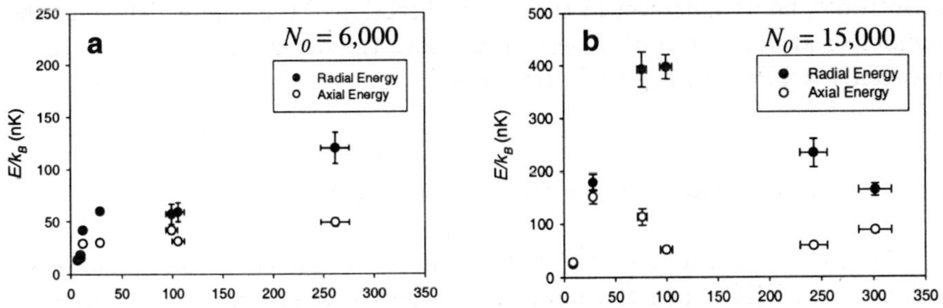

Figure 3. Burst energies. a. and b. The axial and radial burst energies versus $a_{collapse}$ for $N_0 \simeq 6,000$ and $N_0 \simeq 15,000$, respectively. On average, ten burst focuses were measured for each trap direction at each value of $a_{collapse}$ studied. The energies were higher for the larger-N_0 condensates over the full range of $a_{collapse}$ studied. The vertical and horizontal error bars indicate the standard error of the measurements and the uncertainty in $a_{collapse}$ arising from the magnetic-field calibration, respectively. For several of the points, the uncertainties are smaller than the symbol size and the error bars are not visible.

are responsible for stabilizing the remnant since we excite large-amplitude breathing modes. For $N_0 = 6,000$ and $|a_{collapse}| < 10\ a_0$, more atoms were lost than the number required to lower $N_{remnant}$ to below N_{cr}.

Under very specific experimental conditions, we have observed streams of atoms with highly anisotropic velocities emerging from the collapsing condensates. These "jets" are distinguished from the "burst" in that the jets have much lower kinetic energies (on the order of a few nanokelvin), in that their velocity is nearly purely radial, and in that they appear only when the collapse is interrupted (i.e., by jumping to $a_{quench} = 0$) during the period of number loss. Collapse processes that were allowed to evolve to completion (until $N \simeq N_{remnant}$) were not observed to emit jets. The jet size and shape varied from image to image even when all conditions were unchanged, and as many as three jets were sometimes observed to be emitted from the collapse of a single condensate. Also, the jets were not always symmetric about the condensate axis.

We believe that these jets are manifestations of local "spikes" in the condensate density that form during the collapse and expand when the balance of forces is changed by quenching the collapse. We can estimate the size of the spikes using the uncertainty principle. After a jump to $a_{quench} = 0$, the kinetic energy of the atoms in the resulting jet is equal to the confinement energy that the spike had prior to quenching the collapse. The anisotropy of the jets indicates that the spikes from which they originated were also highly anisotropic, being narrower in the radial direction. From the widths and the number of atoms in the jets, we can estimate the density in the spikes to be between 1×10^{13} and 1×10^{14} $atoms/cm^3$.

5 Overview of the current theoretical models.

Several theoretical papers[7,8,9,10,11,12] have considered the problem of the collapse of a BEC when the number of atoms exceeds the critical number. These treatments all use a mean-field approach and describe the condensate dynamics using the Gross-Pitaevskii (GP) equation. In most cases[7,8,10,11], the loss mechanism is three body recombination, but Duine and Stoof[12] propose a different loss mechanism that has a similar density dependence. The comparison with our experimental results is difficult because the calculation of the evolution of the shape of the condensate has only been carried out with either the Gaussian approximation, which the jets show is clearly not appropriate, or assuming a completely isotropic potential. From the anisotropy of the jets and the burst we can see that the anisotropy is clearly important. However, the calculations that have been carried out (Gaussian approximation and/or isotropic)show little resemblance to what we have observed. Until better calculations of the shape are carried out it is difficult to know whether or not the fundamental physics contained in the theory is correct.

There are a number of non-intuitive features of the collapse that a theoretical treatment should be able to explain, and the success or failure to do this will clearly determine the accuracy of any theoretical treatment.

- The decay constant τ_{decay} is independent of both N_0 and $|a_{collapse}|$ for $|a_{collapse}| < 100\ a_0$, and only weakly depends on these quantities for larger $|a_{collapse}|$.

- The burst energy per atom dramatically increases with initial condensate number.

- The number of burst atoms is constant versus $a_{collapse}$.

- The number of cold remnant BEC atoms surviving the violent collapse varies between much less and much more than N_{cr}, depending on N_0 and $a_{collapse}$, but the fraction of remnant atoms, burst atoms, and missing atoms are independent of N_0.

Acknowledgments

We extend our thanks to Stephan Dürr, Gora Shlyapnikov, Henk Stoof, Murray Holland, Masahito Ueda, and Rembert Duine for helpful discussions. This research has been supported by the ONR, NSF, ARO–MURI, and NIST. S. L. C. acknowledges the support of a Lindemann Fellowship.

References

1. F. Dalfovo, S. Giorgini, L.P. Pitaevskii, & S. Stringari, *Rev. Mod. Phys.* **71**, 463–512 (1999).
2. P.A. Ruprecht, M.J. Holland, K. Burnett, & M. Edwards, *Phys. Rev. A* **51**, 4704–4711 (1995).
3. E. Tiesinga, A.J. Moerdijk, B.J. Verhaar, & H.T.C. Stoof, *Phys. Rev. A* **46**, R1167 (1992).
4. E. Tiesinga, B.J. Verhaar, & H.T.C. Stoof, *Phys. Rev. A* **47**, 4114–4122 (1993).
5. S.L. Cornish, N.R. Claussen, J.L. Roberts, E.A. Cornell, & C.E. Wieman, *Phys. Rev. Lett.* **85**, 1795–1798 (2000).
6. J.L. Roberts, *et al. Phys. Rev. Lett.* **86**, 4211–4214 (2001).
7. Y. Kagan, E.L. Surkov, & G.V. Shlyapnikov, *Phys. Rev. Lett.* **79**, 2604–2607 (1997).
8. Y. Kagan, A.E. Muryshev, & G.V. Shlyapnikov, *Phys. Rev. Lett.* **81**, 933–937 (1998).
9. M. Ueda, & K. Huang, *Phys. Rev. A* **60**, 3317–3320 (1999).
10. H. Saito & M. Ueda, *Phys. Rev. A* **63**, 043601–7 (2001).
11. H. Saito & M. Ueda, *Phys. Rev. Lett.* **86**, 1406–1409 (2001).
12. R.A. Duine & H.T.C. Stoof, *Phys. Rev. Lett.* **86**, 2204–2207 (2001).
13. C.J. Myatt, N.R. Newbury, R.W. Ghrist, S. Loutzenhiser, & C.E. Wieman, *Opt. Lett.* **21**, 290–292 (1996).
14. J.L. Roberts, N.R. Claussen, S.L. Cornish, & C.E. Wieman, *Phys. Rev. Lett.* **85**, 728–731 (2000).
15. J. Stenger, *et al.*, *Phys. Rev. Lett.* **82**, 2422–2425 (1999).
16. A.K. Pattanayak, A. Gammal, C.A. Sackett, & R.G. Hulet, *Phys. Rev. A* **63**, 033604–4 (2001).

BOSE-EINSTEIN CONDENSATION OF METASTABLE HELIUM: SOME EXPERIMENTAL ASPECTS

C.I. WESTBROOK, A. ROBERT, O. SIRJEAN, A. BROWAEYS, D. BOIRON, A. ASPECT

Laboratoire Charles Fabry de l'Institut d'Optique, UMR 8501 du CNRS, B.P. 147, F-91403 Orsay Cedex, France
E-mail: christoph.westbrook@iota.u-psud.fr

We describe our recent realization of BEC using metastable helium. All detection is done with a microchannel plate which detects the metastables or ions coming from the trapped atom cloud. This discussion emphasizes some of the diagnostic experiments which were necessary to quantitatively analyze our results.

1 Introduction

Since the announcement of the observation of Bose-Einstein condensation (BEC) in a rubidium vapor at the 12th ICOLS meeting in 1995[1], this fascinating state of matter has occupied central stage in the field of atomic, molecular and optical physics. The many new advances reported at this meeting indicate that the field may continue to do so for some time to come. This paper concerns one of these advances, the condensation of metastable helium atoms (He*), and is intended as a supplement to the recently published Ref. [2]. We shall not repeat the data from that paper, but rather concentrate on some details that were left out of Ref. [2] for lack of space.

Until recently BEC had been observed in 4 different atomic species, H, Li, Na and Rb[3,4,5,6], and the first question to ask before embarking on the quest for BEC of He* was whether a new atom was of interest. During the 1990s, several groups have been working on laser cooling of He*, and of course one answer to the above question is simply that the attainment of BEC is the best cooling one can do, and many of the same justifications for laser cooling apply to BEC. In addition, one might hope that a new atomic species might allow one to observe new phenomena, not accessible to the previously studied cases. In this respect it seemed clear that the metastability of the atoms might be very important. Simple, rapid and efficient detection of the He* atoms is possible using electron multipliers such as microchannel plates (MCP) and these detectors can also be used to observe ions resulting from Penning ionizing collisions, either with residual gas atoms or between the He* atoms themselves. Thus, when BEC was observed in alkali gases, groups working on laser cooling of He* naturally considered the feasibility of He*[7].

A potential impediment to the achievement of the high densities necessary for evaporative cooling and BEC is the Penning ionization reaction:

$$He^* + He^* \rightarrow He + He^+ + e^- .$$

Many experiments have shown that in a magneto-optical trap (MOT), this process is very rapid and limits the density in such a trap[8,9,10,11,12,13]. Thus, when loading atoms from a MOT into a non-dissipative trap, there was a danger that the density and more importantly the elastic collision rate would be too small to efficiently cool by evaporation. In addition, it was known experimentally[9,10,11] and theoretically[14] that even in the absence of resonant light, the rate constant for Penning ionization was on the order of $10^{-10}\mathrm{cm}^3/\mathrm{s}$. Thus, even if it were possible to begin evaporative cooling, such a large inelastic collision rate is likely to prevent evaporative cooling down to BEC.

On the other hand it was predicted[18,19] that the elastic scattering length a would be quite large for He*. This result was very encouraging because it indicated that despite the low densities achievable in a MOT, thermalizing collisions in a MOT-loaded magnetic trap could be rapid enough to allow evaporative cooling.

Even more importantly, it was also known that, for a spin polarized sample, the Penning ionization rate is suppressed by spin conservation. He* has total angular momentum one, therefore only one trapping state exists and magnetically trapped He* atoms are neccesarily 100% polarized. So one might hope to accumulate large densities of He* in a magnetic trap. Experimentally, a suppression of one order of magnitude had already been demonstrated as early as 1972[15], this result was followed up by more recent measurements[16,17]. In the mid 1990's it was predicted that the degree of suppression could be as high as 5 orders of magnitude[18,19], which would easily permit long storage times at densities necessary for BEC. The theoretical predictions for the scattering properties of He* motivated the serious attempt to achieve BEC using a strategy analogous to that used in the alkali gases.

2 Experiment

In our apparatus, a Zeeman cooled atomic beam loads a MOT which, after an optical molasses cooling and an optical pumping stage, loads a magnetic trap. The magnetic trap is of the cloverleaf design[20]. The only unusual feature of the apparatus is an MCP placed 5 cm below the trap center. A grid in front of the MCP allows one to either attract or repel positive ions. The front face of the MCP is at negative high voltage, and so electrons are never detected. We typically trapped 3×10^8 atoms in the MOT, and transferred approximately

50% to the magnetic trap. The MOT temperature was typically of order 1 mK, while after molasses cooling, the atomic temperature reached 300 μK. After loading the magnetic trap, the atomic sample was compressed by lowering the magnetic field bias. During this process the temperature increased again to about 1 mK. Then, an RF-knife was applied and ramped down from 130 MHz to effect the forced evaporation.

The evolution of the temperature, phase space density and elastic collision rate during the evaporation ramp were all derived from measurements of the time of flight distribution of atoms to the MCP after the magnetic trap was turned off. Several examples are shown in Fig. 1. The change in gravitational energy of the atoms in falling 5 cm to the detector corresponds to 240 μK. For the initial temperature of the trapped atoms, this energy is negligible compared to the kinetic energy, and so the atoms expand nearly isotropically after release, and the collection efficiency of the detector corresponds roughly to its solid angle of 0.5%. In this situation, the time of flight distribution is peaked at a value corresponding to the flight time of an atom moving toward the detector with the most probable speed at that temperature. The signature for cooling is a shift of the arrival time toward later times as can be seen in Fig. 1. Unfortunately, as the atoms are evaporatively cooled, their number also diminishes and the detected signal drastically decreases. Indeed when the temperature is of order 100 μK, no signal is visible on the detector. As the temperature decreases further however, the atoms begin to fall down rather than fly away, and the collection efficiency of the detector increases dramatically. Near a temperature of 10 μK we observe a "revival" of the MCP signal, and at 1 μK, close to the BEC threshold, nearly all the atoms remaining in the trap reach the detector and we observe the characteristic structure of an expanding cloud below the BEC temperature: a broad peak whose width corresponds to the temperature and a narrow peak on top of it.

As is discussed in Ref. [2], a careful analysis of the narrow BEC peak reveals that its width increases as the 1/5 power of the number of atoms N_0 in the peak as predicted by the Gross-Pitaevski equation in the Thomas-Fermi approximation. Upon closer examination however, the results were puzzling. First, it was surprising that we detected the atoms at 1 μK at all. A magnetic field gradient as small as 30 mG/cm is enough to deviate the atomic trajectory so as to miss the MCP. We were virtually certain that residual field gradients larger than this were present in the apparatus. Secondly, the study of the expansion of the gas depends on being able to turn the trap off suddenly compared to the inverse of the oscillation frequencies in the trap (50 and 1300 Hz). Here again we were certain from magnetic field measurements that effects such as eddy currents in the reentrant flanges holding the magnetic trap coils,

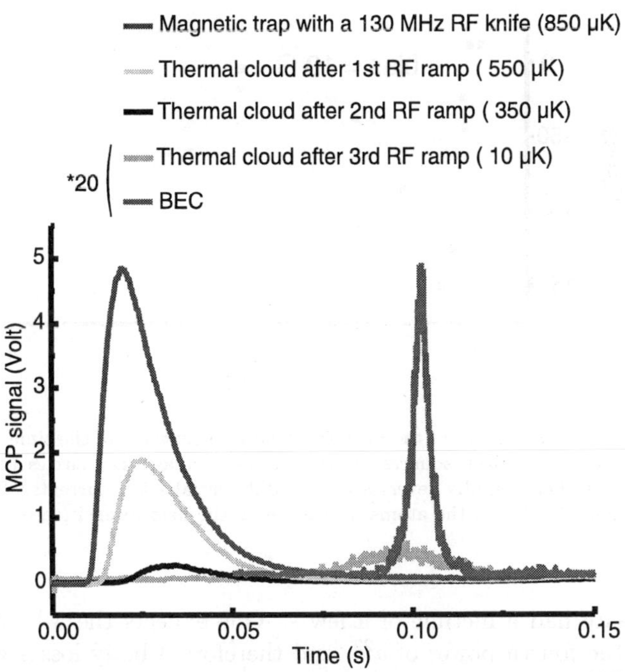

Figure 1. MCP time of flight signals at different stages of the RF evaporation. The vertical scales on last two curves were multiplied by 20. At temperatures above 100 μK, the atomic cloud expands nearly isotropically, and the MCP detects primarily the downward going atoms. At temperatures low compared to 100μK, the atoms fall in the earth's gravititional field and are nearly all detected. The mean arrival time is about 100 ms, the time to fall 5 cm. Because of the loss of atoms during evaporative cooling, and the small solid angle of the detector, very few atoms are detected at temperatures between 300 and 30 μK.

limited our field turn-off time to about 1 ms. Thirdly, we could estimate the number of atoms in the cloud at or near the critical temperature and compare it with the expected number in the ideal gas limit. The predicted number is given by[21]:

$$N_C = 1.202\,(kT_C/\hbar\tilde{\omega})^3. \tag{1}$$

Here $\tilde{\omega}$ denotes the geometric mean of the trap oscillation frequencies. Our calibration of the detector indicated a number of atoms smaller than this by an order of magnitude. Finally a quantitative analysis of the $N_0^{1/5}$ dependence, resulted in an estimate of the scattering length a of order 100 nm. Such a large value of the scattering length seemed to be ruled out by the fact that

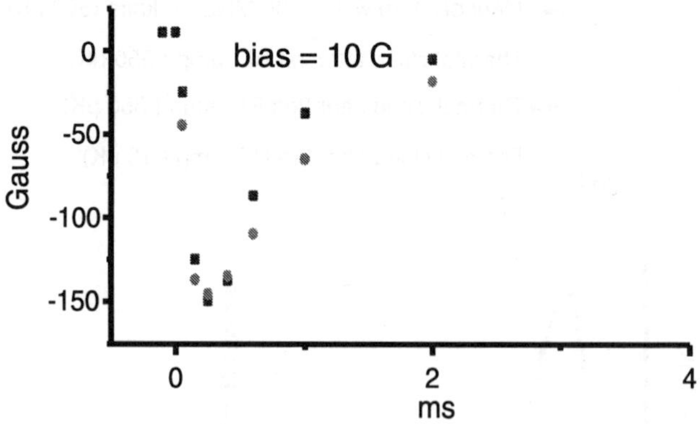

Figure 2. *In situ* measurement of the magnetic field as a function of the time after turning off the magnetic trap. Black squares correspond to circular, grey circles to linear laser polarization. The field rapidly reverses due to differential eddy currents. This reversal causes approximately 10% of the atoms to transfer to the field-insensitive $m = 0$ state.

the condensate had a lifetime of a few s. One expects the 3 body loss rate to scale as the fourth power of a[22], and therefore 3 body losses would have caused the BEC to decay in much less than 1s.

A clue to resolving these difficulties came from earlier, *in situ* measurements of the magnetic fields in our apparatus when the trap was turned off. In Fig. 2 we show the results of such a measurement. We began with atoms in a magnetic trap at 1 mK. While monitoring the MCP signal, we directed a 10 mW/cm^2 laser pulse of 20 μs duration at the cloud at a time t after the magnetic field turnoff. When the laser, which propagated parallel to the bias field, was resonant with the atoms, including the Zeeman shift due to the trapping fields, the atoms scattered the laser light and were pushed from the path to the detector. The laser detuning which minimized the MCP signal therefore corresponded to the Zeeman shift at the time of the pulse.

The magnetic field deduced from this Zeeman shift is plotted in Fig. 2 as a function of t. Negative times correspond to atoms still in the trap, and the value we observe agrees with the calculated magnetic field bias in the trap. One sees that the magnetic field in the trapping region undergoes a violent reversal during the turnoff, reaching a value of more than -150 G in less than 100 μs. This reversal is possibly due to the combined effects of the eddy currents induced by the pinch coils and the compensation coils. These two sets of coils carry currents in opposite directions and have very different sizes. The eddy currents they induce therefore have different signs and probably different time constants, which would explain the observed behavior. An uncompensated trap does not exhibit a magnetic field reversal.

It appears therefore that when the field reverses sign, some of the atoms undergo transitions to the field insensitive $m = 0$ state. We presume that after the transition the atoms adiabatically remain in the $m = 0$ state in the presence of the weak residual fields remaining in the apparatus. They therefore remain insensitive to any further field gradients.

We can test this interpretation by deliberately applying a magnetic field gradient with an external coil, and observing the time of flight spectrum. The results are shown in Fig. 3. The gradient was turned on about $100\,\mu s$ after the currents to the magnetic trap were turned off. The figure shows two peaks, one arriving at approximately 100 ms after the field turnoff, as occurs without an applied gradient. The location and height of this peak does not depend on the magnitude of the gradient. The other peak occurs earlier, corresponding to atoms accelerated by the gradient and the arrival time shifts with the magnitude of the gradient. The height of the peak can be varied by varying the horizontal components of the gradient. The figure shows the largest early peak we were able to produce. These data show that indeed a fraction of the atoms makes the flight to the detector in the $m = 0$ state and that at least 7 times more atoms are in a field sensitive state after the trap turnoff. The applied gradient was produced by a coil above the trap and thus the atoms which are accelerated are weak field seekers ($m = +1$). It is also possible that some atoms are in the strong field seeking state ($m = -1$), but since they are accelerated upwards, they have little chance to reach the MCP.

To get an estimate of the number of atoms in the magnetic trap we can use an analysis similar to that which leads to Eq. 1 leading to the number N_{th} of atoms in the thermal cloud below the critical temperature. By fitting the wings of the time of flight spectra, we are able to determine the temperature T of the atomic cloud, and to infer N_{th}. As discussed in Ref. [21], and experimentally demonstrated in Refs. [20,23,24], this number should be given by: $N_{th} = 1.202\,(kT/\hbar\tilde{\omega})^3$. This relation gives an absolute thermodynamic measurement of the number of atoms. It is higher by a factor $f = 8 \pm 4$ than the value we derive from the calibration of the MCP. Taking this correction into account, the largest condensate we have observed contained about 10^5 atoms, and the number of atoms present at the critical temperature is a few times 10^5.

The magnetic field measurements also help to explain why the analysis of the expansion of the trapped atoms after release works so well. Because of the fast reversal, the atoms which make transitions to the $m = 0$ state are indeed released extremely rapidly. A careful analysis of the expansion may require

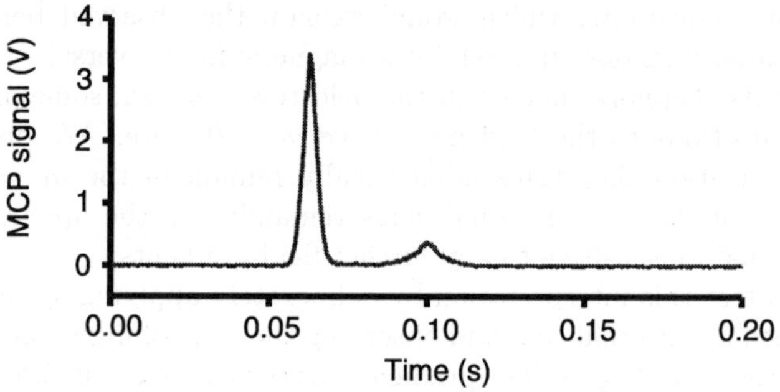

Figure 3. Time of flight spectrum in the presence of a magnetic field gradient of an evaporatively cooled cloud of atoms. The height and arrival time of the small peak are independent of the applied gradient. The large peak's arrival time decreases as the vertical gradient (about 1 G/cm) is increased. A 0.1 G/cm horizontal gradient was also applied in order to maximize the number of atoms in the large peak. The ratio of the peak areas is 7. Thus we believe that the small peak corresponds to atoms in the $m = 0$ state, while the large peak corresponds to atoms in the $m = 1$ state.

taking into account the behavior of the weak field seeking atoms observed in Fig. 3. Here we assume that all atoms expand freely independent of their internal state. In fact the atoms in this state are presumably trapped during the decay of the eddy currents, but since in a cloverleaf trap, the confinement rapidly decreases with increasing bias, it is probably a good approximation to treat the atoms as free on the scale of 1 ms.

An analysis of the mean field expansion of the cloud, using the corrected number of atoms leads to a value of the scattering length, $a = 20 \pm 10$nm. This result is consistent with our elastic rate constant measurements at 1 mK[25], as well as with the observations of Ref. [26].

We have also observed the ions produced by the trapped condensate, by negatively biasing a grid above the MCP. An example of the ion detection rate as a function of time is shown in Fig. 4 of Ref. [2]. These ions are due to Penning ionization of residual gases, to two body collisions within the condensate, or possibly other, more complicated processes. We observe a factor of 5 more ions from the condensate than from a thermal cloud at 1 μK, and we attribute this increase to the larger density in the condensate. The lifetime of the condensate, estimated by observing the ion rate is on the order of a few seconds. This is true both with and without an RF-knife to evacuate hot atoms[20,27], although the lifetime is slightly longer with the knife present. The density of the condensate, deduced from its vertical size measurement and its known aspect ratio, is of order $10^{13}\,\mathrm{cm}^{-3}$, so from the lifetime we can place an upper limit of $10^{-13}\,\mathrm{cm}^3\mathrm{s}^{-1}$ on the relaxation induced Penning ionization rate constant, as well as an upper limit of $10^{-26}\,\mathrm{cm}^{-6}\mathrm{s}^{-1}$ on any three-body loss process.

The achievement of BEC in He* together with a MCP detector, offers many new possibilities for the investigation of BECs. Ion detection allows continuous "non-destructive" monitoring of the trapped condensate. We hope to be able to study the formation kinetics of the condensate using the ion signal. Our ability to count individual He* atoms falling out of the trap should allow us to perform accurate comparisons of correlation functions[27] for a thermal beam of ultracold atoms[28] and for an atom laser, realizing the quantum atom optics counterpart of one of the fundamental experiments of quantum optics.

Acknowledgments

This work was supported by the European Union under grants IST-1999-11055, and HPRN-CT-2000-00125, and by the DGA grant 99.34.050.

References

1. M. H. Anderson, J. R. Ensher, M. R. Matthews, C. E. Wieman, E. A. Cornell, Proceedings of the 12 International Conference on Laser Spectroscopy, Capri, June 1995, M. Ingucio, M. Allegrini and A. Sasso Eds., (World Scientific, Singapore, 1996).
2. A. Robert, O. Sirjean, A. Browaeys, J. Poupard, S. Nowak, D. Boiron, C. I. Westbrook, A. Aspect, *Science*, **292**, 463 (2001), published online 22 March 2001, 10.1126/science.1060622.
3. M. H. Anderson, J. R. Ensher, M. R. Matthews, C. E. Wieman, E. A. Cornell, *Science* **269**, 198 (1995).
4. K. B. Davis *et al.*, *Phys. Rev. Lett.* **75**, 3969 (1995).
5. C. C. Bradley, C. A. Sackett, J. J. Tollet, R. G. Hulet, *Phys. Rev. Lett.* **75** 1687 (1995).
6. D. G. Fried *et al.*, *Phys. Rev. Lett.* **81**, 3811 (1998).
7. W. Vassen, OSA TOPS on Ultracold Atoms and BEC 7, 20, K. Burnett, Ed. (Optical Society of America, Washington, DC, 1996).
8. F. Bardou, O. Emile, J. M. Courty, C. I. Westbrook, and A. Aspect, *Europhys. Lett.* **20**, 681 (1992).
9. H. C. Mastwijk, J. W. Thomsen, P. van der Straten, and A. Niehaus, *Phys. Rev. Lett.* **80**, 5516 (1998).

10. P. J. J. Tol, N. Herschbach, E. A. Hessels, W. Hogervorst, and W. Vassen, *Phys. Rev A.* **60**, 761 (1999).
11. M. Kumakura and N. Morita, *Phys. Rev. Lett.* **82**, 2848 (1999).
12. A. Browaeys, J. Poupard, A. Robert, S. Nowak, W. Rooijakkers, E. Arimondo, L. Marcassa, D. Boiron, C. I. Westbrook, and A. Aspect, *Eur. Phys. J. D* **8**, 199 (2000).
13. F. Pereira Dos Santos, F. Perales, J. Léonard, A. Sinatra, Junmin Wang, F. S. Pavone, E. Rasel, C. S. Unnikrishnan, and M. Leduc, *Eur. Phys. J. D* **14**, 15 (2001).
14. V. Venturi and I. Whittingham, *Phys. Rev. A*, **61**, 060703 (2000).
15. J. C. Hill, L. L. Hatfield, N. D. Stockwell, and G. K. Walters, *Phys. Rev. A* **5**, 189 (1972).
16. N. Herschbach, P. J. J. Tol, W. Hogervorst, and W. Vassen, *Phys. Rev. A* **61**, 050702(R) (2000).
17. S. Nowak, A. Browaeys, J. Poupard, A. Robert, D. Boiron, C. I. Westbrook, and A. Aspect, *Appl. Phys. B* **70**, 455 (2000).
18. P. O. Fedichev, M. W. Reynolds, U. M. Rahmanov, G. V. Shlyapnikov, *Phys. Rev. A* **53**, 1447 (1996); G. V. Shlyapnikov, T. M. Walraven, U. M. Rahmanov, M. W. Reynolds, *Phys. Rev. Lett* **73**, 3247 (1994).
19. V. Venturi, I. B. Whittingham, P. J. Leo, G. Peach, *Phys. Rev. A* **60**, 4635 (1999), P. Leo, V. Venturi, I. Whittingham, J. Babb, preprint arXiv:physics/0011072.
20. M. O. Mewes, M. R. Andrews, N. J. van Druten, D. M. Kurn, D. S. Durfee, and W. Ketterle, *Phys. Rev. Lett.* **77**, 416 (1996).
21. F. Dafolvo, S. Giorgini, L. P. Pitaevskii, Rev. Mod. Phys. **71**, 463 (1999), and references therein.
22. P. Fedichev, M. Reynolds, G. Shlyapnikov, *Phys. Rev. Lett.*, **77**, 2921 (1996).
23. L. V. Hau *et al.*, *Phys. Rev. A* **58**, R54 (1998).
24. J. R. Ensher, D. S. Jin, M. R. Matthews, C. E. Wieman, E. A. Cornell, *Phys. Rev. Lett.* **77**, 4984 (1996).
25. A. Browaeys, A. Robert, O. Sirjean, J. Poupard S. Nowak, D. Boiron, C.I. Westbrook, A. Aspect, *Phys. Rev. A* **64**, 034703 (2001).
26. F. Pereira dos Santos, J. Leonard, J. Wang, C. Barrelet, F. Perales, E. Rasel, C.S. Unnikrishnan, M. Leduc, C. Cohen-Tannoudji, *Phys. Rev. Lett.* **86**, 3459 (2001).
27. E. A. Burt *et al.*, *Phys. Rev. Lett.* **79**, 337 (1997).
28. Y. Yasuda, F. Shimizu, *Phys. Rev. Lett.* **77**, 3090 (1996).

COHERENT DYNAMICS OF BOSE-EINSTEIN CONDENSATES IN A 1D OPTICAL LATTICE

S. BURGER, F. S. CATALIOTTI, C. FORT, P. MADDALONI, F. MINARDI, AND M. INGUSCIO

Laboratorio Europeo di Spettroscopia Nonlineare (LENS),
Istituto Nazionale per la Fisica della Materia,
Dipartimento di Fisica dell' Università di Firenze,
Largo Enrico Fermi, 2, I-50125 Firenze, Italia

Bose-Einstein condensates in the combined potential of a harmonic trap and a 1D optical lattice allow us to investigate the coherent atomic current in an array of Josephson junctions and to study the density-dependent critical velocity for the breakdown of superfluid motion.

1 Introduction

Bose-Einstein condensates (BECs) are quantum systems which can be easily manipulated and characterized due to their macroscopic nature[1]. The employment of BECs in atomic physics has stimulated a wealth of new experiments which is often compared to the rapid development in optics and spectroscopy after the invention of the laser.

Atoms confined in a periodic potential exhibit quantum effects known from solid state physics, like Bloch oscillations and Wannier-Stark ladders, which have been observed by exposing cold atoms to the dipole potential of far detuned optical lattices[2,3]. Meanwhile, the achievement of BEC has given the possibility to explore also macroscopic quantum effects in this context.

In a first experiment with BECs loaded into a 1D optical lattice, quantum interference could be observed, leading to the formation of the first "mode-locked" atom laser[4]. Recently, also the squeezing of matter waves[5] and the decoherence of BECs in 2D optical lattices[6] have been investigated. Experiments, in which optical lattices are applied to the BEC on much shorter time scales have investigated Bragg-diffraction as a tool for interferometry and spectroscopy[7,8], Bloch oscillations[9], and dynamical tunnelling[10].

The coherent nature of BECs is of great importance for its dynamics in optical lattices. From the well defined phase of the macroscopically occupied wavefunction describing the BEC it follows that at low fluid velocities the BEC is performing a superfluid motion in the lattice[11]. In a regime of a greater potential height of the optical lattice, the system is also ideally suited to study the Josephson effect: At a potential depth of the lattice sites exceeding the thermal energy BECs do collectively tunnel from one site to the next, at a rate which depends on the difference in phase between the sites, while thermal clouds of atoms are fixed to the wells of the optical lattice, [12].

In this paper we discuss recent experiments on macroscopic quantum effects of BEC dynamics in optical lattices. The paper is organized as follows: In the following chapter we briefly introduce the setup of our BEC experiment and the implementation of the optical standing wave creating the periodic lattice potential. In chapter 3 we report on the superfluid motion of a BEC with a changed effective mass in an optical lattice and on the density-dependent breakdown of superfluidity, allowing to measure the spectrum of critical velocities in the inhomogeneous BEC. Chapter 4 concentrates on the direct observation of a coherent atomic current in an array of Josephson junctions. Chapter 5 concludes the paper with an outlook on future directions.

2 Experimental Setup

BECs of ^{87}Rb in the (F=1,$m_F=-1$) state are produced by the combination of laser cooling in a double magneto-optical trap system and evaporative cooling in a static magnetic trap of the Ioffe-type[13]. The condensates are cigar-shaped with the long axis oriented horizontally; with an atom number of $N = 4 \times 10^5$ their typical dimensions (Thomas-Fermi radii) are $R_x = 55\,\mu$m and $R_\perp = 5.5\,\mu$m.

We create a 1D optical lattice by superimposing to the long axis of the magnetic trap a far detuned, retroreflected laser beam with wavelength λ (see Fig. 1a). The resulting potential is given by the sum of the magnetic (V_B)

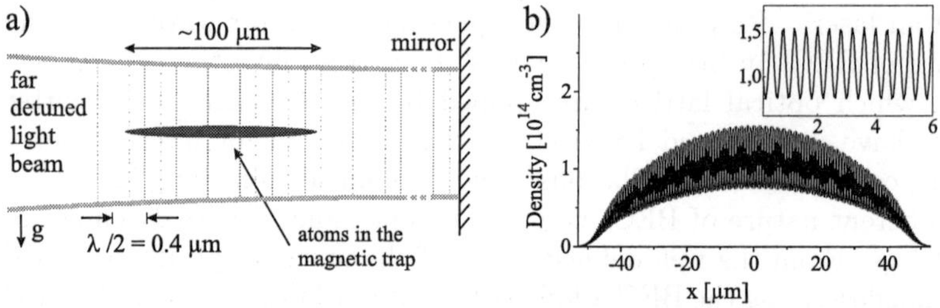

Figure 1. a) Schematic set-up of the experiment. b) Density distribution of the ground state from a numerical simulation for parameters $N = 3 \times 10^5$ and $V_0 = 1.5\,E_R$.

and the optical potential (V_{opt}):

$$V = V_B + V_{opt} = \frac{1}{2}m\left(\omega_x^2 x^2 + \omega_\perp^2 (y^2 + z^2)\right) + V_0 \cos^2 kx \ , \qquad (1)$$

where m is the atomic mass, $\omega_x = 2\pi \times 8.7$ Hz and $\omega_\perp = 2\pi \times 90$ Hz are the axial and radial frequencies of the magnetic harmonic potential, and $k = 2\pi/\lambda$ is the modulus of the wavevector of the optical lattice. By varying the intensity of the laser beam (detuned typically $\Delta = 150\,\text{GHz}$ to the blue of the D_1 transition at $\lambda = 795$ nm) up to 14 mW/mm^2 we can vary the optical lattice potential height V_0 from 0 to $V_0 \sim 5E_R$, where E_R is the recoil energy, $E_R = \hbar^2 k^2/2m$. We calibrate the optical potential measuring the Rabi frequency of the Bragg transition between the momentum states $-\hbar k$ and $+\hbar k$ induced by the standing wave[14]. Due to the large detuning of the optical lattice, spontaneous scattering can be neglected for the experiments on BEC-dynamics which are performed typically on a timescale of $\tau \sim 2\pi/\omega_x$; nevertheless, spontaneous scattering leads to a reduction of the total atom number during the preparation of the BEC.

Bose-Einstein condensates in the combined magnetic trap and optical lattice are prepared by superimposing the optical lattice to the trapping potential already during the last hundreds of ms of the RF-evaporation ramp. Figure 1b shows the density distribution along the x-axis, as obtained by numerical propagation of the Gross-Pitaevskii equation in imaginary time. In the experiment, the density modulation on the length scale of $\lambda/2$ cannot be directly resolved, due to the limited resolution ($\simeq 7\,\mu$m) of the absorption-imaging system.

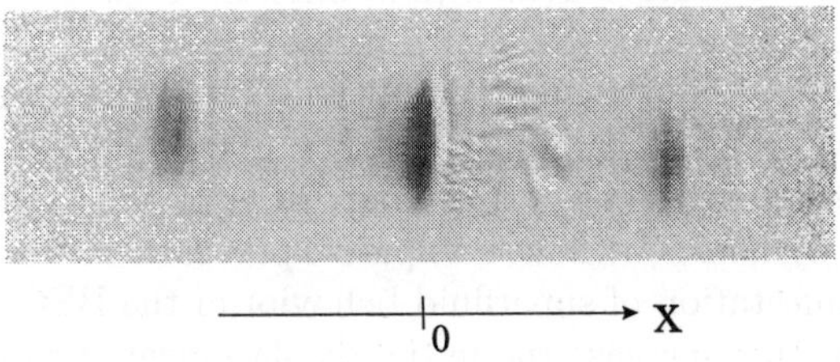

Figure 2. Absorption image of a BEC released from the combined magnetic trap and optical lattice ($V_0 \sim 4\,E_R$, $T < 50\,\text{nK}$).

In order to assure experimentally that the state reached by the atoms is the ground state of the system we use the fact that the periodically modulated density distribution of the BEC in real space corresponds to a comb of equally spaced peaks in momentum space. In a time-of-flight measurement (see Fig. 2) we check that the fraction of atoms in the different momentum components of the ground state does not depend on t_l but only on the depth of the dipole-potential wells[15].

3 Superfluid dynamics

Superfluidity of BECs is a direct consequence of their coherent nature[16]. It is manifested in the appearance of vortices[17,18] and scissors modes[19] as well as in a critical velocity for the onset of dissipative processes[20]. A far detuned optical lattice at a low potential height, $V_0 < 2\,E_R$, is well suited to study in detail the critical velocity because it acts like a medium with a microscopic roughness on the BEC moving through it, being velocity-dependent compressed and decompressed as it propagates.

In order to investigate the dynamics in the combined trap we translate the magnetic trapping potential in the x-direction by a variable distance Δx ranging up to 300 μm in a time $t \ll 2\pi/\omega_x$. Therefore, the BEC finds itself out of equilibrium and is subject to a potential gradient which forces it into motion. After an evolution time t_{ev} in the displaced trap, both the magnetic trapping and the optical lattice are switched off and after a free expansion of 26.5 ms the cloud is imaged, giving information on the momentum distribution of the system.

In the magnetic trap the center-of-mass motion of the BEC in the displaced trap is an undamped oscillation with frequency $\omega_x = 2\pi \times 8.7$ Hz and amplitude Δx. In the combined trap formed by the magnetic and the optical lattice potentials we observe dynamics in different regimes:

For small displacements, $\Delta x < 50\,\mu$m, the dynamics of the BEC resembles the "free oscillation" at the same amplitude but with a significant shift in frequency which can be explained in terms of an effective atomic mass[11]. By varying the potential height V_0 we are able to tune this effective mass. The undamped dynamics without dissipative processes in the small-amplitude regime is a manifestation of superfluid behavior of the BEC.

When we further increase the initial displacement Δx and hence the velocity of the BEC, it enters a regime of dissipative dynamics. We observe a damped oscillation in the trap and dissipative processes heating the cloud.

The critical velocity in a superfluid is proportional to the local speed of sound, c_s, which depends on the density n, $c_s(r) = \sqrt{n(r)/m\,(\delta\mu/\delta n)}$, with

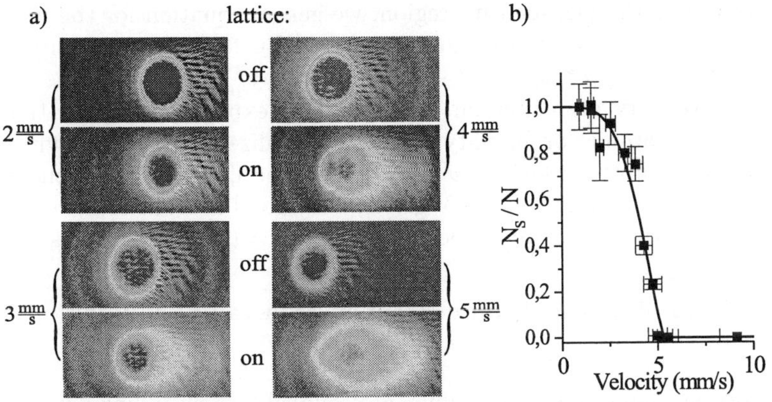

Figure 3. a) False color representation of absorption images of atomic clouds after evolution with different maximum velocities (indicated) in the magnetic trap ("lattice off") and in the combined trap ("lattice on"). b) Fraction of atoms in the superfluid component vs. maximum propagation velocity.

the chemical potential μ. Therefore superfluidity breaks down first in the wings of the BEC where the density is lowest.

In order to measure the velocity- and density-dependent onset of dissipation and thereby the spectrum of critical velocities in the BEC, we have varied the displacement Δx and recorded atomic distributions after a fixed evolution time $t_{ev} = 40\,\mathrm{ms}$. For low velocities, $v < 2\,\mathrm{mm/s}$, the sample follows the position of a freely moving BEC ("lattice off" in Fig. 3a); no thermal component appears.

Upon increasing the velocity of the BEC, we observe a retardation of a part of the cloud, leading to a well detectable separation from the superfluid component after free evolution (see Fig. 3a). The spatial separation from the thermal component allows a clear demonstration of the superfluid properties of inhomogeneous Bose-Einstein condensates.

For velocities $v \sim 4\,\mathrm{mm/s}$ we observe that only the central part of the fluid is moving without retardation; for even higher velocities all of the atoms are retarded and form a heated cloud with a Gaussian density distribution.

Figure 3b shows the ratio of atom number in the non-retarded component (parabolic density-profile, "superfluid component"), N_s, and the total atom number, N, in dependence of the maximum velocity attained during the evolution in the optical lattice[21]. The envelope function of the density distribution of the BEC is an inverted parabola in 3D (see Fig. 1b and hence, by integration over the high-density region, we get an equation for the relative number of atoms in the superfluid part of the BEC for a given velocity v, $N_s(v)/N = \left[5/2 \times (1 - v^2/v_{max}^2)^{3/2} - 3/2 \times (1 - v^2/v_{max}^2)^{5/2}\right]$, where v_{max} is the critical velocity at maximum density. This expression implies that about 90% of the atomic probability density is localized in a region which remains superfluid up to velocities $v \simeq v_{max}/2$. The line in Fig. 3b shows that the above expression for $N_s(v)/N$ gives a very good account of the data, the fitted value of the maximum velocity being $v_{max} = (5.3 \pm 0.5)$ mm/s.

4 Observation of a Josephson current in an array of coupled BECs

Two macroscopic quantum systems which are coupled by a weak link produce the flow of a supercurrent I between them, driven by their relative phase $\Delta\phi$,

$$I = I_c \sin \Delta\phi \,, \qquad (2)$$

where I_c is the critical Josephson current[22,23]. The relative phase evolves in time proportionally to the difference in chemical potential between the two quantum fluids.

The first experimental evidence of a current-phase relation was already observed in superconducting systems soon after Josephson's proposal[24]. We realize a one-dimensional array of bosonic Josephson junctions (JJs) by preparing an array of BECs in the sites of the optical lattice with an interwell barrier energy V_0 which is high compared to the chemical potential of the BECs[12]. Every two condensates in neighbouring wells overlap slightly with each other due to a finite tunnelling probability, and therefore constitute a JJ, with the possibility to adjust the critical current I_c by tuning the laser intensity. By driving the system with the external harmonic potential, we investigate the current-phase dynamics and measure the critical Josephson current as a function of the interwell potential V_0.

Although the system in its ground state consists of spatially separated condensates, tunneling between adjacent wells leads to a constant phase over the whole array. As a result when the condensates are released from the combined trap they show an interference pattern (see Fig. 2). This provides us with information about the relative phase of the different condensates.

To observe a Josephson current in the array we nonadiabatically displace the magnetic trap along the lattice axis by a distance of $\sim 30\,\mu m$. The potential energy the atoms gain in this process is much smaller than the interwell potential barrier, but the relative phases of the BECs in the different wells are driven by this process. Therefore, according to equation 2 we expect a Joseph-

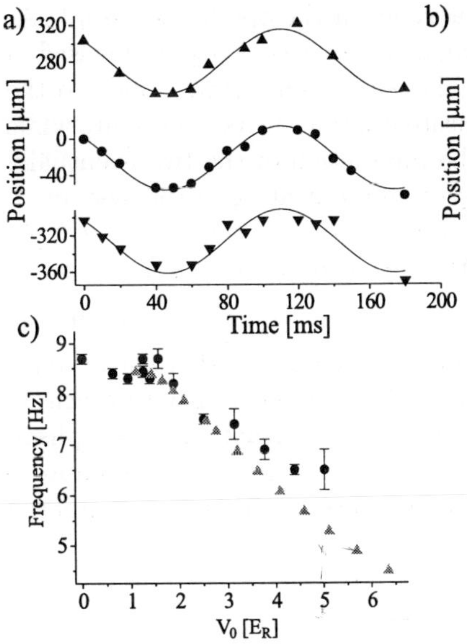

Figure 4. a) The three peaks of the interference pattern of an array BECs, expanded for 28 ms after the propagating in the optical lattice. b) Center-of-mass positions of thermal clouds after expansion from the magnetic trap (filled circles) and from the combined magnetic trap and optical lattice (open circles) as a function of evolution time in the displaced respective trap. c) Frequency of the oscillation of a coherent ensemble in the JJ-array in dependence of the interwell potential height.

son current. Indeed, we observe a collective oscillation of the ensemble. A collective motion can be established only at the price of a well definite phase coherence among the condensates – the relative phases among all adjacent sites should remain locked together in order to preserve the ordering of the collective motion. This locking of the relative phases shows up in the expanded cloud interferogram. In Fig. 4a the positions of the three peaks in the interferogram are plotted as a function of time spent in the combined trap after the displacement of the magnetic trap. The motion performed by the center of mass of the condensate is an undamped oscillation at a frequency $\omega < \omega_x$.

The coherent nature of the oscillation is also proven by repeating the same experiment with a thermal cloud. In this case – although atoms can individually tunnel through the barriers – no macroscopic phase is present in the cloud and no motion of the center of mass is observed. The center-of-mass positions of thermal clouds in the optical lattice are shown in Fig. 4b, together with the oscillation of thermal clouds in absence of the optical potential. As can clearly be seen, the movement of thermal clouds is strongly suppressed in presence of the optical lattice. We have also subjected mixed clouds to the displaced potential, where only the condensate fraction starts to oscillate while the thermal component remains static; the interaction of the two eventually leads to a damping of the condensate motion and a heating of the system.

As can be derived from the phase-current relation of the JJ array[12] the critical Josephson current is related to the small amplitude oscillation frequency ω of the JJ array by the simple relation $I_c = \frac{4\hbar}{m\lambda^2}\left(\frac{\omega}{\omega_x}\right)^2$. Figure 4c shows experimental values of the oscillation frequency ω together with the result of a variational calculation of the JJ array[12]. The possibility to precisely adjust the critical Josephson current presents a major advantage of Josephson junctions in Bose-Einstein condensates, where due to the elaborate manipulation tools of atomic physics a variety of parameters can be tuned, compared to systems realized in solid-state physics.

5 Conclusions

The phase coherence of Bose-Einstein condensates has important consequences on their behaviour in periodic potentials. We have experimentally investigated macroscopic quantum effects in the dynamics of BECs in a 1D optical lattice: We could observe superfluid motion, a density-dependent critical velocity for the onset of dissipation, and – in a regime of higher lattice potentials – an oscillating Josephson current.

Future directions of this work are the study of lower dimensional physics in the strongly confining lattice sites, the investigation of bright solitons formed by BECs in optical lattices[25], and the exploration of collective tunnelling of the ground state fraction of a mixed cloud as an atom-optical filtering technique for the "purification" of condensates.

Acknowledgments

Our understanding of the various experiments performed at LENS benefitted very much from collaborations with M. L. Chiofalo, L. Pitaevskii, A. Smerzi, S. Stringari, M. Tosi, and A. Trombettoni. We also acknowledge stimulating discussions with M. Artoni and G. Ferrari, and support by the EU under contracts HPRI-CT 1999-00111 and HPRN-CT-2000-00125 and by MURST through the PRIN 1999 and PRIN 2000 Initiatives.

References

1. See, e.g., *Bose-Einstein Condensation in Atomic Gases*, ed. M. Inguscio *et al.*, (IOS Press, Amsterdam, 1999); *Bose-Einstein Condensates and Atom Lasers*, ed. S. Martellucci *et al.*, (Kluwer, New York, 2000).
2. M. B. Dahan, E. Peik, J. Reichel, Y. Castin, and C. Salomon. *Phys. Rev. Lett.*, **76**, 4508 (1996).
3. S. R. Wilkinson, C. F. Bharucha, K. W. Madison, Q. Niu, and M. G. Raizen. *Phys. Rev. Lett.*, **76**, 4512 (1996).
4. B. P. Anderson and M. A. Kasevich. *Science*, **282**, 1686 (1998).
5. C. Orzel, A. K. Tuchman, M. L. Fenselau, M. Yasuda, and M. A. Kasevich. *Science*, **291**, 2386 (2001).
6. M. Greiner, I. Bloch, O. Mandel, T. W. Hänsch, and T. Esslinger. e-print cond-mat/0105105 (2001).
7. M. Kozuma, L. Deng, E. W. Hagley, J. Wen, R. Lutwak, K. Helmerson, S. L. Rolston, and W. D. Phillips. *Phys. Rev. Lett.*, **82**, 871 (1999).
8. J. Stenger, S. Inouye, A. P. Chikkatur, D. M. Stamper-Kurn, D. E. Pritchard, and W. Ketterle. *Phys. Rev. Lett.*, **82**, 4569 (1999).
9. O. Morsch, J. H. Müller, M. Cristiani, and E. Arimondo. e-print cond-mat/0103466 (2001).
10. W. K. Hensinger, H. Häffner, A. Browaeys, N. R. Heckenberg, K. Helmerson, C. McKenzie, G. J. Milburn, W. D. Phillips, S. L. Rolston, H. Rubinsztein-Dunlop, and B. Upcroft. *Nature*, **412**, 52 (2001).
11. S. Burger, F. S. Cataliotti, C. Fort, F. Minardi, M. Inguscio, M. L. Chiofalo, and M. Tosi. *Phys. Rev. Lett.*, **86**, 4447 (2001).
12. F. S. Cataliotti, S. Burger, C. Fort, P. Maddaloni, F. Minardi, M. Inguscio, A. Trombettoni, and A. Smerzi. Science **293**, (2001).
13. C. Fort, M. Prevedelli, F. Minardi, F. S. Cataliotti, L. Ricci, G. M. Tino, and M. Inguscio. *Europhys. Lett.*, **49**, 8 (2000).
14. E. Peik, M. B. Dahan, I. Bouchoule, Y. Castin, and C. Salomon, *Phys. Rev. A*, **55**, 2989 (1997).
15. P. Pedri, L. Pitaevskii, S. Stringari, C. Fort, S. Burger, F. S. Cataliotti, P. Maddaloni, F. Minardi, and M. Inguscio, to be published.
16. F. Dalfovo, S. Giorgini, L. P. Pitaevskii, and S. Stringari. *Rev. Mod. Phys.*, **71**, 463 (1999).
17. M. R. Matthews, B. P. Anderson, P. C. Haljan, D. S. Hall, C. E. Wieman, and E. A. Cornell. *Phys. Rev. Lett.*, **83**, 2498 (1999).
18. K. W. Madison, F. Chevy, W. Wohlleben, and J. Dalibard. *Phys. Rev. Lett.*, **84**, 806 (2000).
19. O. M. Maragó, S. A. Hopkins, J. Arlt, E. Hodby, G. Heckenblaikner, and C. J. Foot. *Phys. Rev. Lett.*, **84**, 2056 (2000).
20. C. Raman, M. Köhl, R. Onofrio, D. S. Durfee, C. E. Kuklewicz, Z. Hadzibabic, and W. Ketterle. *Phys. Rev. Lett.*, **83**, 2502 (1999).
21. Due to spontaneous scattering during the preparation of the BEC in the combined trap the total atom number is reduced by a factor ~ 0.7.
22. A. Barone and G. Paterno, *Physics and Applications of the Josephson Effect* (Wiley, New York 1982).
23. A. Smerzi, S. Fantoni, S. Giovannazzi, S. R. Shenoy, *Phys. Rev. Lett.*, **79**, 4950 (1997).
24. B. D. Josephson, *Phys. Lett.*, **1**, 251 (1962).
25. O. Zobay, S. Pötting, P. Meystre, and E. M. Wright. *Phys. Rev. A*, **59**, 643 (1999).

^{6}LI AND ^{7}LI: NON-IDENTICAL TWINS

RANDALL G. HULET, KEVIN E. STRECKER, ANDREW G. TRUSCOTT, AND GUTHRIE B. PARTRIDGE

Department of Physics and Astronomy, Rice University, Houston, Texas 77005, USA
E-mail: randy@atomcool.rice.edu

We have cooled a magnetically trapped gas of bosonic ^{7}Li and fermionic ^{6}Li atoms into the quantum degenerate regime. The atoms are loaded from a laser-slowed atomic beam into a magneto-optical trap, and then transferred to a magnetic trap. Forced evaporation is used to cool the ^{7}Li atoms, and the ^{6}Li are cooled "sympathetically" via collisional interaction with the ^{7}Li. As the temperature of the Fermi gas is reduced below the Fermi temperature, we observe that its spatial size is greater than that of the Bose gas. This effect is quite pronounced at the lowest temperature achieved of 0.25 times the Fermi temperature, and is a manifestation of the Fermi pressure resulting from the Pauli exclusion principle.

1 Introduction

The development of the techniques of atomic laser cooling and trapping culminated several years ago with the achievement of Bose-Einstein condensation (BEC) of a weakly interacting gas. This development has greatly expanded our ability to probe and understand bosonic matter in a regime dominated by quantum statistics rather than interactions. In contrast, prior to the results presented here, there has been only one realization of quantum degeneracy of a trapped Fermi gas, that being the work of Demarco and Jin with ^{40}K [1]. In this paper, we report the first realization of quantum degeneracy in a mixed Bose/Fermi gas of trapped lithium atoms [2]. Of the two stable isotopes of lithium, ^{7}Li is a composite boson, while ^{6}Li is a fermion. We have succeeded in cooling the gas to a temperature of 240 nK, which corresponds to 0.25 times the Fermi temperature T_F for the fermions. At this temperature, the spatial distribution of the fermions is strongly affected by Fermi pressure, in exact analogy with the stabilization of white dwarf and neutron stars, and is in stark contrast with the behavior of the Bose gas. Our experiment is very similar to one in Paris, and they report similar results [3].

Trapping and cooling fermions is similar to bosons except for one major difference: identical fermions are symmetry-forbidden to undergo *s*-wave rethermalization collisions needed to evaporatively cool. This obstacle can be circumvented by evaporative cooling using a two-component Fermi gas [1,4] or by sympathetic cooling with a Bose/Fermi mixture [2,5]. We have chosen the latter approach because any two magnetically trappable spin-states in ^{6}Li will rapidly undergo spin-exchange collisions [6]. In our experiment, ^{6}Li is cooled by thermal contact with the evaporatively cooled ^{7}Li. Ultimately, it may be possible to cause a BCS-like Cooper pairing of a two-spin state mixture of the ^{6}Li atoms [7].

2 Experimental Approach

We have constructed an entirely new apparatus for this experiment, although the techniques and apparatus are similar to those used to achieve BEC in ^{7}Li [8]. The apparatus consists of an ultrahigh vacuum chamber containing a magnetic trap. The trap is loaded from a dual-species magneto-optical trap (MOT). Approximately 3×10^{10} atoms of ^{7}Li are loaded into the MOT by laser slowing a thermal atomic beam using the Zeeman slower technique. The ^{6}Li MOT is loaded using the same Zeeman slower, but for only 20 ms to minimize interference with the ^{7}Li MOT. The interference arises because of a near coincidence of the D1-line transition frequency of ^{7}Li with the D2-line transition in ^{6}Li. Approximately 10^7 atoms of ^{6}Li are loaded into the MOT. Both isotopes are then optically pumped into the "stretched" low-field seeking Zeeman sub-level, corresponding to F=2, m_F = 2 for ^{7}Li, and F = 3/2, m_F = 3/2 for ^{6}Li. After further cooling, and compressing, approximately 10% of the atoms of each isotope are transferred to the magnetic trap.

The magnetic trap has an Ioffe-Pritchard field configuration, and was built using the "clover-leaf" design of MIT [9]. The trap produces an axial curvature of 75 G/cm^2 and a radial gradient of 110 G/cm, which at a 2 G bias field, correspond to measured axial and radial trapping frequencies for ^{7}Li of 39 Hz and 433 Hz, respectively. The trapped atom lifetime is limited by collisions with background gas, and has been measured to be in excess of 3 minutes.

Cooling to degeneracy is accomplished by microwave-induced evaporative cooling to an untrapped spin-state of ^{7}Li. The ^{6}Li atoms are cooled sympathetically through their elastic interactions with the ^{7}Li and are not themselves ejected. The triplet *s*-wave scattering lengths determine the elastic scattering cross sections for thermalization. For ^{7}Li/^{7}Li collisions the scattering length is –1.5 nm, whereas for ^{6}Li/^{7}Li it is 2.2 nm [10,11]. Although neither of these correspond to particularly large cross sections, they are sufficient to cool both species to quantum degeneracy in ~60 s. Similar methods of sympathetic cooling have been previously used in a two-species ion trap [12] and in a two-component Bose gas of ^{87}Rb cooled to BEC [13].

Once the sympathetic cooling cycle is complete, the atoms are held in the trap for at least 2 s to ensure complete thermalization. The ^{7}Li atoms are then probed with a near-resonant laser beam, and their absorption shadows imaged onto a CCD camera with a magnification corresponding to 5 μm/pixel. A high-intensity on-resonant laser pulse is applied for 10 μs to quickly remove all remaining ^{7}Li atoms from the magnetic trap. Because this pulse is detuned by more than 10 GHz from any ^{6}Li resonance, it has no measurable effect on the ^{6}Li atom cloud. The ^{6}Li are then imaged in a similar manner.

The number of atoms and their temperature are obtained from the images by fitting them to the appropriate quantum statistical density distribution functions. Unlike the fermions, the shape of the density distribution for bosons changes significantly as quantum degeneracy is approached. For this reason, the bosons prove to be a sensitive thermometer for determining the common temperature, which reduces the uncertainty in both the number and temperature of the Fermi gas. It is assumed that the interactions have a negligible effect on the density. This is a good approximation because ^{7}Li has attractive interactions that limit the number of condensate atoms, and hence the magnitude of the mean field [8,14]. This effect also constrains the magnitude of the mean-field experienced by the ^{6}Li as a result of the ^{7}Li, while the self-interaction between the fermions is identically zero in the *s*-wave limit.

3 Results

We have cooled the gas to temperatures as low as T = 240 nK, corresponding to T/T_F = 0.25, where the Fermi temperature $T_F = \hbar\varpi\,(6N_F)^{1/3}/k_B$, and ϖ is the geometric mean of the trap frequencies for ^{6}Li, N_F is the number of ^{6}Li atoms, and k_B is Boltzmann's constant. Density profiles for two pairs of images, corresponding to two different temperatures, are shown in Fig. 1. At high temperature, where classical statistics are a good approximation, the spatial distributions of the bosons and fermions show little difference, as can be seen in Fig. 1A. As the gas is cooled further, the ^{6}Li distribution is observed to be broader than that of the ^{7}Li. This difference is clearly visible at the lowest temperatures, as shown in Fig. 1B. The broadening is the result of Fermi pressure and is a direct manifestation of quantum statistics.

The square of the axial radius of the ^{6}Li clouds is plotted versus T/T_F in Fig. 2, where it can be seen that at relatively high temperatures, the radius decreases as $T^{1/2}$, as expected for a classical gas (dashed line). At a temperature near 0.5 T_F, however, the radius deviates from the classical prediction, and at the lowest temperatures, it plateaus to a value near the Fermi radius. At T = 0, every trap state is singly occupied up to the Fermi energy, giving rise to a nonzero mean energy and a resulting Fermi pressure. Fermi pressure is responsible for the minimum radius and is a striking manifestation of Fermi-Dirac statistics. In white dwarf and neutron stars, which are essentially "dead" due to the depletion of their nuclear fuel, it is the Fermi pressure that stabilizes the star against gravitational collapse. The stabilization of the size of the atom cloud with decreasing temperature is another manifestation of the same physics.

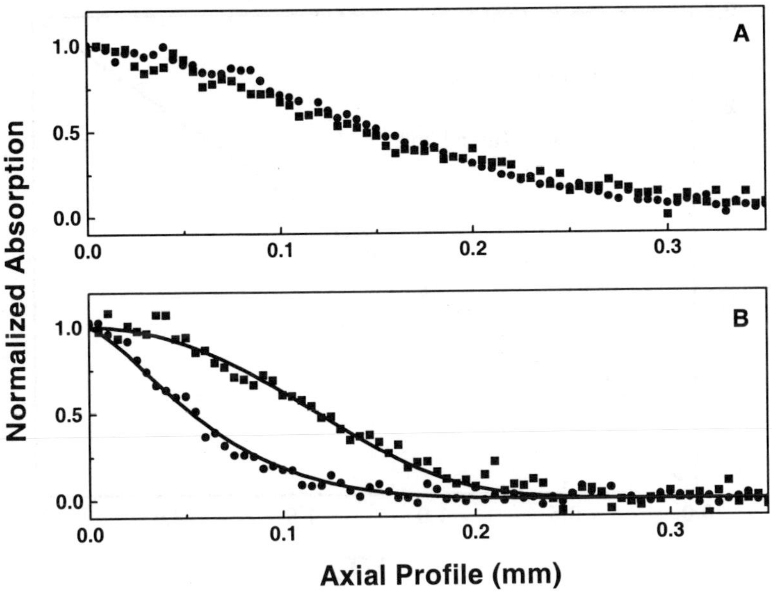

Figure 1. Axial profiles obtained from absorption images. The squares correspond to ^{6}Li, and the circles to ^{7}Li. (A) T = 810 nK, corresponding to T/T_F = 1.0 for the fermions and T/T_c = 1.5 for the bosons. (B) T = 240 nK, corresponding to T/T_F = 0.25 and T/T_c = 1.0. The fits to the data are shown as solid lines. (Reprinted from Ref. [2]).

4 Discussion

In the current experiment, ^{6}Li in the F = 3/2, m_F = 3/2 state is sympathetically cooled by allowing it to thermalize with evaporatively cooled ^{7}Li in the F = 2, m_F = 2 state. We found that it was not possible to cool below ~0.25 T_F with this method. We believe that this is a fundamental limit of sympathetic cooling of fermions with bosons. For sympathetic cooling to work, the heat capacity of the evaporatively cooled gas, the bosons in this case, must exceed that of the sympathetically cooled gas [2]. For a quantum degenerate Fermi gas, the heat capacity $C_F \approx \pi^2 N_F k_B (T/T_F)$ [15]. For a harmonically confined Bose gas at the critical temperature T_c for Bose-Einstein condensation, the heat capacity $C_B \approx 10.86\, N_B k_B$, where N_B is the number of bosons [16]. By equating these heat capacities, we find that $C_B > C_F$, only for $T/T_F > 0.3$ [2]. The key aspect of this argument is that the heat capacity of the bosons takes the value at T_c. This is clearly true for bosons with attractive

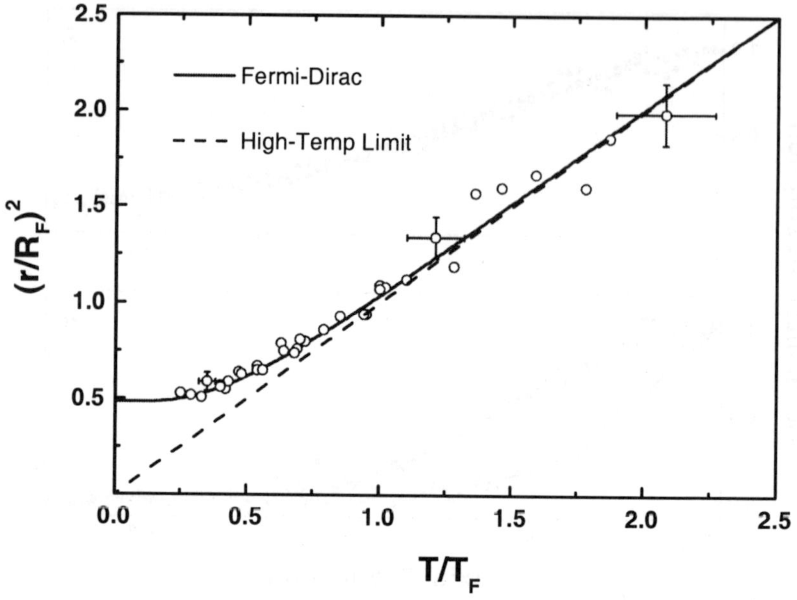

Figure 2. Square of the 1/e axial radius vs. T/T_F. The radius is normalized by $R_F = (2k_BT_F/m\omega_a^2)^{1/2}$, where ω_a is the axial trap frequency, and m is the atomic mass of ^{6}Li. The solid line is the prediction for an ideal Fermi gas, whereas the dashed line is the high-temperature limit. The divergence of the data from the classical prediction is the result of Fermi pressure. Several representative errors bars are shown. (Reprinted from Ref. [2]).

interactions, as for the F = 2, m_F = 2 state of ^{7}Li. In this case, the condensate number is limited to values much less than N_B, and T is therefore restricted to values only incrementally below T_c. Although not as obvious, the same restriction on C_B also applies to condensates with repulsive interactions. Only the thermal atoms in a Bose gas contribute to the heat capacity as the condensate itself has none. Therefore, below T_c, the total heat capacity is the heat capacity of the gas at T_c, 10.86 N_c k_B, where N_c is the critical number at temperature T. Again, the same limit on sympathetic cooling, $T/T_F > 0.3$, is found. There are several possible ways to achieve lower temperatures, but the most straightforward is simply to evaporate both isotopes simultaneously, so that C_F is lowered at the same rate as C_B.

A strong, attractive interaction in a two spin-state gas will be required to induce *s*-wave Cooper pairing. The best candidate states for ^{6}Li seem to be the energetically lowest pair of Zeeman sublevels, the F = ½, m_F = ½ and the F = ½, m_F = -½ states. These states are predicted to exhibit an enormous Feshbach resonance, for which the interaction may be arbitrarily tuned [6]. Because these states are energetically the lowest states, there are no open two-body inelastic collision channels. Furthermore, three-body recombination should also be suppressed, since there is no way to produce a totally anti-symmetric three-body state from a gas with only two-spin states.

5 Acknowledgements

The Office of Naval Research, NASA, the National Science Foundation, and the R. A. Welch Foundation supported this work. We are grateful to B. Ghaffari and D. Homan for their help in constructing the apparatus.

References

1. DeMarco B. and Jin D. S., Onset of Fermi degeneracy in a trapped atomic gas. *Science* **285**, (1999) pp. 1703.
2. Truscott A. G., Strecker K. E., McAlexander W. I., Partridge G. B. and Hulet R. G., Observation of Fermi Pressure in a Gas of Trapped Atoms. *Science* **291**, (2001) pp. 2570.
3. Schreck F. *et al.*, Quasipure Bose-Einstein Condensate Immersed in a Fermi Sea. *Phys. Rev. Lett.* **87**, (2001) pp. 080403.
4. O'Hara K. M., Gehm M. E., Granade S. R., Bali S. and Thomas J. E., Stable, Strongly attractive, Two-State Mixture of Lithium Fermions in an Optical Trap. *Phys. Rev. Lett.* **85**, (2000) pp. 2092.
5. Schreck F. *et al.*, Sympathetic Cooling of Bosonic and Fermionic Lithium Gases Towards Quantum Degeneracy. *Phys. Rev. A* **64**, (2000) pp. 011402.
6. Houbiers M., Stoof H. T. C., McAlexander W. I. and Hulet R. G., Elastic and inelastic collisions of ^{6}Li atoms in magnetic and optical traps. *Phys. Rev. A* **57**, (1998) pp. 1497-1500.
7. Stoof H. T. C., Houbiers M., Sackett C. A. and Hulet R. G., Superfluidity of Spin-Polarized ^{6}Li. *Phys. Rev. Lett.* **76**, (1996) pp. 10.
8. Sackett C. A., Bradley C. C., Welling M. and Hulet R. G., Bose-Einstein Condensation of Lithium. *Appl. Phys. B* **65**, (1997) pp. 433-440.
9. Mewes M.-O. *et al.*, Bose-Einstein Condensation in a Tightly Confining dc Magnetic Trap. *Phys. Rev. Lett.* **77**, (1996) pp. 416-419.
10. Abraham E. R. I. *et al.*, Triplet s-wave resonance in ^{6}Li collisions and scattering lengths of ^{6}Li and ^{7}Li. *Phys. Rev. A* **55**, (1997) pp. R3299-R3302.
11. van Abeelen F. A., Verhaar B. J. and Moerdijk A. J., Sympathetic cooling of ^{6}Li atoms. *Physical Review A.* **55**, (1997) pp. 4377.

12. Larson D. J., Bergquist J. C., Bollinger J. J., Itano W. M. and Wineland D. J., Sympathetic Cooling of Trapped Ions: A Laser-Cooled Two-Species Nonneutral Ion Plasma. *Phys. Rev. Lett.* **57**, (1986) pp. 70-73.

13. Myatt C. J., Burt E. A., Ghrist R. W., Cornell E. A. and Wieman C. E., Production of Two Overlapping Bose-Einstein Condensates by Sympathetic Cooling. *Phys. Rev. Lett.* **78**, (1997) pp. 586-589.

14. Bradley C. C., Sackett C. A. and Hulet R. G., Bose-Einstein Condensation of Lithium: Observation of Limited Condensate Number. *Phys. Rev. Lett.* **78**, (1997) pp. 985-989.

15. Butts D. A. and Rokhsar D. S., Trapped Fermi Gases. *Physical Review A* **55**, (1997) pp. 4346-4350.

16. Bagnato V., Pritchard D. E. and Kleppner D., Bose-Einstein Condensation in an External Potential. *Phys. Rev. A* **35**, (1987) pp. 4354-4358.

Quantum Degenerate Bosonic and Fermionic Gases: A ^{7}Li Bose-Einstein Condensate Immersed in a ^{6}Li Fermi Sea

L. Khaykovich, F. Schreck, K. L. Corwin, G. Ferrari *, T. Bourdel, J. Cubizolles and C. Salomon

Laboratoire Kastler Brossel, Ecole Normale Suprieure, 24 rue Lhomond, 75231 Paris CEDEX 05, France

** : LENS-INFM, Largo E.Fermi, 2 Firenze 50125 Italy*

Using sympathetic cooling between fermionic and bosonic lithium atoms in a magnetic trap, we have reached quantum degeneracy for both species [1]. In a first set of experiments we trap both isotopes in their upper hyperfine states where bosonic ^{7}Li atoms have a negative scattering length. We observe a common temperature of $T \sim 1\mu$K, which corresponds to the Bose-Einstein condensation critical temperature T_C while $T/T_F = 0.22(5)$ where T_F is the Fermi temperature of ^{6}Li. In the second set of experiments both isotopes are trapped in their lower hyperfine states. We produce a stable ^{7}Li condensate because of the positive scattering length in this state. With a very small fraction of thermal atoms, the condensate is quasi-pure and coexists with the ^{6}Li Fermi sea. The lowest common temperature is $0.28\,\mu$K $\simeq 0.2(1)\,T_C = 0.2(1)\,T_F$. The ^{7}Li condensate has a one-dimensional character.

1 Introduction

Bose-Einstein condensation of atomic gases has been very actively studied in recent years [2,3]. The dilute character of the samples and the ability to control the atom-atom interactions allowed a detailed comparison with the theories of quantum gases. Atomic Fermi gases, on the other hand, have only been investigated experimentally for two years [4,5,6]. They are predicted to possess intriguing properties and may offer an interesting link with the behavior of electrons in metals and semi-conductors, and the possibility of Cooper pairing [7] such as in superconductors and neutron stars. Mixtures of bosonic and fermionic quantum systems, with the prominent example of ^{4}He - ^{3}He fluids, have also stimulated intense theoretical and experimental activity [8]. This has led to new physical effects including phase separation, influence of the superfluidity of the Bose system on the Fermi degeneracy and to new applications such as the dilution refrigerator [8,9,10].

In this paper, we present a new mixture of bosonic and fermionic systems, a stable Bose-condensed gas of ^{7}Li atoms in internal state $|F = 1, m_F = -1\rangle$ immersed in a Fermi sea of ^{6}Li atoms in $|F = 1/2, m_F = -1/2\rangle$ (fig. 1). Confined in the same magnetic trap, both atomic species are in thermal equilibrium with a temperature of $0.2(1)\,T_F \ll T_C$. All previous experiments performed with ^{7}Li in $|F = 2, m_F = 2\rangle$ had condensate numbers limited to $N \leq 1400$ because

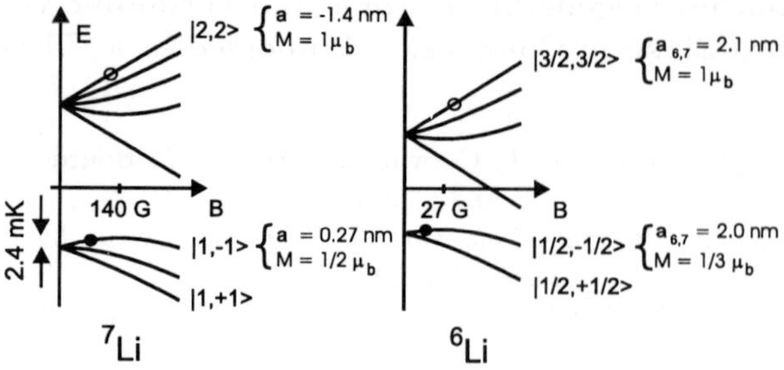

Figure 1: Energy levels of ^{7}Li and ^{6}Li ground states in a magnetic field. Relevant scattering lengths,a, and magnetic moments, M, are given. μ_b is the Bohr magneton. The $|1,-1\rangle$ state (resp. $|1/2,-1/2\rangle$) is only trapped in fields weaker than 140 Gauss (resp. 27 Gauss). Open circles: first cooling stage; black circles: second cooling stage.

of the negative scattering length, $a = -1.4\,\text{nm}$, in this state [11,12]. Our condensate is produced in a state which has a positive, but small, scattering length, $a = +0.27\,\text{nm}$ [13]. The number of condensed atoms is typically 10^4, and BEC appears unambiguously both in the position distribution in the trap and in the standard time of flight images. An interesting feature is the one-dimensional character of this condensate, behaving as an ideal gas in the transverse direction of the trap[14,15].

Because of the symmetrization postulate, colliding fermions have no *s*-wave scattering at low energy. In the low temperature domain of interest, the *p*-wave contribution vanishes. Our method for producing simultaneous quantum degeneracy for both isotopes of lithium is sympathetic cooling [5,6]; *s*-wave collisions between two different atomic isotopes are allowed and RF evaporation selectively removes from the trap high energy atoms of one species. Elastic collisions subsequently restore thermal equilibrium of the two-component gas at a lower temperature.

Our experimental setup has been described in detail previously [5,16]. A mixture of ^{6}Li and ^{7}Li atoms is loaded from a magneto-optical trap into a strongly confining Ioffe-Pritchard trap at a temperature of about $2\,\text{mK}$. As depicted in fig. 1, this relatively high temperature precludes direct magnetic trapping of the atoms in their lower hyperfine state because of the shallow magnetic trap depth, $2.4\,\text{mK}$ for ^{7}Li in $|F=1, m_F=-1\rangle$ and $0.2\,\text{mK}$ for ^{6}Li in $|F=1/2, m_F=-1/2\rangle$. Therefore we proceed in two steps. Both isotopes are first trapped and cooled in their upper hyperfine states. The ^{7}Li $|F=2, m_F=2\rangle$ and ^{6}Li $|F=3/2, m_F=3/2\rangle$ states have no energy maximum as a function of magnetic field (fig. 1). Thus the trap depth can be large. Evaporation is performed selectively on ^{7}Li using a microwave field near 803

MHz that couples $|F=2, m_F=2\rangle$ to $|F=1, m_F=1\rangle$. When both gases are cooled to a common temperature of about $9\,\mu$K, atoms are transferred using a combination of microwave and RF pulses into states $|F=1, m_F=-1\rangle$ and $|F=1/2, m_F=-1/2\rangle$ with an energy far below their respective trap depths. Evaporative cooling is then resumed until ^{7}Li reaches the BEC threshold.

2 Experiments in the upper hyperfine states

In the first series of experiments, both Li isotopes are trapped in their higher HF states. ^{6}Li is sympathetically cooled to Fermi degeneracy by performing 30 seconds of evaporative cooling on ^{7}Li [5]. Trap frequencies for ^{7}Li are $\omega_{\rm rad} = 2\pi * 4000(10)s^{-1}$ and $\omega_{\rm ax} = 2\pi * 75.0(1)s^{-1}$ with a bias field of 2 G. Absorption images of both isotopes are recorded on a single CCD camera with a resolution of $10\,\mu$m. Images are taken quasi-simultaneously (only 1 ms apart) in the trap or after a time of flight expansion. Probe beams have an intensity below saturation and a common duration of $30\,\mu$s. Typical in-situ absorption images in the quantum regime are reported in [1]. The temperature T is typically $1.4(1)\,\mu$K and $T/T_{\rm F} = 0.33(5)$, where the Fermi temperature $T_{\rm F}$ is $(\hbar\bar{\omega}/k_{\rm B})(6N_{\rm F})^{1/3}$, with $\bar{\omega}$ the geometric mean of the three oscillation frequencies in the trap and $N_{\rm F}$ the number of fermions. For images recorded in the magnetic trap, the common temperature is measured from the spatial extent of the bosonic cloud in the axial direction since the shape of the Fermi cloud is much less sensitive to temperature changes when $T/T_{\rm F} < 1$ [17]. The spatial distributions of bosons and fermions are recorded after a 1 sec thermalization stage at the end of the evaporation. As the measured thermalization time constant between the two gases, 0.15 s, is much shorter than 1 s, the two clouds are in thermal equilibrium [a]. Both isotopes experience the same trapping potential. Thus the striking difference between the sizes of the Fermi and Bose gases [6] is a direct consequence of Fermi pressure. The measured axial profiles in [1] are in excellent agreement with the calculated ones for a Bose distribution at the critical temperature $T_{\rm C}$. In our steepest traps, Fermi temperatures as high as $11\,\mu$K with a degeneracy of $T/T_F = 0.36$ are obtained. This $T_{\rm F}$ is a factor 3 larger than the single photon recoil temperature at 671 nm, opening interesting possibilities for light scattering experiments [18].

Our highest Fermi degeneracy in ^{6}Li $|F=3/2, m=3/2\rangle$, achieved by cooling with ^{7}Li $|F=2, m=2\rangle$, is $T/T_{\rm F} = 0.25(5)$ with $T_{\rm F} = 4\,\mu$K, very similar to [6]. We observe that the boson temperature cannot be lowered below

[a] In this measurement, we abruptly prepare an out of equilibrium ^{7}Li energy distribution in the trap using the microwave evaporation knife and measure the time needed to restore thermal equilibrium through the evolution of the axial size of the ^{7}Li and ^{6}Li clouds.

T_C. Indeed, because of the negative scattering length in ^{7}Li $|F=2, m=2\rangle$, for our trap parameters, collapse of the condensate occurs when its number reaches ~ 300 [11]. Since sympathetic cooling stops when the heat capacity of the bosons becomes lower than that of the fermions, this limits the Fermi degeneracy to about 0.3 [6].

3 Experiments in the lower hyperfine states

In order to explore the behavior of a Fermi sea in the presence of a BEC with a temperature well below T_C, we perform another series of experiments with both isotopes trapped in their lower HF states where the positive ^{7}Li scattering length (fig. 1) allows the formation of a stable BEC with high atom numbers. To avoid large dipolar relaxation, ^{6}Li must also be in its lower HF state [19]. First, sympathetic cooling down to $\sim 9\,\mu K$ is performed on the ^{7}Li $|F=2, m=2\rangle$, ^{6}Li $|F=3/2, m=3/2\rangle$ mixture as before. Then, to facilitate state transfer, the trap is adiabatically opened to frequencies $\omega_{\text{rad}} = 2\pi * 100\,s^{-1}$ and $\omega_{\text{ax}} = 2\pi * 5\,s^{-1}$ (for ^{7}Li, F=2). The transfer of each isotope uses two

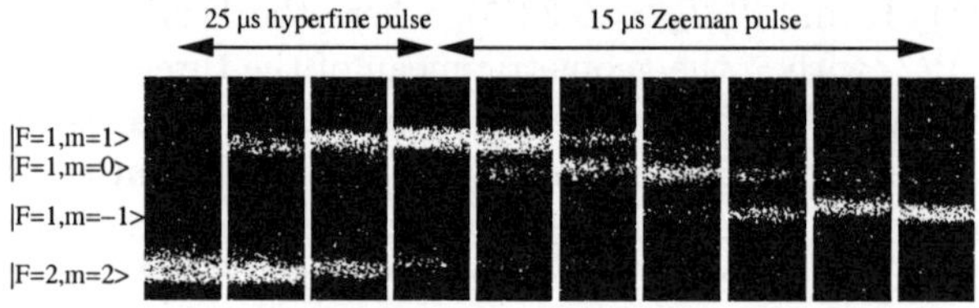

Figure 2: Transfer of ^{7}Li atoms from $|F=2, m=2\rangle$ to $|F=1, m=-1\rangle$. The different hyperfine states are spatially separated in a Stern and Gerlach experiment. Up to 70 % of the atoms are transferred.

microwave π pulses (see fig. 2). The first pulse at 803 MHz for ^{7}Li (228 MHz for ^{6}Li) transfers the bosons from $|2,2\rangle$ to $|1,1\rangle$ (the fermions from $|3/2,3/2\rangle$ to $|1/2,1/2\rangle$). These states are magnetically untrapped states (see fig. 1). The second π pulse at 1 MHz for ^{7}Li (1.3 MHz for ^{6}Li) transfers the bosons to $|1,-1\rangle$, a magnetically trapped state (the fermions to $|1/2,-1/2\rangle$). Adiabatic opening of the trap cools the cloud. It decreases the energy broadening of the resonance and gives more time for the passage through untrapped states. The duration of the π pulses are 17 μs and 13 μs and more than 70% of each isotope are transferred. Finally the trap is adiabatically recompressed to the steepest confinement giving $\omega_{\text{rad}} = 2\pi * 4970(10)s^{-1}$ and $\omega_{\text{ax}} = 2\pi * 83(1)s^{-1}$ for ^{7}Li $|F=1, m=-1\rangle$, compensating for the reduced magnetic moment.

Because of the very large reduction of the ^{7}Li s-wave scattering cross section from the F=2 to the F=1 state (a factor ~ 27 [19]), we were unable to reach

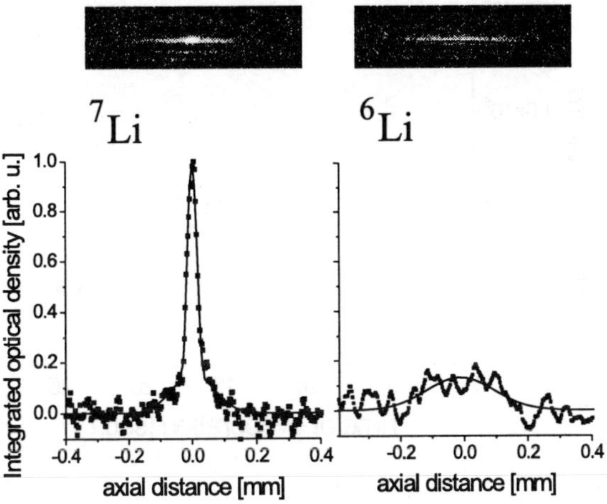

Figure 3: Mixture of Bose and Fermi gases. In situ spatial distributions after sympathetic cooling. The characteristic Bose condensed peak is surrounded by very few thermal atoms : the condensate is *quasi-pure*. The Fermi distribution is wider because of the smaller magnetic moment and Fermi pressure. The barely detectable thermal cloud indicates a temperature of $\simeq 0.28\,\mu\mathrm{K} \simeq 0.2(1)\,T_{\mathrm{C}} = 0.2(1)\,T_{\mathrm{F}}$.

runaway evaporation with ^{7}Li atoms alone in $F = 1$. In contrast, the ^{6}Li/^{7}Li cross section is $\sim$ 27 times higher than the ^{7}Li/^{7}Li one [19,13]. We therefore use ^{6}Li atoms as a mediating gas to increase the thermalization rate of both gases. Two different methods were used to perform the evaporation. The first consists in using two RF ramps on the HF transitions of ^{6}Li (from $|1/2, -1/2\rangle$ to $|3/2, -3/2\rangle$) and ^{7}Li (from $|1, -1\rangle$ to $|2, -2\rangle$), which we balanced to maintain roughly equal numbers of both isotopes. After 10 s of evaporative cooling, Bose-Einstein condensation of ^{7}Li occurs together with a ^{6}Li degenerate Fermi gas (fig. 3). Surprisingly, a single 25 s ramp performed only on ^{6}Li achieved the same results. In this case the equal number condition was fulfilled because of the reduced lifetime of the ^{7}Li cloud that we attribute to dipolar collisional loss [19]. The duration of the RF evaporation was matched to this loss rate (see fig. 4). In the following we concentrate on this second, and simpler, evaporation scheme, sympathetic cooling of ^{7}Li by evaporative cooling of ^{6}Li.

In fig. 3 in-situ absorption images of bosons and fermions at the end of the evaporation are shown. The bosonic distribution shows the typical double structure: a strong and narrow peak forms the condensate at the center, surrounded by a much broader distribution, the thermal cloud. As the Fermi distribution is very insensitive to temperature, this thermal cloud is a very useful

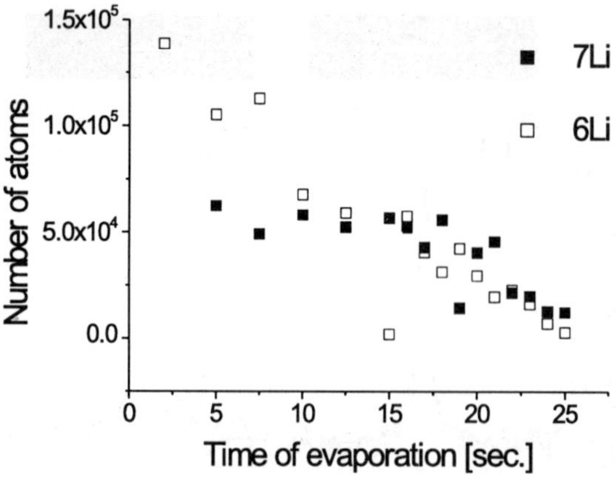

Figure 4: Evolution of the number of atoms of each isotope during the evaporation ramp. The ^{6}Li atoms are removed by a microwave knife coupling $|1/2, -1/2\rangle$ and $|3/2, -3/2\rangle$, while the loss of ^{7}Li is due to dipolar relaxation[19].

tool for the determination of the common temperature. Note that, as cooling was only performed on ^{6}Li atoms, the temperature measured on ^{7}Li cannot be lower than the temperature of the fermions. Measuring N_B, N_F, the condensate fraction N_0/N_B, and $\bar{\omega}$, we determine the quantum degeneracy of the Bose and Fermi gases. In fig. 3, the condensate is quasi-pure; $N_0/N_B = 0.77$; the thermal fraction is near our detectivity limit, indicating a temperature of $\simeq 0.28\,\mu\text{K} \leq 0.2\,T_C = 0.2(1)\,T_F$ with $N_B = 10^4$ bosons and $4\,10^3$ fermions. The condensate fraction N_0/N_B as a function of T/T_C is shown in fig. 5 (a), while the size of the Fermi gas as a function of T/T_F is shown in fig. 5 (b). With the strong anisotropy ($\omega_{\text{rad}}/\omega_{\text{ax}} = 59$) of our trap, the theory including anisotropy and finite number effects differs significantly from the thermodynamic limit[3], in agreement with our measurements even though there is a 20% systematic uncertainty on our determination of T_C and T_F. We have also obtained samples colder than those presented in fig. 5, for which the ^{7}Li thermal fraction is below our detectivity floor, indicating $T < 0.2\,T_C \simeq 0.2\,T_F$. Clearly a more sensitive thermal probe is required now to investigate this temperature domain. An elegant method relies on the measurement of thermalization rates with impurity atoms including Pauli blocking[20,21].

Because of the small scattering length, this ^{7}Li condensate has interesting properties. Time of flight images, performed after expansion times of 0-10 ms

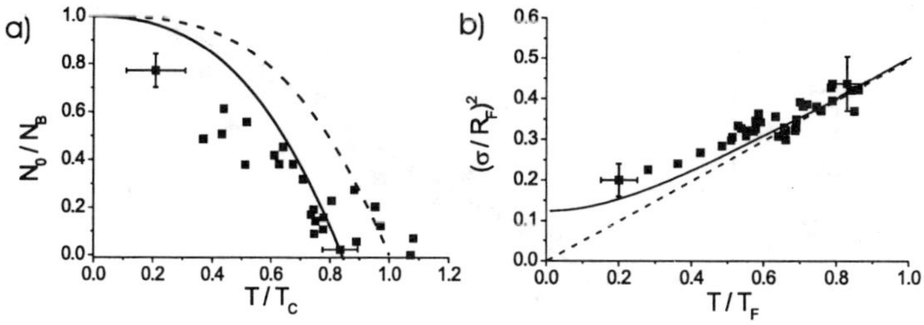

Figure 5: Temperature dependence of mixtures of quantum gases: a) normalized BEC fraction as a function of T/T_C. Dashed line: theory in the thermodynamic limit. Solid line: theory including finite size and trap anisotropy ; b) fermion cloud size: variance of gaussian fit divided by the square of Fermi radius $R_F^2 = 2k_B T_F / M\omega_{ax}^2$ as a function of T/T_F. Solid line: theory. Dashed line: Boltzmann gas.

with $N_0 = 10^4$ condensed atoms, reveal that the condensate is one-dimensional (1D). In contrast to condensates in the Thomas-Fermi (TF) regime, where the release of interaction energy leads to a fast increase in radial size, our measurements agree to better than 5% with the time development of the radial ground state wave function in the harmonic magnetic trap (fig. 6). This behavior is expected when the chemical potential μ satisfies $\mu < \hbar\omega_{rad}$ [14]. Searching for the ground state energy of the many-body system with a Gaussian ansatz radially and TF shape axially [14], we find that the mean-field interaction increases the size of the Gaussian by $\approx 3\%$. The calculated TF radius is 28 μm or 7 times the axial harmonic oscillator size and is in good agreement with the measured radius, 30 μm in fig. 3. Thus with $\mu = 0.45\,\hbar\omega_{rad}$, the gas is described as an ideal gas radially but is in the TF regime axially. This 1D situation has been also realized recently in sodium condensates[15]. As $\mu < \hbar\omega_{rad}$ implies that the linear density of a 1D condensate is limited to $\simeq 1/a$, the 1D regime is much easier to reach with ^{7}Li (small a) than with Na or ^{87}Rb which have much larger scattering lengths.

What are the limits of this BEC-Fermi gas cooling scheme? First, the $1/e$ condensate lifetime of about 3 s in this steep trap will limit the available BEC-Fermi gas interaction time. Second, the boson-fermion mean field interaction can induce a spatial phase separation [9] that prevents thermal contact between ^{7}Li and ^{6}Li. Using the method of [9] developed for $T = 0$, we expect, for the parameters of fig. 3 (top), that the density of fermions is only very slightly modified by the presence of the condensate in accordance with our observations. Third, because of the superfluidity of the condensate, impurity atoms

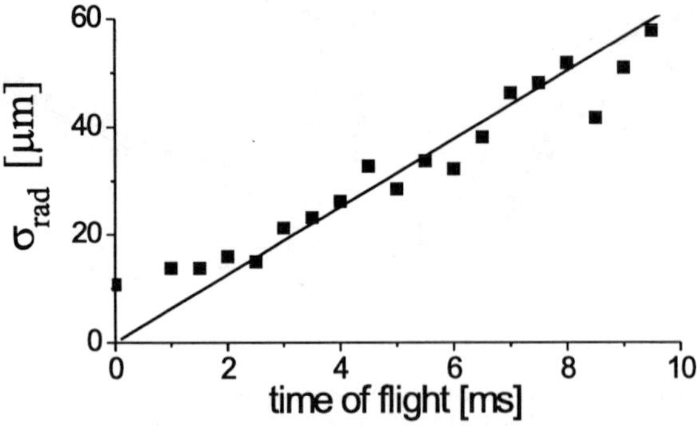

Figure 6: Signature of 1D condensate. Radial size of expanding condensates with 10^4 atoms as a function of time of flight. The straight line is the expected behavior for the expansion of the ground state radial harmonic oscillator.

(such as ^{6}Li), which move through the BEC slower than the sound velocity v_c, are no longer scattered [10,22]. When the Fermi velocity v_F becomes smaller than v_c, cooling occurs only through collisions with the bosonic thermal cloud, thus slowing down drastically. With 10^4 condensed atoms, $v_c \simeq 0.9\,\mathrm{cm/s}$. The corresponding temperature where superfluid decoupling should occur is $\simeq 100\,\mathrm{nK}$, a factor 3 lower than our currently measured temperature.

In summary, we have produced a new mixture of Bose and Fermi quantum gases. Future work will explore the degeneracy limits of this mixture. Phase fluctuations of the 1D ^{7}Li condensate should also be detectable via density fluctuations in time of flight images, as recently reported [23]. The transfer of the BEC into $|F = 2, m = 2\rangle$ with negative a should allow the production of bright solitons and large unstable condensates where interesting and still unexplained dynamics has been recently observed [12,24]. Finally, the large effective attractive interaction between ^{6}Li $|F = 1/2, m_F = +1/2\rangle$ and $|F = 1/2, m_F = -1/2\rangle$ makes this atom an attractive candidate for searching for BCS pairing at lower temperatures [7].

We are grateful to Y. Castin, J. Dalibard, C. Cohen-Tannoudji, and G. Shlyapnikov for useful discussions. F. S., and K. C. were supported by a fellowship from the DAAD and by MENRT. Work supported by CNRS, Collège de France and Région Ile de France. Laboratoire Kastler Brossel is *Unité de recherche de l'Ecole Normale Supérieure et de l'Université Pierre et Marie Curie, associée au CNRS.*

Bibliography

1. F. Schreck et. al., (2001), *Phys. Rev. Let.* **87**, 80403.
2. Proc. of the Int. School of Phys. "Enrico Fermi", M. Inguscio, S. Stringari and C. E. Wieman eds, (IOS Press, Amsterdam 1999).
3. Dalfovo, F., Giorgini, S., Pitaevskii, L. P., (1999), *Rev. Mod. Phys.* **71**, 463.
4. DeMarco, B. and Jin, D. S., (1999), *Science* **285**, 1703.
5. Schreck, F., et. al., (2001), *Phys. Rev. A*, **64**, 011402R.
6. Truscott, A. G. et. al., (2001), *Science*, **291**, 2570.
7. Stoof, H. T. C. and Houbiers, M in ref. 2, p.537; Micnas, R et. al., (1990), *Rev. Mod. Phys.*, **62**, 113.
8. Ebner, C. and Edwards, D. O., (1970), *Phys. Rep.*, **2C**, 77.
9. Mølmer, K., (1998), *Phys. Rev. Lett.*, **80**, 1804.
10. Timmermans, E. and Côté, R., (1998), *Phys. Rev. Lett.*, **80**, 3419.
11. Bradley, C. C., Sackett, C. A. and Hulet, R. G. ,(1997), *Phys. Rev. Lett.*, **78**, 985.
12. Gerton, J. M. et. al., (2000), *Nature*, **408**, 692.
13. R.G. Hulet, private communication.
14. Petrov, D. S., Shlyapnikov, G. V. and Walraven, J. T. M., (2000), *Phys. Rev. Lett.*, **85**, 3745.
15. Görlitz, A. et. al., (2001) *cond-mat/0104549.*
16. Mewes, M. -O, et. al., (2000), *Phys. Rev. A*, **61**, 011403R.
17. Butts, D. A. and Rokhsar, D. S., (1997) *Phys. Rev. A*, **55**, 4346.
18. Busch, T. et. al., (1998), *Europhys. Lett.*, **44**, 755, DeMarco, B. and Jin, D. S., (1998) *Phys. Rev. A*, **58**, 4267.
19. Van Abeelen,F. A., Verhaar, B. J. and Moerdijk, A. J., (1997, *Phys. Rev. A*, **55**, 4377.
20. Ferrari, G. (1999), *Phys. Rev. A.*, **59**, R4125.
21. DeMarco, B., Papp, S. B. and Jin, D. S., (2001), *Phys. Rev. Lett.*, **86**, 5409.
22. Chikkatur, A. P. et. al., (2000), *Phys. Rev. Lett.*, **85**, 483.
23. Dettmer, S. et. al., (2001), *cond-mat/0105525.*
24. Roberts, J. L. et. al., (2001), *Phys. Rev. lett.*, **86**, 4211.

OPTICAL TRAPPING OF A TWO-COMPONENT FERMI GAS

J. E. THOMAS, S. R. GRANADE,
M. E. GEHM, M.-S. CHANG, AND K. M. O'HARA
Physics Department, Duke University, Durham, NC 27708-0305, USA
E-mail: jet@phy.duke.edu

Stable, strongly attractive, two-component mixtures of lithium fermions are confined and evaporatively cooled in an ultrastable optical trap. The optical trap has a lifetime of 370 seconds with a measured residual heating rate of 6 nK/sec, and is the first optical trap to achieve a background gas limited lifetime at 10^{-11} Torr. After 60 seconds of evaporation, a final temperature corresponding to 2.2 T_F is obtained, where T_F is the Fermi temperature. We describe the physics of evaporation in time-dependent optical traps and our progress toward achieving degeneracy in a two-component mixture which is suitable for studies of superfluidity in a Fermi gas.

1 Introduction

Trapping and cooling of neutral fermionic atoms offers exciting prospects for precision fundamental studies of interactions and collective behavior in Fermi gases for which the interaction strength, density, and temperature can be experimentally controlled. However, at low temperatures, where s-wave scattering is dominant, the Pauli exclusion principle forbids scattering in a single-component Fermi gas. Hence, to study interactions and to exploit evaporative cooling, it is necessary to trap a two-component gas. Recently, evaporation to degeneracy has been accomplished in fermionic ^{40}K by magnetically trapping two different spin components. [1] Sympathetic cooling of fermionic ^{6}Li to degeneracy also has been accomplished by using mixtures with bosonic ^{7}Li contained in the same magnetic trap. [2,3] These experiments have been used to dramatically illustrate the effects of Fermi statistics in a degenerate one-component Fermi gas.

In certain two-component Fermi gases with strongly attractive interactions, i.e., large and negative scattering lengths, a cold gas analog of superfluidity is predicted to occur at experimentally accessible temperatures. [4] Especially interesting are certain stable two-state mixtures of ^{6}Li and ^{40}K which have very strong, magnetically tunable, attractive interactions. Unfortunately, none of the stable attractive mixtures can be confined in a magnetic trap, such as used to study Bose gases, since the required spin states are repelled. For this reason, optical traps are essential for studies of superfluidity in Fermi gases.

2 Ultrastable CO_2 Laser Trap

To enable trapping of arbitrary mixtures of fermions, we have developed an all-optical trap based on an ultrastable CO_2 laser. [5] This trap consists of a single focused beam with $1/e^2$ intensity radii of 47 μm and 67 μm respectively. With 36 W at the focus of the trap, the intensity is $\simeq 1$ MW/cm^2, producing a trap depth of 0.3 mK for ^{6}Li. This trap confines atoms in a nearly spin-state-independent potential with a trap lifetime of 370 seconds. [5,6]

In our experiments, a 50-50 mixture of the two lowest spin states of ^{6}Li is directly loaded into the trap from a ^{6}Li magneto-optical-trap (MOT). Typically, a total of $5 \times 10^5 - 8 \times 10^5$ atoms are loaded into the CO_2 laser trap at a temperature of $150 - 200\,\mu$K. Mixtures of the two lowest spin states have a negligible scattering length at magnetic fields up to a few gauss, but the scattering length increases to more than -300 a_0 for magnetic fields of a few hundred gauss and exhibits a Feshbach resonance near 800 gauss. Another stable mixture can be created from a mixture of the two lowest states by using a pair of optical fields to induce a Raman π-pulse which transfers one spin component to a higher lying state. The stable two-component gas so produced exhibits a scattering length of up to -1600 a_0 at low magnetic field and enables rapid evaporative cooling. [6] These mixtures are very well suited for fundamental studies of interaction dynamics and superfluidity in degenerate Fermi gases. Direct evaporative cooling in the CO_2 laser trap can be used to produce two-state degenerate mixtures.

2.1 Residual Heating

For evaporative cooling over long time periods, it is important that residual heating be minimized. Optical heating is negligible, since the optical scattering rate scales as the cube of the laser frequency. For our CO_2 laser trap, which operates at a wavelength of $10.6\,\mu$m, the scattering rate at the trap focus is only 1 photon/hour, corresponding to a heating rate of 8 pK/sec which is entirely negligible. We have investigated additional sources of heating arising from laser intensity and pointing noise, [7,8] as well as from diffractive background gas collisions. [9]

To measure the residual heating rate, we employ evaporative cooling in a mixture of the two lowest spin states of ^{6}Li. These states are well suited for trap diagnostics, since elastic scattering can be turned on and off simply by applying or not applying a bias magnetic field. By reversing the polarity of one of the MOT gradient coils, we obtain a bias magnetic field of 130 G, producing a scattering length of $-130\,a_0$. In this case, rapid evaporation

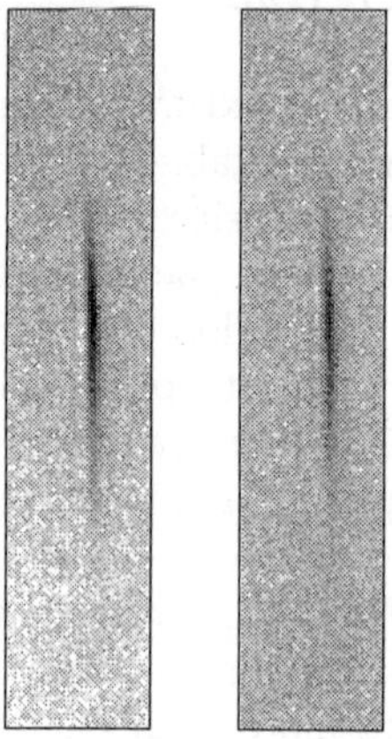

Figure 1. Absorption images of atoms in the CO_2 laser trap at 0 seconds (left) and 200 seconds (right) after evaporation is halted at an initial temperature of 30 μK.

occurs until the thermal energy drops to about 1/10 of the trap depth, or about 30 μK. Forced evaporation is achieved by lowering the trap depth as described below.

To determine the net residual heating rate at the full trap depth of 0.33 mK, we simply use evaporation to lower the temperature of the trapped atoms to 30 μK. Then, the bias magnetic field is turned off to stop the evaporation. Fig. 1 shows absorption images of the spatial distribution of the trapped atoms at 0 seconds and 200 seconds after evaporation is halted. From the measured 1/e widths of these spatial distributions, we estimate that the residual heating rate is 6 nK/sec. Since the elastic scattering cross section at low magnetic field is nearly zero, evaporation is negligible. Hence, the measured very small heating rate shows that trap loss arises from background gas collisions in the ultrahigh vacuum system. To our knowledge, this is the first demonstration of an all-optical atom trap which achieves a background gas limited lifetime at 10^{-11} Torr. Parameters for our trap are summarized in Table 1.

3 Forced Evaporative Cooling in Optical Traps

Evaporation in an optical trap of fixed depth is strongly suppressed when the thermal energy kT is approximately 1/10 of the trap depth U. At such low temperatures, the number of colliding pairs of atoms with enough energy for

Measured Lifetime	370 sec
Well depth	0.33 mK
Background pressure	$< 10^{-11}$ Torr
Measured residual heating rate	< 6 nK/sec

Table 1. CO_2 Laser Trap Parameters

one to escape the trap is suppressed by the Boltzmann factor $\exp(-U/kT)$, which is quite small. Further cooling is achieved by forced evaporation.

Forced evaporative cooling is accomplished by lowering the depth of the optical trap by reducing the trap laser intensity. To estimate the increase in phase-space density as the trap depth is lowered, we have developed a simple scaling law model based on energy conservation which includes the energy losses arising both from evaporation and from the decrease in potential energy as the trap is lowered. The latter accounts for the decrease in temperature by adiabatic cooling which preserves the phase-space density in the absence of evaporation. Previous scaling law models for evaporation in magnetic traps using radio-frequency-knife methods [10] have not considered the adiabatic energy loss. This is appropriate for magnetic traps, where the trap strength is constant and the adiabatic contribution is negligible, as shown by a Boltzmann equation model with a time-dependent evaporation threshold. [11]

Scaling laws for the number of trapped atoms N, the elastic collision rate γ, and the phase-space density ρ are derived by considering the total energy loss from the trap. Defining the zero of energy at the bottom of the trap, an evaporating atom carries away an energy between U and $U + kT$, where the energy spread is a consequence of the Boltzmann factor. Hence, the energy loss rate by evaporation can be conservatively estimated as $\dot{N}U$, where $\dot{N}$ is the evaporation rate of the atoms. An additional energy loss occurs as the trap is lowered by reducing the trap laser intensity. Assuming that the thermal energy is small compared to the trap depth, i.e., $\eta \equiv U/kT >> 1$, the atoms vibrate near the bottom of the trap in a potential which is approximately harmonic. In this case, the average potential energy is $E/2$, where E is the sum of the kinetic and potential energy. Then, the rate of potential energy loss is $(\dot{U}/U)E/2$, where $\dot{U}$ is the rate at which the trap depth is lowered. Hence, the net the rate of energy loss from the trap obeys the approximate evolution equation,

$$\dot{E} = U\dot{N} + \frac{\dot{U}}{U}\frac{E}{2}. \tag{1}$$

It can be shown that Eq. 1 follows with the same approximations from the s-wave Boltzmann equation with a time-dependent harmonic potential, assuming sufficient ergoticity and neglecting background gas collisions. [12]

In the classical limit, $E = 3NkT$ is the total energy of the trapped gas so that $\dot{E} = 3Nk\dot{T} + 3\dot{N}kT$. Then, the contribution to $\dot{T}$ from evaporation is proportional to $\dot{N}(U - 3kT)$. Hence, the corresponding cooling rate is proportional to the difference between the average energy carried away per particle ($\simeq U$) and the average thermal energy $3kT$, as it should be. Solving Eq.1 with fixed value of $U/kT = \eta$, we obtain

$$\frac{N_f}{N_i} = \left(\frac{U_f}{U_i}\right)^{\frac{3}{2(\eta-3)}} \tag{2}$$

The phase space density in the classical regime is $\rho = N(h\nu)^3/(kT)^3$, where $\nu \propto \sqrt{U}$ is the geometric mean of the trap oscillation frequencies. Using Eq. 2, it is easy to show that ρ scales with number and trap depth as

$$\frac{\rho_f}{\rho_i} = \left(\frac{N_i}{N_f}\right)^{\eta-4} = \left(\frac{U_i}{U_f}\right)^{\frac{3}{2}\frac{\eta-4}{\eta-3}}, \tag{3}$$

Similarly, for an energy-independent scattering cross section σ, the elastic collision rate $\gamma \propto N\sigma\nu^3/(kT)$ scales as

$$\frac{\gamma_f}{\gamma_i} = \left(\frac{U_f}{U_i}\right)^{\frac{\eta}{2(\eta-3)}} \tag{4}$$

An important feature of Eq. 3 is that the increase in phase space density with decreasing number is identical to that obtained using an radio-frequency-knife method with a fixed trap depth and the same value of η. This is a consequence of the adiabatic energy loss which ensures that the phase-space density does not change as the trap depth is lowered in the absence of evaporation. For an optical trap with $\eta = 10$, only a modest decrease in number is needed to greatly increase the phase space density. Unlike evaporation from a fixed well, however, the collision rate for a fixed cross section decreases as $(U/U_i)^{5/7}$ for $\eta = 10$, and runaway evaporation is not achieved. Nevertheless, for ^{6}Li, where the scattering length is anomalously large, evaporation is still rapid and background collision induced loss can be minimized despite the reduced collision rate. This is confirmed by detailed modeling using a Boltzmann equation which includes Fermi statistics and background gas collisions. [12] In some cases, where the scattering cross section is unitarity limited and inversely proportional to the relative kinetic energy, one expects an average collision cross section $\propto 1/U$, and runaway evaporation may be achievable.

4 Evaporative Cooling of Fermions

It might appear that the evaporation will become inefficient as the temperature is lowered toward degeneracy, since Fermi statistics leads to a well-known suppression of the classical collision rate. Neglecting Fermi statistics, at low temperatures, collisions occur at the classical rate γ_{cl}, where γ_{cl} is evaluated at the Fermi energy. Including statistics, for $T << T_F$, the collision rate within the trap is reduced to $\Gamma \propto \gamma_{cl}\,(T/T_F)^2$, where T is the temperature and T_F is the Fermi temperature. The suppression factor $(T/T_F)^2$ is a consequence of Pauli blocking, which forbids collisions into occupied energy states as observed recently [13]. However, in evaporation, one of the final states is essentially unoccupied, since it is outside the trap. Hence, the evaporation rate is suppressed by only T/T_F, i.e.,

$$\Gamma_{Evap} \propto \gamma_{cl}\,\frac{T}{T_F}\,\exp\left(-\frac{U - k_B T_F}{k_B T}\right).$$

The exponential factor describes the high energy tail of the Fermi distribution which is responsible for evaporation when the trap depth $U >> k_B T$. This is essentially the same factor that appears in the evaporation of a classical gas. Since the heat capacity also scales as T/T_F, [14] the efficiency of evaporative cooling in lowering the temperature is not seriously compromised in a two-component Fermi gas. Further, the collision rate within the trap is always fast compared to the evaporation rate when $U - k_B T_F >> k_B T$, i.e., $T/T_F >> \exp\left(-\frac{U-k_B T_F}{k_B T}\right)$ so that rethermalization is faster than evaporation. This physical picture explains why Pauli blocking does not appear to strongly affect the rate of increase of phase space density in a recent theoretical model of evaporation for a two-component Fermi gas. [15]

5 ^{6}Li Evaporative Cooling Experiments

In our experiments, the CO_2 laser trap is loaded and then evaporatively cooled in two stages. To increase the number of atoms loaded, the intensity of the trap is increased by recollimating and retroreflecting the trap laser beam. The retroreflected beam is cross-polarized to minimize the standing-wave component and to aid in decoupling the back-reflected beam from the trap laser. An acousto-optic modulator is used to control the overall trap laser power and to suppress feedback of the back-reflected beam into the trap laser.

The ^{6}Li MOT is loaded from a Zeeman-slowed beam for 5 seconds, after which the MOT-trapping beams are lowered in intensity and tuned close to resonance to achieve Doppler-limited cooling to a temperature between 150 and 200 μK. The retroreflected beam increases the trap depth by approximately 50%, increasing the number of atoms loaded by a factor of 3. In this way, up to 800,000 atoms are loaded in a 50-50 mixture of the two lowest hyperfine states. Immediately after the loading stage, a bias magnetic field of 130 G is introduced to initiate evaporation as described above. Rapid evaporation occurs over a second or two during which the retroreflected beam is adiabatically blocked using an air-dashpot controlled pick-off mirror which diverts power smoothly into a 100 W power meter. After this first stage of evaporation, up to 400,000 atoms are confined in the focus of a single CO_2 laser beam, at a trap depth of 0.3 mK, and an initial temperature of 30 μK. The phase-space density for each spin state is $\simeq 3 \times 10^{-3}$ at this point.

Further evaporation is accomplished by adiabatically lowering the trap laser intensity via the acousto-optic modulator. In the initial experiments, we employ an exponential reduction of the intensity with a time constant of 35 seconds. Absorption images of the trap density distribution as a function of time are shown in Fig. 2. Note that the images show the spatial distributions after recompression of the trap to the maximum depth of 0.3 mK. The phase-space density increases for the first 60 seconds reaching a maximum of 2×10^{-2} at 12.6 μK in the recompressed well. This temperature corresponds to $T/T_F =$ 2.2. We believe that residual noise in the trap lowering electronics is causing an increase in the fractional intensity fluctuation as the trap beam intensity is further lowered. Currently, we are eliminating this source of residual heating.

6 Conclusions

In conclusion, we have developed a stable CO_2 laser trap which confines arbitrary spin states of neutral atoms for hundreds of seconds. This is the first all-optical trap to achieve background gas limited lifetime at 10^{-11} Torr. A new scaling law for evaporative cooling in optical traps has been presented which determines how the number, collision rate, and phase-space density scale with trap depth as the trap is lowered. Evaporative cooling of a mixture of the two lowest spin states of ^{6}Li in the CO_2 laser trap has been demonstrated, and a temperature of $2.2\,T_F$ is achieved. By applying a much larger bias magnetic field, this system will be used to explore superfluidity in this two-state fermionic mixture in the region of the Feshbach resonance.

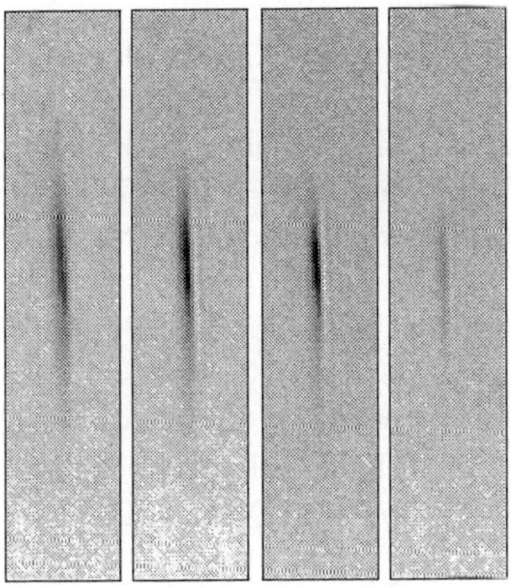

Figure 2. Absorption images of evaporatively cooled atoms in the CO_2 laser trap between 0 and 90 seconds after trap lowering is initiated.

Acknowledgments

This research was supported by the Army Research Office, the National Science Foundation, and the National Aeronautics and Space Administration.

References

1. B. DeMarco and D. S. Jin, *Science* **285**, 1703 (1999).
2. A. G. Truscott, K. E. Strecker, W. I. McAlexander, G. B. Patridge, and R. G. Hulet, *Science*, **291**, 2570-2572 (2001).
3. F. Schreck, G. Ferrari, K. L Corwin, J. Cubizolles, L. Khaykovich, M.-O. Mewes, and C. Salomon, Phys. Rev. A **64**, 011402 (R), (2001).
4. H. T. C. Stoof, M. Houbiers, C. A. Sackett, and R. G. Hulet,Phys. Rev. Lett. **76**, 10 (1996); See also, M. Houbiers, R. Ferwerda, H. T. C. Stoof, W. I. McAlexander, C. A. Sackett, and R. G. Hulet, Phys. Rev. A **56**, 4864 (1998).
5. K. M. O'Hara, S. R. Granade, M. E. Gehm, T. A. Savard, S Bali, C. Freed and J. E. Thomas, Phys. Rev. Lett. **82**, 4204 (1999).
6. K. M. O'Hara, M. E. Gehm, S. R. Granade, S. Bali, and J. E. Thomas, Phys. Rev. Lett. **85**, 2092 (2000).
7. T. A. Savard, K. M. O'Hara, and J. E. Thomas, Phys. Rev. A **56**, R1095

(1997).
8. M. E. Gehm, K. M. O'Hara, T. A. Savard, and J. E. Thomas, Phys. Rev. A f58, 3914 (1998).
9. S. Bali, K. M. O'Hara, M. E. Gehm, S. R. Granade, and J. E. Thomas, Phys. Rev. A Rapid Comm. **60**, R29 (1999).
10. K. B. Davis, M.-O. Mewes, and W. Ketterle, Appl. Phys. B **60**, 155 (1995).
11. K. Berg-Sorensen, Phys. Rev. A **55**, 1281 (1997); Erratum: Phys. Rev. A **56**, 3308 (1997).
12. For a discussion of evaporative cooling in an optical trap, see K. M. O'Hara, Ph.D. Thesis, (Duke University, 2000), unpublished.
13. B. DeMarco, S. B. Papp, and D. S. Jin, Phys. Rev. Lett. **86**, 5409 (2001).
14. Charles Kittel, *Thermal Physics* (Wiley, New York, 1969), pp. 230-234.
15. M. J. Holland, B. DeMarco, and D. S. Jin, Phys. Rev. A **61**, 053610 (2000).

ATOMIC COLLISIONS IN TIGHTLY CONFINED ULTRA-COLD GASES

D.S. PETROV[1,2], M.A. BARANOV[1,2], AND G.V. SHLYAPNIKOV[1,2,3]

[1]*FOM Institute for Atomic and Molecular Physics, Kruislaan 407, 1098 SJ Amsterdam, The Netherlands*

[2]*Russian Research Center Kurchatov Institute, Kurchatov Square, 123 182 Moscow, Russia*

[3]*Laboratoire Kastler Brossel,* 24 rue Lhomond, F-75231 Paris Cedex 05, France*

tel 31-20-608-13-38, fax 31-20-668-41-06, e-mail shlyapnikov@amolf.nl

We discuss pair interatomic collisions in an ultracold gas tightly confined in one (axial) direction and identify two regimes of scattering: the quasi2D regime and the confinement-dominated 3D regime. The latter case is realized at temperatures $T \sim \hbar\omega_0$, where ω_0 is the frequency of the tight confinement. In this regime the confinement can change the momentum dependence of the scattering amplitudes. The quasi2D regime requires temperatures $T \ll \hbar\omega_0$, and in this case the scattering rates exhibit 2D features of the particle motion. We then show that there are promising prospects for achieving the BCS superfluid phase transition in quasi2D two-component Fermi gases.

1 Introduction

The discovery of Bose-Einstein condensation in trapped gases of alkali atoms [1] stimulated a search for non-trival trapping geometries for neutral atoms. Present facilities allow one to (tightly) confine the motion of particles in one direction to zero point oscillations and thus create a gas which is kinematically two-dimensional (2D). For example, in the Stanford [2,3] and ENS [4,5] experiments with optically trapped Cs about 90% of atoms were accumulated in the ground state of the harmonic oscillator potential in the direction of the tight confinement. Collisional properties of such 2D gases and quantum-degenerate regimes in these systems are expected to be drastically different from those in ordinary 3D samples [6]. Recently, several groups [7,8,9] have realized 2D and 1D regimes for trapped condensates.

One can also think of achieving the 2D and 1D regimes for quantum degenerate Fermi gases. For example, optical confinement of the gas in one direction can be made so tight that the corresponding spacing between the trap levels exceeds the Fermi energy ε_F. Then the particles of the degenerate Fermi gas predominantly occupy only the ground state of the tightly confining potential and the gas becomes kinematically 2D. This option seems realistic in view of the recent success in cooling 3D fermionic gases of K [10] and Li [11,9] to well below the temperature of quantum degeneracy.

In this paper we study pair interatomic collisions in an ultracold gas tightly confined in one (axial) direction. We then show that such fermionic (two-component) gases are promising for achieving the superfluid BCS transition.

2 Regimes of scattering in a tightly confined gas

We will confine ourselves to the case of the s-wave scattering, which is generally the leading scattering channel in the ultracold limit. At temperatures T of the order of or smaller than the frequency ω_0 of the tight axial confinement, the harmonic oscillator length $l_0 = (\hbar/m\omega_0)^{1/2}$ (m is the atom mass) plays the role of the axial de Broglie wavelength. We assume that l_0 greatly exceeds the characteristic radius of interatomic interaction R_e, which ensures the 3D character of the relative motion of particles at distances $\sim R_e$. Then the (s-wave) scattering problem involves two distance scales: l_0 and the 3D scattering length a.

We identify two regimes of scattering. At $T \ll \hbar\omega_0$ only the ground state of the axial harmonic oscillator is occupied, and one has a quasi2D regime. In this case, the 2D character of the relative motion of particles at large separation between them, manifests itself in a logarithmic energy dependence of the scattering amplitude. For $a < 0$, we observe resonances in the dependence of the elastic scattering rate on a. This is quite different from the 3D case where the scattering rate always increases with a^2. These resonances originate from the presence of the quasi2D bound state. The corresponding binding energy ϵ_0 depends exponentially on the frequency ω_0 and is given by

$$\epsilon_0 = 0.915(\hbar\omega_0/\pi)\exp\left(-\sqrt{2\pi}l_0/|a|\right). \tag{1}$$

Our calculations show that in order to get to the quasi2D regime of scattering one should at least reach temperatures $T < 0.1\hbar\omega_0$.

At temperatures $T \sim \hbar\omega_0$ we have a confinement-dominated 3D regime of scattering, where the 2D character of the particle motion is no longer pronounced in the scattering process. However, the axial confinement can strongly influence the energy (temperature) dependence of the scattering rate. Treating collisions as three-dimensional, the wavevector p of the relative motion of colliding atoms does not decrease with T at $T < \hbar\omega_0$. The atoms undergo zero-point oscillations in the axial direction and this corresponds to $p \sim 1/l_0$. This means that for a large 3D scattering length, i.e. at $|a| > l_0$, the tight axial confinement suppresses a resonant enhancement of the collisional

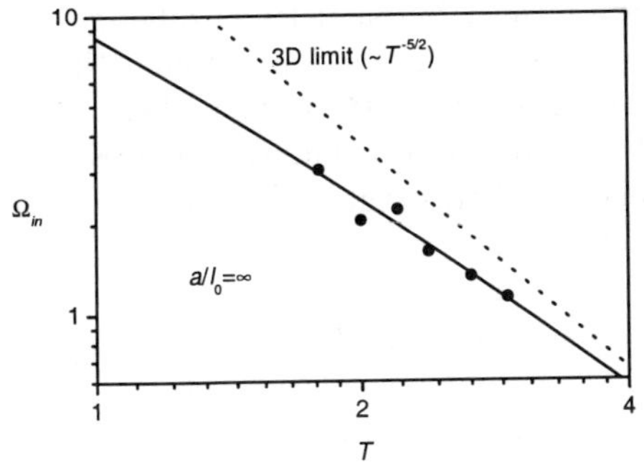

Figure 1. Inelastic rate Ω_{in} normalized to Ω_{in} at $T = 3\hbar\omega_0$, versus temperature (in units of $\hbar\omega_0$) in the unitarity limit ($a \to \infty$). The solid curve shows the result of our numerical calculations, and the dotted line the 3D limit. Circles show the data of the Stanford experiment.

rate at low temperatures. We obtain a similar suppression of resonances for inelastic collisions, where the resonant temperature dependence in 3D is related to the energy dependence of the initial wavefunction of colliding atoms.

In many of the current experiments with ultra-cold gases one tunes a to large positive or negative values by varying the magnetic field and achieving Feshbach resonances. In the unitarity limit ($|a| \to \infty$), the 3D rate constant of pair inelastic collisions $\alpha_{in} \propto 1/T$. This originates from the fact that the initial wavefunction of colliding atoms at interatomic distances $\sim R_e$ is inversly proportional to the relative momentum of the collision, whereas the final-state wavefunction is independent of the initial momenta of particles. In a harmonically trapped 3D Boltzmann gas the density $n \propto 1/T^{3/2}$ and, hence, the frequency of inelastic collisions $\Omega_{in} \approx \alpha_{in} n \propto 1/T^{5/2}$. In a tightly (axially) confined gas at T smaller than $\hbar\omega_0$, the rate constant α_{in} is temperature independent, and the density $n \propto 1/T$. We thus have $\Omega_{in} \propto 1/T$. In Fig.1 we display our results for the temperatute dependence of the inelastic rate Ω_{in} and compare them with the data of the Stanford experiment [3] on spin relaxation in a tightly confined gas of Cs atoms. The Stanford results agree fairly well with our calculations and show deviations from the 3D behavior.

3 Towards the BCS transition in quasi2D Fermi gases

We now show that there are promising prospects for achieving the superfluid BCS phase transition in two-component Fermi gases in quasi2D geometries.

The Cooper pairs are formed by particles of different components, and this requires an attractive s-wave interaction between them. In the purely 2D case the mean-field interaction is attractive if there is a weakly bound state for two atoms and the energy of colliding particles (Fermi energy ε_F) exceeds the binding energy of this state. This result follows directly from the solution of the 2D scattering problem (see, e.g. [12]). Then the interaction strength depends logarithmically on the collision energy, i.e. the Fermi energy $\varepsilon_F = \pi\hbar^2 n/m$, and hence on the 2D density n. We have established that this logarithmic density dependence is retained for the quasi2D regime of scattering in tightly confined Fermi gases: the interaction strength is inversly proportional to $\ln(\epsilon_0/\varepsilon_F)$, where the binding energy of the quasi2D weakly bound state ϵ_0 is given by Eq.(1). The interaction is attractive if the density is sufficiently high and the inequality

$$\epsilon_0/\varepsilon_F \ll 1 \qquad (2)$$

is satisfied. As the quasi2D regime is realized for $\varepsilon_f \ll \hbar\omega_0$, the condition (2) automatically requires $\hbar\omega_0 \gg \epsilon_0$. From Eq.(1) one can see that this is the case if the 3D scattering length a is negative and $|a|$ is significantly smaller than l_0.

Similarly to the purely 2D case [13], the logarithmic dependence of the interaction strength on the density changes the exponential density dependence of the BCS transition temperature to power law. For finding this critical temperature T_c, below which the formation of Cooper pairs becomes favorable, we go beyond the simple BCS approach. Along the lines of the Gor'kov and Melik-Barkhudarov theory developed for the 3D case [14], we take into account the modification of the interparticle interaction due to the presence of other particles (a particle interacts with a particle-hole pair virtually created by another particle). Then for the critical temperature we obtain

$$T_c = 0.16\sqrt{\epsilon_F\hbar\omega_0}\exp(-\sqrt{2\pi}l_0/|a|). \qquad (3)$$

The relative correction to this equation is of the order of

$$1/|\ln(\epsilon_0/\varepsilon_F)| \ll 1. \qquad (4)$$

Note that Eq. (3) predicts a value for the critical temperature smaller by a factor of e as compared to a simple BCS calculation. This means that the attractive interaction between particles becomes weaker once we take into account the polarization of the medium.

The exponent in Eq.(3) is assumed to be large. Nevertheless, one can have a comparatively large ratio T_c/ε_F. For example, using Feshbach resonances the scattering length for ^{40}K atoms can be tuned over a wide interval of negative values [15]. Keeping the exponential term in Eq.(3) equal to 0.05, with $\omega_0 \sim 100$ kHz we then obtain $T_c/\varepsilon_F \sim 0.1$ for 2D densities $\sim 10^9$ cm^{-2} ($T_c \sim 40$ nK).

As in the 2D case, the transition temperature $T_c \propto n^{1/2}$, whereas $\varepsilon_F \propto n$. In the presence of the radial confinement, one can decrease the ratio T/T_c for the quantum-degenerate Fermi gas by adiabatically expanding the quasi2D trap in the radial direction(s). As the ratio T/ε_F plays a role of the degeneracy parameter, it will be conserved in the course of the adiabatic expansion. Hence, the temperature will decrease as n and the ratio T/T_c as $n^{1/2}$. Eq.(3) also shows that one can increase T_c by adiabatically tuning $|a|$ to larger values or by making the tight confinement stronger and thus decreasing l_0.

However, the problem of achieving the superfluid transition by doing the described manipulations is more subtle. Our result for T_c assumes that we have $T_c \ll \varepsilon_F$. Generally speaking, one can think of T_c/T_F of the order of or smaller than 0.1. The discussed manipulations preserve the ratio T/T_F. Therefore, in order to achieve the BCS transition one should initially create a non-superfluid degenerate Fermi gas at a temperature $T < 0.1\varepsilon_F$. Note that for a non-superfluid fermionic gases in 3D traps the currently achieved minimum ratio T/T_F is 0.2 [10,11,9], and there are serious difficulties in making this ratio much smaller.

One can raise the question of what happens if the binding energy ϵ_0 becomes comparable with ε_F. In this case Eq.(3) leads to $T_c \sim \varepsilon_F$ and is no longer valid. A related problem has been discussed for submonolayers of ^{3}He on thin ^{4}He films and for copper-oxide 2D superconducting metals [13,16,17]. Actually, if the ratio ϵ_0/ε_F increases and approaches unity, one expects the formation of bound bosonic molecules. In our case these will be quasi2D molecules. For a large ratio ϵ_0/ε_F one encounters the problem of molecular BEC, and the stability of the expected molecular condensate will depend on the interaction between molecules. For the repulsive interaction one will have a stable molecular BEC, and the attractive interaction should cause a sort of a collapse. These problems require a separate analysis.

References

1. M.H. Anderson, J.R. Ensher, M.R. Matthews, C.E. Wieman, E.A. Cornell, Science **269**, 198 (1995); K.B. Davis, M.-O. Mewes, M.R. Andrews, N.J. van Druten, D.S. Durfee, D.M. Kurn, and W. Ketterle, Phys. Rev. Lett. **75**, 3969 (1995); C.C. Bradley, C.A. Sackett, J.J. Tollett, and R.G. Hulet, Phys. Rev. Lett. **75**, 1687 (1995).

2. V. Vuletić, C. Chin, A.J. Kerman, and S. Chu, Phys. Rev. Lett. **81**, 5768 (1998).
3. V. Vuletić, A.J. Kerman, C. Chin, and S. Chu, Phys. Rev. Lett. **82**, 1406 (1999).
4. I. Bouchoule, H. Perin, A. Kuhn, M. Morinaga, and C. Salomon, Phys. Rev. A **59**, R8 (1999); M. Morinaga, I. Bouchoule, J.-C. Karam, and C. Salomon, Phys. Rev. Lett. **83**, 4037 (1999).
5. I. Bouchoule, PhD Thesis, LKB ENS, Paris, 2000; I. Bouchoule, M. Morinaga, C. Salomon, and D.S. Petrov, (to be published).
6. D.S. Petrov, M. Holzmann, and G.V. Shlyapnikov, Phys. Rev. Lett. **84**, 2551 (2000).
7. A. Goerlitz, J.M. Vogels, A.E. Leanhardt, C. Rahman, T.L. Gustavson, J.R. Abo-Shaeer, A.P. Chikkatur, S. Gupta, S. Inouye, T.P. Rosenband, D. Pritchard, and W. Ketterle, Phys. Rev. Lett. **87**, 130402 (2001).
8. S. Burger, F.S. Cataliotti, C. Fort, Maddaloni, F. Minardi, and M. Inguscio, cond-mat/0108037.
9. F.S. Schreck, L. Khaykovich, K.L. Corwin, G. Ferrari, T. Bourdwel, J. Cubizolles, and C. Salomon, cond-mat/0107442.
10. B. DeMarco and D.S. Jin, Science **285**, 1703 (1999); B. DeMarco, S.B. Papp, and D.S. Jin, Phys. Rev. Lett. **86**, 5409 (2001).
11. A.G. Truscott, K.E. Strecker, W.I. McAlexander, G.B. Partridge, and R.G. Hulet, Science **291**, 2570 (2001).
12. L.D. Landau and E.M. Lifshitz, *Quantum Mechanics*, (Butterworth-Heinemann, Oxford, 1999).
13. K. Miyake, Progr. Theor. Phys. **69**, 1794 (1983).
14. L.P. Gor'kov and T.K. Melik-Barhudarov, Sov. Phys. JETP **13**, 1018 (1961).
15. J.L. Bohn, Phys. Rev. A **61**, 053409 (2000).
16. Mohit Randeria, Ji-Min Duan, and Lih-Yir Shieh, Phys. Rev. Lett. **62**, 981 (1989).
17. S. Schmitt-Rink, C.M. Varma, and A.E. Ruckenstein, Phys. Rev. Lett. **62**, 981 (1989).

Nucleation of vortices in a rotating Bose-Einstein condensate

Subhasis Sinha and Yvan Castin
Laboratoire Kastler Brossel, Ecole normale supérieure,
24 rue Lhomond, 75 231 Paris Cedex 5, France

For the precise experimental procedure used at ENS to stir a Bose-Einstein condensate, we have shown that the vortex nucleation is driven by a dynamic, hydrodynamic type instability rather than by a thermodynamic mechanism. The universal value of the rotation frequency, $\Omega = 0.7\omega_\perp$, leading to vortex formation in a very slightly anisotropic rotating trap corresponds simply to a resonance between the stirring harmonic potential and the condensate, at the origin of the dynamic instability. The fact that a resonance is involved also explains simply why vortex nucleation takes place in a narrow interval of rotation frequencies.

1 Introduction

The concept of quantum vortex applies to a quite general class of coherent macroscopic systems, that is systems which have a so-called *macroscopic wavefunction* $\psi(\vec{r})$. Examples of coherent macroscopic systems are typically N-body bosonic systems, like superfluid ^{4}He (ψ is the order parameter), the laser (ψ is the electric field amplitude in the lasing mode), and the dilute Bose condensed gases ($\psi(\vec{r}) = N_0^{1/2}\phi(\vec{r})$, where N_0 is the number of condensate particles and ϕ is the condensate wavefunction normalized to unity). Fermionic systems can also exhibit a macroscopic wavefunction when the fermions are paired, as for example in superconductors with Cooper pairs, ψ being a possibly position dependent gap parameter Δ.

Restricting to the case where $\psi(\vec{r})$ is a complex field (rather than a spinor), we say that ψ supports a vortex if (i) there exists a line in space along which ψ vanishes, and (ii) the phase of ψ changes by a non-zero integer multiple of 2π, that is $2q\pi$, on any closed contour around this nodal line. The nodal line is called a vortex line. Although the vortex line is a phase singularity, the field ψ itself is not singular on the line, it simply vanishes like $(X \pm iY)^{|q|}$, where $\pm$ is the sign of q and (X, Y, Z) is a local system of coordinates with Z axis tangent to the vortex line.

The presence of a vortex is at the origin of quantum phenomena at the macroscopic level. The superfluid velocity field for example has a circulation $2q\pi\hbar/m$ around the vortex line, where m is the mass of the particles.

We are concerned here with the fundamental problem of understanding how quantum vortices form in an atomic Bose-Einstein condensate subject to a stirring potential. This nucleation problem is difficult for a general coherent macroscopic system if one wants to reach understanding at the microscopic level. For example the issue of vortex nucleation in superfluid helium II in a rotating container has long been the subject of intense work[1]. With the recent production of gaseous Bose-Einstein condensates [2] the subject has gained a renewed interest. Quantum gases have the advantage of being clean systems (very weakly coupled to their environment), dilute systems with well known interactions (so that the theory is hopefully tractable), and atomic physics has powerful experimental techniques allowing almost perfect control and imaging. One can therefore hope to reach a full understanding of vortex nucleation in these systems.

2 Experimental results at ENS and early theoretical predictions

On the experimental side, several groups have succeeded in obtaining vortices in atomic condensates, with two different techniques: a phase imprinting technique at JILA [3] and the equivalent of the helium rotating bucket experiment at ENS [4], at MIT [5] and at Oxford [6]. The precise features of vortex nucleation actually depend on the precise experimental technique used to transfer angular momentum to the gas. We shall restrict here to the case of the ENS experiment.

At ENS the condensate atoms are stored in an (almost) cylindrically symmetric magnetic trap, and one suddenly applies a laser beam creating a rotating potential with adjustable anisotropy ϵ and rotation frequency Ω. In the frame rotating at frequency Ω around the trap eigenaxis z the trapping potential can be written as

$$U(\vec{r}) = \frac{1}{2}m\omega_\perp^2 \left\{ [1-\epsilon]x^2 + [1+\epsilon]y^2 + \left(\frac{\omega_z}{\omega_\perp}\right)^2 z^2 \right\} \tag{1}$$

where m is the mass of an atom, ϵ is the trap anisotropy. The parameters $\omega_\perp$ and ω_z are the oscillation frequencies of the atoms in transverse and axial directions for vanishing anisotropy of the stirring potential. For a well chosen range of variation for Ω one or several vortices are nucleated, and then detected as holes in the density profile of the gas after ballistic expansion [4] or by a measurement of the angular momentum of the condensate[7]. A striking feature of the ENS experimental results is that, for a very weak anisotropy ϵ, nucleation of vortices takes place in a narrow interval of rotation frequencies $[\Omega_{\rm min}, \Omega_{\rm max}]$ around $0.7\omega_\perp$ whatever the number of atoms or the oscillation frequency ω_z along z in the experiment[8]. An example of experimental results is shown in the figure of §4 (disks with error bars) giving the angular momentum per particle in the condensate as function of the rotating frequency Ω for an anisotropy parameter $\epsilon = 0.01$. A non-vanishing angular momentum indicates that one or several vortices have been nucleated [9].

Several theoretical articles, inspired by the case of superfluid helium, have tried to predict the value of the lower vortex nucleation frequency $\Omega_{\rm min}$ from purely thermodynamic arguments [10,11,12,13,14,15,16], most of them before the ENS experiment was completed. The gas is supposed to reach thermal equilibrium in the rotating frame, and the energy $E_{\rm rf}$ per particle of the gas in the rotating frame is related to the energy $E_{\rm lab}$ per particle of the gas in the lab frame as

$$E_{\rm rf} = E_{\rm lab} - \Omega L_z \tag{2}$$

where L_z is the angular momentum per particle of the gas. This formula immediately suggests that configurations with vortices (which have a non-vanishing angular momentum) are energetically favored for a large enough rotation frequency, so that vortices should form for Ω greater than some critical value $\Omega_{\rm nuc}$. Several values for $\Omega_{\rm nuc}$ have been proposed: Ω_c such that the one-vortex configuration becomes the absolute minimum of energy, Ω_L such that the zero-vortex configuration is not a local minimum of energy (Landau criterion), etc. The proposed values for $\Omega_{\rm min}$ are significantly different from the observed value of $0.7\omega_\perp$, or depend on the trap aspect ratio $\omega_z/\omega_\perp$ or on the atom number, in contradiction with the observations at ENS. Also thermodynamic reasonings are not able to predict the upper vortex nucleation frequency $\Omega_{\rm max}$, which is also close to $0.7\omega_\perp$ for low anisotropy ϵ.

3 A time dependent approach to understand vortex nucleation

As thermodynamic reasoning fails, we take as a way-out a purely time dependent approach, not relying on the assumption of thermal equilibrium in the rotating frame. Within the mean field approximation, the time evolution of the condensate field or macroscopic wavefunction $\psi(\vec{r}, t)$ can be described by the time dependent Gross-Pitaevskii equation [17]:

$$i\hbar\frac{\partial\psi}{\partial t} = \left[-\frac{\hbar^2}{2m}\vec{\nabla}^2 + U(\vec{r}, t) + g|\psi|^2 - \Omega(t)\hat{L}_z\right]\psi, \tag{3}$$

where $g = 4\pi\hbar^2 a/m$ is the coupling constant, proportional to the s-wave scattering length a of the atoms, here taken to be positive, and where the inertial term proportional to the angular momentum operator $\hat{L}_z$ along z-axis accounts for the frame rotation.

A first way to tackle with the Gross-Pitaevskii equation is to solve it numerically, as done in [14,18,19]. The results are encouraging: for example in [19] vortices appear for $\Omega > 0.7\omega_\perp$, but numerics do not explain why!

So we wish to obtain analytical results. We use the classical hydrodynamic approximation to the time dependent Gross-Pitaevskii equation (GPE), an approximation well justified for the ENS parameters [20]. We are then able to reformulate the partial differential hydrodynamic equations in terms of ordinary differential equations, which allows an almost analytical solution [21]. As we explain now, our main result is the discovery of dynamical instabilities in the evolution of the condensate for a certain range of the rotation frequency and of the trap anisotropy. These instabilities will invalidate the classical hydrodynamic approximation after some evolution time, but we have checked with a numerical solution of the Gross-Pitaevskii equation that vortices then enter the condensate.

The condensate field ψ can be written in terms of density ρ and phase S,

$$\psi(\vec{r},t) = \sqrt{\rho(\vec{r},t)}e^{iS(\vec{r},t)/\hbar}. \tag{4}$$

The equation obtained from the GPE for ρ is just the continuity equation. The equation for S contains the so-called quantum pressure term $\hbar^2\vec{\nabla}^2\sqrt{\rho}/2m\sqrt{\rho}$ that we neglect here as compared to the mean-field term ρg in the Thomas-Fermi approximation [20]. We obtain the so-called classical hydrodynamics equation for a superfluid, equivalent to Euler equation for a potential flow. Finally here are the equations of motion:

$$\frac{\partial\rho}{\partial t} = -\text{div}\left[\rho\left(\frac{\vec{\nabla}S}{m} - \vec{\Omega}(t)\times\vec{r}\right)\right] \tag{5}$$

$$-\frac{\partial S}{\partial t} = \frac{(\vec{\nabla}S)^2}{2m} + U(\vec{r},t) + g\rho - (\vec{\Omega}\,(t)\times\vec{r})\cdot\vec{\nabla}S. \tag{6}$$

In the absence of vortex the approximation of neglecting the quantum pressure term is valid when the chemical potential μ of the gas is much larger than $\omega_\perp, \omega_z$: in the ENS experiment one has $\mu \simeq 10\hbar\omega_\perp \simeq 150\hbar\omega_z$. In presence of a vortex this approximation fails within the vortex core radius: the radius is on the order of the healing length $\xi = (\hbar^2/2m\mu)^{1/2}$, so that the quantum pressure term is on the order of $\hbar^2/2m\xi^2 = \mu \simeq \rho g$ inside the vortex core. The healing length is considered as infinitely small in the Thomas-Fermi approximation so that the apparition of a vortex due to the stirring should manifest itself as the emergence of singularities in the classical hydrodynamics equation!

A very fortunate feature of the harmonic trap is that these superfluid hydrodynamic equations (5, 6) can be solved exactly for a condensate initially at equilibrium in the non-rotating trap with the following quadratic ansatz for the condensate density and phase [22]:

$$\rho_c(\vec{r},t) = \rho_0(t) + \frac{m\omega_\perp^2}{g}\sum_{i,j=1}^{3} x_i A_{ij}(t) x_j, \tag{7}$$

$$S_c(\vec{r},t) = s_0(t) + m\omega_\perp \sum_{i,j=1}^{3} x_i B_{ij}(t) x_j, \tag{8}$$

where x_1, x_2 and x_3 are the coordinates along x, y and z axes respectively. The time dependent dimensionless coefficients $A_{i,j}$ and $B_{i,j}$ form 3×3 symmetric matrices A and B which from Eqs.(5,6) obey the evolution equations:

$$\omega_\perp^{-1}\frac{dA}{dt} = -2A\,\mathrm{Tr}B - 2\{A,B\} + \frac{\Omega}{\omega_\perp}[R,A], \tag{9}$$

$$\omega_\perp^{-1}\frac{dB}{dt} = -2B^2 - W - A + \frac{\Omega}{\omega_\perp}[R,B] \tag{10}$$

where $\{,\}$ stands for the anti-commutator, $[,]$ stands for the commutator of two matrices, the matrix W is diagonal, with components $W_{11} = (1-\epsilon)/2$, $W_{22} = (1+\epsilon)/2$, and $W_{33} = (\omega_z/\omega_\perp)^2/2$, and the matrix R, originating from the vectorial product in $\hat{L}_z$, has vanishing elements except for $R_{12} = -R_{21} = 1$ [23]. These equations do not depend on the number of atoms nor on the coupling constant g.

We solve the equations for A and B for the stirring scenario used at ENS. We find that the stirring potential can communicate angular momentum to the condensate. For a given rotation frequency Ω the angular momentum is an oscillating function of time, oscillating around a non zero mean $\bar{L}_z$. We then plot this mean transferred angular momentum $\bar{L}_z$ as function of the rotation frequency Ω in the figure of §4 (dashed line). But we note that clearly a condensate with a vortex cannot be described within the quadratic ansatz (7,8) as the phase S_c corresponds to an irrotational velocity flow, so that the transferred angular momentum is not associated yet to vortex formation. These results raise two important questions.

First, how can a vortex free condensate store angular momentum ? Our results seem to be in conflict with the common saying that a superfluid cannot rotate in the absence of vortices. A look at matrix A shows that the condensate, round in the $x-y$ plane at the beginning of the stirring procedure, takes after some time the shape of a stick. Such a stick, even superfluid, can rotate in the lab frame, and the common saying is correct for round superfluids only.

Second, how to explain the peak structure of the curve in the figure ? It is due to a resonance of the stirring potential with a quadrupole mode of the condensate, as is most easily seen in the lab frame. The stirring potential is a periodic function of pulsation 2Ω in the lab frame, as after a time π/Ω the axes have rotated by an angle π and the quadratic potential (1) is mapped back onto itself. The lowest quadrupole mode frequency is $\sqrt{2}\omega_\perp$ [20,24] so that the resonance condition is $\Omega = (\sqrt{2}/2)\omega_\perp \simeq 0.7\omega_\perp$. We have recovered the universal value 0.7 obtained in the ENS experiment! The peak of $\bar{L}_z$ is not exactly located at $\Omega = 0.7\omega_\perp$ because of non-linear effects in Eqs.(9, 10).

4 Looking for dynamic instability

The previous section has given some clue about the magic number 0.7 but lets us with the puzzle that the exact solutions ρ_c, S_c to the hydrodynamics equations have no vortex! So the precise scenario for the vortex nucleation that we put forward is the following: initially very small deviations $\delta\rho(\vec{r},t)$ of the condensate density and $\delta S(\vec{r},t)$ of the condensate phase from the quadratic shapes ρ_c and S_c may grow exponentially fast in the course of time evolution, eventually leading the condensate to a structure very different from Eqs.(7,8). This may happen when a dynamic instability is present.

To reveal such an instability we obtain from the evolution equations (5, 6) linearized equations of motion for initially small deviations $\delta\rho$ and δS from ρ_c and S_c:

$$\frac{D\delta\rho}{Dt} = -\mathrm{div}\left(\rho_c\frac{\vec{\nabla}\delta S}{m}\right) - \delta\rho\frac{\vec{\nabla}^2 S_c}{m}, \tag{11}$$

$$\frac{D\delta S}{Dt} = -g\delta\rho. \tag{12}$$

In these equations, we have introduced the convective derivative $\frac{D}{Dt} \equiv \frac{\partial}{\partial t} + \vec{v}_c(\vec{r},t)\cdot\vec{\nabla}$ where $\vec{v}_c = \vec{\nabla}S_c/m - \vec{\Omega}\times\vec{r}$ is the velocity field of the condensate in the rotating frame. A polynomial ansatz for δS and $\delta\rho$ of an arbitrary total degree n in the coordinates x, y and z solves these linear equations exactly [25]. This is another nice consequence of the harmonicity of the trap. Note that after rescaling of the variables, Eqs.(11,12) become independent of the number of atoms and of the coupling constant g, in a way similar to Eqs.(9,10).

In practice, we calculate the evolution operator $\mathcal{U}_n(t)$ mapping the coefficients of the polynomials at time zero onto their values after a time evolution t. Dynamic instability takes place when one or several eigenvalues of $\mathcal{U}_n$ grow exponentially fast with time t. So we calculate $Z_{\max}(t)$, the eigenvalue of $\mathcal{U}_n(t)$ with the largest modulus. Then we define the mean instability exponent $\mathrm{Re}\,\langle\lambda\rangle$ as the 'mean' slope of $\ln|Z_{\max}(t)|$ as function of time, the mean being taken over an integer number of periods of oscillations of L_z.

This reveals that within certain range of rotation frequency the system becomes dynamically unstable, see the solid line in the figure. In the limit of a low anisotropy ϵ the instability sets in when the rotation frequency Ω is close to the value $\simeq 0.7\omega_\perp$: as pointed out in the previous section, the stirring potential of frequency 2Ω in the lab frame is then resonant with a quadrupole mode of the condensate of frequency $\sqrt{2}\omega_\perp$, and induces large amplitude oscillations of the condensate, resulting in a dynamic instability.

To show the connection between the dynamic instability and this resonance effect more quantitatively we plot the mean angular momentum $\bar{L}_z$ of the vortex free condensate ρ_c, S_c as function of the rotation frequency Ω, in dashed line in the figure. The peak structure of the instability exponent in the figure is alike the peak structure of $\bar{L}_z$, with a narrower width as dynamic instability of the vortex free solution ρ_c, S_c sets in for the higher values of $\bar{L}_z$ only. For values of Ω significantly above or below $0.7\omega_\perp$ the stirrer is out of resonance with the quadrupole mode and induces only small and stable oscillations of the condensate. For larger values of ϵ, the instability interval in Ω broadens. We have also checked that the instability interval depends weakly on $\omega_z/\omega_\perp$, as observed in the ENS experiment.

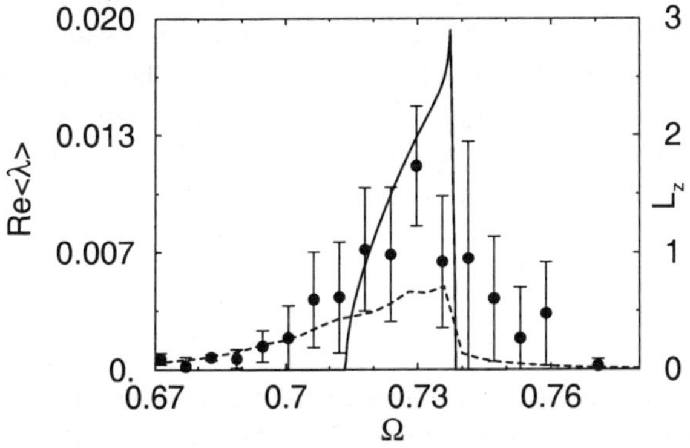

Figure 1: For the ENS stirring procedure, with $\epsilon = 0.01$ and $\omega_z/\omega_\perp = 0.1$: Solid line: mean instability exponent $\mathrm{Re}\,\langle\lambda\rangle$ (see text) of the vortex free classical hydrodynamic solution ρ_c, S_c as function of Ω, for a degree $n = 3$. Dashed line: mean angular momentum per particle $\bar{L}_z$ (see text) obtained from ρ_c, S_c. Filled disks: experimentally measured angular momentum L_z per particle in the condensate after vortices have possibly entered the condensate [8]. The initial steady state condensate in the calculation of $\bar{L}_z$ has a chemical potential $\mu = 10\hbar\omega_\perp$, close to the experimental value. $\mathrm{Re}\,\langle\lambda\rangle$ and Ω are in units of $\omega_\perp$, and L_z, $\bar{L}_z$ are in units of $\hbar$.

How to prove the connection between the dynamic instabilities found here and the nucleation of vortices ? To obtain a theoretical answer to this question, one has to go beyond a linear stability analysis to determine the evolution of the condensate in the long run: for a few values of the rotation frequency Ω we have checked by a numerical integration of the time dependent GPE in three dimensions, that vortices are indeed nucleated in the predicted instability domains: after some evolution time, the angular momentum in the numerical solution suddenly becomes larger than the classical hydrodynamic prediction, as vortices enter in the condensate. An experimental answer to this question is obtained by comparing the experimental data to our instability exponent $\langle\lambda\rangle$, see the figure: our instability domain in Ω coincides with the experimental vortex nucleation interval within a few percent.

To test our scenario of dynamic instability we have suggested another procedure to stir the condensate: the stirrer is set into rotation very slowly, so that the condensate follows adiabatically a branch of steady state calculated in [26]. We have predicted that this branch becomes dynamically unstable at some rotation frequency Ω, where the first vortex should appear [27], a prediction confirmed experimentally at ENS [28].

Another stirring procedure has been used by the Oxford group [6], to which we plan to extend our analysis. Also understanding why the thermodynamics arguments failed for the ENS experiment is a relevant question to address!

We thank S. Rica, V. Hakim, G. Shlyapnikov, F. Chevy, K. Madison and J. Dalibard for helpful discussions. We acknowledge financial support from Ministère de la Recherche et de la Technologie. LKB is a unité de recherche de l'Ecole normale supérieure et de l'Université Pierre et Marie Curie, associée au CNRS.

References

1. R. J. Donnelly, *Quantized vortices in Helium II* (Cambridge, 1991).
2. For a review see e.g. *Bose-Einstein Condensation in Atomic gases*, eds M. Inguscio, S. Stringari, and C.E. Wieman (IOS Press, Amsterdam, 1999).
3. M. R. Matthews, B. P. Anderson, P. C. Haljan, D. S. Hall, C. E. Wieman, and E. A. Cornell, Phys. Rev. Lett. **83**, 2498 (1999).
4. K. W. Madison, F. Chevy, W. Wohlleben, and J. Dalibard, Phys. Rev. Lett. **84**, 806 (2000).
5. J.R. Abo-Sheer, C. Raman, J.M. Vogels, and W. Ketterle, Science **292**, 476 (2001).
6. G. Hechenblaikner, private communication at this ICOLS conference,

and preprint cond-mat/0106262.
7. F. Chevy, K.W. Madison, and J. Dalibard, Phys. Rev. Lett. **85**, 2223 (2000).
8. F. Chevy, K. Madison, V. Bretin, J. Dalibard, to be published in *Proceedings of Trapped particles and fundamental physics Workshop* (Les Houches 2001), edited by S. Atutov, K. Kalabrese, L. Moi.
9. In this experiment the stirring potential is applied for some time then one lets the gas relax in the (almost) cylindrically symmetric magnetic trap before measuring the angular momentum. As the condensate is round in the $x-y$ plane at this stage it cannot have a non vanishing angular momentum without having vortices (see end of §3).
10. G. Baym and C. J. Pethick, Phys. Rev. Lett. **76**, 6 (1996).
11. F. Dalfovo and S. Stringari, Phys. Rev. A **53**, 2477 (1996).
12. S. Sinha, Phys. Rev. A **55**, 4325 (1997).
13. E. Lundh, C. J. Pethick, and H. Smith, Phys. Rev. A **55**, 2126 (1997).
14. D. L. Feder, A. A. Svidzinsky, A. L. Fetter, and C. W. Clark, Phys. Rev. Lett. **86**, 564 (2001).
15. Y. Castin and R. Dum, Eur. Phys. J. D **7**, 399 (1999).
16. T. Isoshima, K. Machida, Phys. Rev. A **60**, 3313 (1999).
17. F. Dalfovo, S. Giorgini, L. P. Pitaevskii, and S. Stringari, Rev. Mod. Phys. **71**, 463 (1999).
18. B. M. Caradoc-Davies, R. J. Ballagh, and K. Burnett, Phys. Rev. Lett. **83**, 895 (1999).
19. David L. Feder, Charles W. Clark, Barry I. Schneider, Phys. Rev. A **61**, 011601 (2000).
20. S. Stringari, Phys. Rev. Lett. **77**, 2360 (1996).
21. This reformulation is possible here because the stirring potential is harmonic. In presence of an obstacle the situation is much more involved, see e.g. C. Josserand, Y. Pomeau, and S. Rica, Physica D **134**, 111 (1999), and C. Huepe, M.-É. Brachet, Physica D **140**, 126 (2000).
22. ρ_c is set to zero where the ansatz is negative.
23. A scaling formulation of the ansatz is given by P. Storey and M. Olshanii, Phys. Rev. A **62**, 033604 (2000).
24. F. Dalfovo and S. Stringari, Phys. Rev. A **63**, 011601(R) (2000).
25. We take 'degree' of the modes in the strict sense: the degree of the highest degree monomial with non vanishing coefficient.
26. A. Recati, F. Zambelli and S. Stringari, Phys. Rev. Lett. **86**, 377 (2001).
27. S. Subhasis and Y. Castin, to appear in Phys. Rev. Lett. (2001).
28. F. Chevy, K.W. Madison, and J. Dalibard, cond-mat/0101051, to appear in Phys. Rev. Lett. (2001).

RESONANCE SUPERFLUIDITY IN A QUANTUM DEGENERATE FERMI GAS

SERVAAS KOKKELMANS, MURRAY HOLLAND, AND REINHOLD WALSER
JILA, University of Colorado and National Institute of Standards and Technology, Boulder, Colorado 80309-0440, USA
E-mail: servaas.kokkelmans@jila.colorado.edu

MARILU CHIOFALO
Scuola Normale Superiore,Piazza dei Cavalieri 7,I-56126 Pisa, Italy

We consider the superfluid phase transition that arises when a Feshbach resonance pairing occurs in a dilute Fermi gas. This is related to the phenomenon of superconductivity described by the seminal Bardeen-Cooper-Schrieffer (BCS) theory. In superconductivity, the phase transition is caused by a coupling between pairs of electrons within the medium. This coupling is perturbative and leads to a critical temperature T_c which is small compared to the Fermi temperature T_F. Even high-T_c superconductors typically have a critical temperature which is two orders of magnitude below T_F. Here we describe a resonance pairing mechanism in a quantum degenerate gas of potassium (^{40}K) atoms which leads to superfluidity in a novel regime—a regime that promises the unique opportunity to experimentally study the crossover from the BCS phase of weakly-coupled fermions to the Bose Einstein condensate of strongly-bound composite bosons. We find that the transition to a superfluid phase is possible at the high critical temperature of about $0.5\,T_F$. It should be straightforward to verify this prediction, since these temperatures can already be achieved experimentally.

The study of superfluid phase transitions in fermion and boson systems has played an important role in the development of many areas of quantum physics. Their characteristics determine the observed properties of some of the most distinct systems imaginable, including the cosmology of neutron stars, the non-viscous flow of superfluid liquid Helium, the non-resistive currents in superconductors, and the structure and dynamics of microscopic elemental nuclei. Recently, physicists have succeeded in demonstrating the creation of weakly interacting quantum fluids by cooling dilute gases to temperatures in the nanokelvin scale. For these near ideal gases, reaching such incredibly low temperatures is required in order to cross the threshold for superfluid properties to emerge. These systems offer great opportunities for study since they can be created in table-top experiments, manipulated by laser and magnetic fields which can be controlled with high precision, and directly observed using conventional optics. Furthermore their microscopic behavior can be understood theoretically from first principles. Observations of Bose-Einstein condensation (BEC) [1], and demonstrations of the near ideal degenerate Fermi gas [2], are becoming fairly routine in atomic physics—something which would have been hard to foresee even ten years ago.

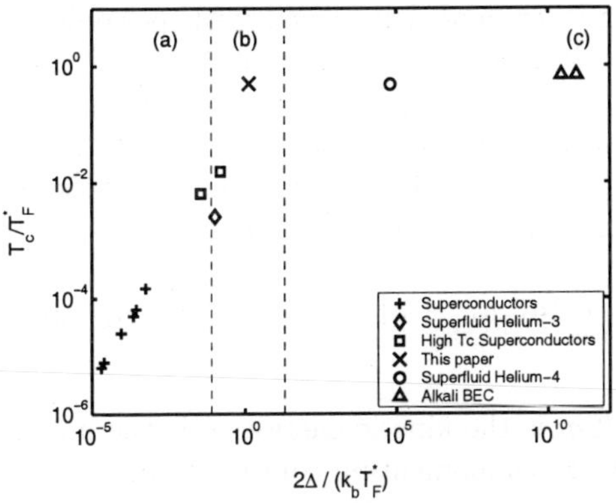

Figure 1. A log-log plot showing six distinct regimes for quantum fluids. The transition temperature T_c is shown as a function of the relevant gap energy 2Δ. Both quantities are normalized by an effective Fermi temperature T_F^*. For the BCS systems in region (a), and the systems in the cross-over region (b), 2Δ is the energy needed to break up a fermion pair, and T_F^* is the Fermi energy. For the systems in region (c), which are all strongly bound composite bosons and exhibit BEC phenomenology, 2Δ is the smallest energy needed to break the composite boson up into two fermions, i.e. ionization to a charged atomic core and an electron, and T_F^* is the ionic Fermi temperature.

The phenomenology of superfluid dilute gases can be quite distinct from that of condensed matter systems. In this letter, we present a striking illustration of this point by predicting the existence of a Feshbach resonance superfluidity in a gas of fermionic potassium atoms. This system has an ultrahigh critical phase transition temperature in close proximity to the Fermi temperature. This is a novel regime for quantum fluids, as illustrated in Fig. 1 where our system and others which exhibit superfluidity or BEC are compared. Simply by modifying a control parameter, in this case the strength of magnetic field, the system we consider can potentially explore the crossover regime between the Bardeen-Cooper-Schrieffer (BCS) [3] transition of weakly-coupled fermion pairs and the Bose-Einstein condensation of strongly-bound composite particles [4]. This is an intriguing regime for quantum fluids as it bridges the physics of superconductors and superfluid ^{3}He, and the physics of superfluid ^{4}He and bosonic alkali gases. Non-resonant pairing applied to a dilute gas yields a T_c that depends exponentially on the inverse scattering length [5], as will be pointed out in the following.

The BCS theory of superconductivity applied to a dilute gas considers binary interactions between particles in two distinguishable quantum states, say $|\uparrow\rangle$ and $|\downarrow\rangle$. For a uniform system, the fermionic field operators may be Fourier-expanded in a box with periodic boundary conditions giving wavevector-$\boldsymbol{k}$ dependent creation and annihilation operators $a^\dagger_{\boldsymbol{k}\sigma}$ and $a_{\boldsymbol{k}\sigma}$ for states $|\sigma\rangle$. At low energy, the binary scattering processes are assumed to be completely characterized by the s-wave scattering length a in terms of a contact quasipotential $U = 4\pi\hbar^2 an/m$, where n is the number density. The Hamiltonian describing such a system is given by

$$H = \sum_{\boldsymbol{k}} \epsilon_k (a^\dagger_{\boldsymbol{k}\uparrow} a_{\boldsymbol{k}\uparrow} + a^\dagger_{\boldsymbol{k}\downarrow} a_{\boldsymbol{k}\downarrow}) + U \sum_{\boldsymbol{k}_1 \ldots \boldsymbol{k}_3} a^\dagger_{\boldsymbol{k}_1\uparrow} a^\dagger_{\boldsymbol{k}_2\downarrow} a_{\boldsymbol{k}_3\downarrow} a_{\boldsymbol{k}_4\uparrow}, \tag{1}$$

where $\epsilon_k = \hbar^2 k^2/2m$ is the kinetic energy, m is the mass, and the constraint $\boldsymbol{k}_4 = \boldsymbol{k}_1 + \boldsymbol{k}_2 - \boldsymbol{k}_3$ gives momentum conservation.

For a negative scattering length, the thermodynamic properties of the gas show a superfluid phase transition at a critical temperature T_c which arises due to an instability towards the formation of Cooper-pairs. When the gas is dilute, as characterized by the inequality $n|a|^3 \ll 1$ (or equivalently $k_F|a| \ll 1$ where k_F is the Fermi wavenumber), the application of mean-field theory gives a well-known solution for the ratio of T_c to the Fermi temperature T_F [5]:

$$\frac{T_c}{T_F} \sim \exp(-\frac{\pi}{2|a|k_F}). \tag{2}$$

The exact prefactor to the exponential depends on the precise form of the analytic integral approximations made in the derivation. Several papers have pointed out that the presence of a scattering resonance in dilute alkali gases can be used to obtain a very large negative value for the scattering length [6]. This promises the opportunity for the system to enter the high-T_c superfluidity regime as the ratio in Eq. (2) approaches unity. However, direct application of the BCS theory close to resonance then becomes speculative due to the potential breakdown of a number of underlying assumptions:

1. Exactly on resonance the theory fails as the scattering length passes through $\pm\infty$ and the Hamiltonian in Eq. (1) can not be defined.

2. For the mean field approach to be accurate it is required that there be many particles inside a volume associated with the spatial scale of a Cooper-pair. This condition begins to break down as T_c approaches T_F.

3. The theory of the dilute gas is formulated on a perturbation approach based on an expansion in the small parameter $n|a|^3$. When this parameter approaches unity the perturbation theory fails to converge.

These points show that care should be taken in applying Eq. (2) near the point of resonance where the basis for the conventional mean-field theory is not well-founded.

Despite these limitations, on general grounds, one would expect to be able to derive a renormalizable low-energy effective field theory even in close proximity to a resonance. This statement is based on the identification that at relevant densities the range of the interparticle potential is always orders of magnitude smaller than the interparticle spacing. Here we present a theory of superfluidity in a gas of dilute fermionic atoms which handles correctly the scattering resonance and places the transition temperature to the superfluid state in the experimentally accessible range.

While the scattering length a usually characterizes the range of the interatomic potential for a collision, this is a poor approximation in the vicinity of a scattering resonance. The scattering properties are completely determined by the positions of the bound states in the interaction potentials. In a multichannel system, a bound state may cross the threshold as a function of magnetic field and enter the continuum, resulting in a field-dependent Feshbach scattering resonance [7]. As this occurs, the scattering length becomes strongly dependent on the field, and exactly at threshold it changes sign by passing through $\pm\infty$.

When such resonance processes occur, it is necessary to formulate the Hamiltonian by separating out the resonance state and treating it explicitly. This is motivated by the microscopic identification of two types of scattering contributions: one from the scattering resonance, and one from the background non-resonant processes that includes the contributions from all the other bound states. The non-resonant contributions give rise to a background scattering length $a_{\rm bg}$ which is a good characterization of the potential range. The corresponding quasipotential in that case is given by $U_{\rm bg} = 4\pi\hbar^2 a_{\rm bg} n/m$. The Feshbach resonance occurs due to a coupling with a molecular state, that is long-lived in comparison with characteristic non-resonant collision timescales. This state is a composite boson which is described by bosonic annihilation operators $b_{\boldsymbol{k}}$. It is parameterized by a detuning energy from threshold, denoted by 2ν, that is dependent on the value of the magnetic field. The coupling strength of $b_{\boldsymbol{k}}$ to the two-particle continuum is well characterized by a single coupling constant g, independent of $\boldsymbol{k}$. These considerations imply that the Hamiltonian given in Eq. (1) is not sufficient to account for the important resonance processes and must be extended to incorporate explicitly the coupling between the atomic and molecular gases:

$$
\begin{aligned}
H = 2\nu \sum_{\boldsymbol{k}} b^{\dagger}_{\boldsymbol{k}} b_{\boldsymbol{k}} &+ \sum_{\boldsymbol{k}} \epsilon_k \left(a^{\dagger}_{\boldsymbol{k}\uparrow} a_{\boldsymbol{k}\uparrow} + a^{\dagger}_{\boldsymbol{k}\downarrow} a_{\boldsymbol{k}\downarrow}\right) \\
&+ U_{\text{bg}} \sum_{\boldsymbol{k}_1 \ldots \boldsymbol{k}_3} a^{\dagger}_{\boldsymbol{k}_1\uparrow} a^{\dagger}_{\boldsymbol{k}_2\downarrow} a_{\boldsymbol{k}_3\downarrow} a_{\boldsymbol{k}_4\uparrow} \\
&+ g \sum_{\boldsymbol{k},\boldsymbol{q}} b^{\dagger}_{\boldsymbol{q}} a_{\frac{\boldsymbol{q}}{2}+\boldsymbol{k}\uparrow} a_{\frac{\boldsymbol{q}}{2}-\boldsymbol{k}\downarrow} + b_{\boldsymbol{q}} a^{\dagger}_{\frac{\boldsymbol{q}}{2}-\boldsymbol{k}\downarrow} a^{\dagger}_{\frac{\boldsymbol{q}}{2}+\boldsymbol{k}\uparrow} .
\end{aligned}
\tag{3}
$$

Evolution generated by this Hamiltonian conserves the particle number $N = \sum_{\boldsymbol{k}} (a^{\dagger}_{\boldsymbol{k}\uparrow} a_{\boldsymbol{k}\uparrow} + a^{\dagger}_{\boldsymbol{k}\downarrow} a_{\boldsymbol{k}\downarrow}) + 2\sum_{\boldsymbol{k}} b^{\dagger}_{\boldsymbol{k}} b_{\boldsymbol{k}}$. Note that the Hamiltonian does not contain a explicitly, and that the field dependence of the scattering is completely characterized by the parameters: g, ν, and U_{bg}. The magnitude of g is derived in the following way. We define κ as the product of the magnetic field width of the resonance and the magnetic moment difference of the Feshbach state and the continuum state. For large values of ν, the boson field $b_{\boldsymbol{k}}$ can be adiabatically eliminated from the theory, and then $g = \sqrt{\kappa U_{\text{bg}}}$ is required in order for the scattering properties to have the correct dependence on magnetic field [8].

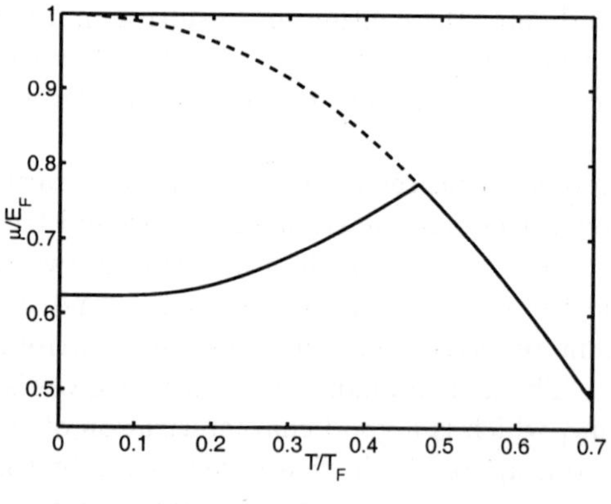

Figure 2. Chemical potential as a function of temperature for the system of resonance pairing (solid line). The second order phase transition occurs at $T_c \approx 0.5\, T_F$ where a clear cusp is visible. The dashed line shows the chemical potential of a non-interacting Fermi gas.

The essential point is that this Hamiltonian, founded on the microscopic basis of resonance scattering, is well-behaved at all detunings ν; even for the pathological case of exact resonance. The diluteness criterion is now given by constraints which require both the potential range and the spatial extent of the Feshbach resonance state, to be much smaller than the interparticle spacing (e.g. $n|a_{\rm bg}|^3 \ll 1$).

We apply this Hamiltonian to derive the self-consistent mean-fields for given thermodynamic constraints by formulating a Hartree-Fock-Bogoliubov theory [9,10,11]. The mean-fields present include the fermion number $f = \sum_{\boldsymbol{k}} \langle a^\dagger_{\boldsymbol{k}\uparrow} a_{\boldsymbol{k}\uparrow} \rangle$, the molecule field $\phi_m = \langle b_{\boldsymbol{k}=0} \rangle$ taken to be a classical field, and the pairing-field $p = \sum_{\boldsymbol{k}} \langle a_{\boldsymbol{k}\uparrow} a_{-\boldsymbol{k}\downarrow} \rangle$. It is well-known that such a theory must be renormalized in order to remove the ultraviolet divergence which arises from the incorporation of second-order vacuum contributions. This implies replacing the physical parameters in the Hamiltonian, U, g, and ν, by renormalized values so that observables are independent of a high momentum cut-off used in the formulation of the effective field-theory [12]. In order to diagonalize the Hamiltonian, we construct Bogoliubov quasiparticles according to the general canonical transformation

$$\begin{pmatrix} \alpha_{\boldsymbol{k}\uparrow} \\ \alpha^\dagger_{-\boldsymbol{k}\downarrow} \end{pmatrix} = \begin{pmatrix} \cos\theta & -e^{i\gamma}\sin\theta \\ e^{i\gamma}\sin\theta & \cos\theta \end{pmatrix} \begin{pmatrix} a_{\boldsymbol{k}\uparrow} \\ a^\dagger_{-\boldsymbol{k}\downarrow} \end{pmatrix}. \tag{4}$$

Given single particle energies, $U_k = \epsilon_k - \mu + Uf$, where μ is the chemical potential, and the gap parameter in the quasiparticle spectrum $\Delta = Up - g\phi_m$, the two transformation angles are specified as $\tan(2\theta) = |\Delta|/U_k$ and $\phi_m = |\phi_m|\exp(i\gamma)$. The corresponding quasiparticle spectrum is $E_k = \sqrt{U_k^2 + \Delta^2}$. Dropping terms of higher order than quadratic in the fermion operators, gives the resulting many-body Hamiltonian

$$\begin{aligned} H - \mu N &= 2(\nu - \mu)|\phi_m|^2 \\ &\quad + \sum_{\boldsymbol{k}} \Big(U_k + E_k \big(\alpha^\dagger_{\boldsymbol{k}\uparrow}\alpha_{\boldsymbol{k}\uparrow} + \alpha^\dagger_{\boldsymbol{k}\downarrow}\alpha_{\boldsymbol{k}\downarrow} - 1 \big) \Big). \end{aligned} \tag{5}$$

which is now in diagonal form.

The next task is to calculate the thermodynamic solutions. Equilibrium populations for the quasiparticles are given by the Fermi-Dirac distribution. The fermion number and pairing field are not only inputs to the Hamiltonian, but also determine the quasiparticle spectrum. Therefore, they must be self-consistent with the values derived by summing the relevant equilibrium density matrix elements over all wave numbers. In practice, at a given temperature, chemical potential, and molecule number ϕ_m, this requires an iterative method to locate self-consistent values for f and p. The value of ϕ_m is calculated by minimizing the grand potential $\Phi_G = -k_b T \ln \Xi$ at fixed temperature and chemical potential, with k_b denoting Boltzmann's constant. The partition function $\Xi = \text{Tr}[\exp(-(H - \mu N)/k_b T)]$ is found from Eq. (5). This procedure is mathematically equivalent to minimizing the Helmholtz free-energy at fixed temperature and density and corresponds uniquely to the maximum entropy solution. This solution has an associated particle number, $\langle N \rangle = -\partial\Phi_G/\partial\mu$ taken at constant temperature and volume, which must match the actual particle density of the gas, so that the final step is to adjust the chemical potential until this condition is satisfied. The whole procedure is repeated over a range of temperatures to determine the locus of thermodynamic equilibrium points. For large positive detunings, where the molecule field could be eliminated from the theory entirely, regular BCS theory emerges. For this case, when the scattering length a is negative the behaviour of the critical temperature on $1/a$ is given by the usual exponential law [5].

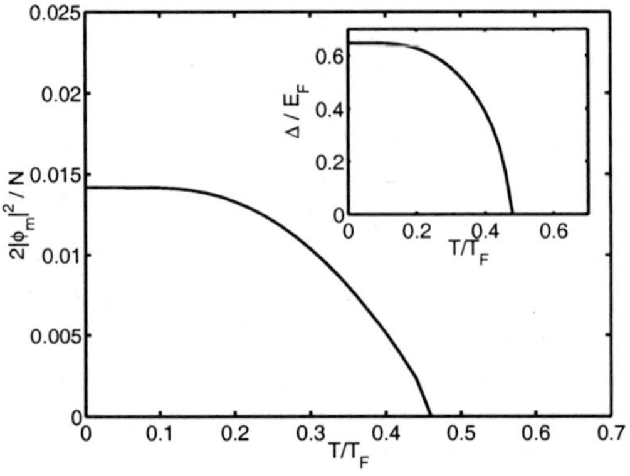

Figure 3. The temperature at the phase transition is also visible from the amplitude of the molecular field. This amplitude is non-zero only when the broken symmetry exists in the region $T < T_c$. For $T = 0$, the molecules form a Bose condensed fraction of 1.5% of the total gas sample. The inset shows the behaviour of the gap $\Delta = Up - g\phi_m$. The critical temperature T_c can be related to the value of the gap at $T = 0$. For comparison, in superconductors the analagous gap is simply the binding energy of a fermion pair.

In this paper, we use fermionic ^{40}K atoms as an example of the application of this theory. The values of our interaction parameters $a_{\text{bg}} = 176$ a$_0$ and $\kappa/k_b = 657$ μK are obtained from [14]. We fix the total density to be $n = 10^{14}$ cm^{-3}, a typical experimental value expected for this quantum degenerate gas in an optical trap. We set the detuning to be $\nu = +E_F$ so that the quasi-bound state is detuned slightly above the atomic resonance. For a temperature above T_c, the grand potential surface is shaped like a bowl, and the value of

ϕ_m which minimizes the grand potential is $\phi_m = 0$, associated with the self-consistent solution $p = 0$. For $T < T_c$,the grand potential surface is shaped like a Mexican hat, and its minimum is given by a ϕ_m with non-zero amplitude and an undetermined phase. The superfluid phase transition therefore leads to a spontaneously broken symmetry. The value of T_c can be clearly found from Figs. 2 and 3, where we show the chemical potential, the molecular density, and the gap as a function of temperature. We find for our parameter set for ^{40}K and almost zero detuning a remarkably high value for the critical temperature $T_c \approx 0.5\,T_F$, i.e. $T_c \approx 0.6$ μK. Furthermore we find a weak dependence of $T_c \approx 0.5\,T_F$ on the density, so that the value of T_c has more-or-less the same density behavior as T_F. When we increase the detuning to $\nu = +17.6\,E_F$ (this corresponds to a magnetic field detuning of 0.5 G away from the Feshbach resonance), the value of T_c drops to approximately $0.25\,T_F$.

The system of ^{40}K atoms, equally distributed among the two lowest hyperfine states, is a good candidate for demonstrating the superfluid phase transition. It not only exhibits a Feshbach resonance, but also, the inelastic binary collision events are energetically forbidden. Three-body interactions are highly suppressed, since the asymptotic three-body wave function should consist of a product of three s-wave two-body scattering wave functions. In a three-body interaction, two-particles are always in the same initial hyperfine state, and therefore the corresponding s-wave state is forbidden. The only three body relaxation could come from asymptotic p-waves, but these have very little contribution at the low temperatures considered. Although the detailed three-body collision problem is an intricate one, this asymptotic statistical effect should lead to a large suppression of the vibrational relaxation of quasi-bound molecules.

Current experimental techniques for ultracold gases do not produce samples which are spatially uniform. An optical dipole trap may be needed to confine the high field seeking atoms, and the conditions for the superfluid phase transition would be satisfied first in the trap center where the density is highest. The presence of the quasi-bound molecules may be a very useful aspect allowing direct observation of the phase transition through imaging the molecular field.

Acknowledgements

We thank J. Cooper, E. Cornell, D. Jin, C. Wieman, and B. DeMarco for very stimulating discussions. Support is acknowledged for M.H. from the National Science Foundation, for S.K. from the U.S. Department of Energy, Office of Basic Energy Sciences via the Chemical Sciences, Geosciences and Biosciences

Division, for M.C. from the INFM and SNS, Pisa (Italy), and for R.W. from the Austrian Academy of Sciences.

References

1. M. H. Anderson *et al.*, Science **269**, 198 (1995); K. B. Davis *et al.*, Phys. Rev. Lett. **75**, 3969 (1995); C. C. Bradley *et al.*, Phys. Rev. Lett. **75**, 1687 (1995); **79**, 1170(E) (1997).
2. B. DeMarco and D. S. Jin, Science **285**, 1703 (1999); A. G. Truscott, K. E. Strecker, W. I. McAlexander, G. B. Partridge, and R. G. Hulet, Science **291**, 2570 (2001); F. Schreck, L. Khaykovich, K. L. Corwin, G. Ferrari, T. Bourdel, J. Cubizolles, and C. Salomon, Phys. Rev. Lett. **87**, 080403 (2001).
3. J. Bardeen, L. N. Cooper, and J. R. Schrieffer, Phys. Rev. **108**, 1175 (1957); J. R. Schrieffer, *Theory of Superconductivity*, Perseus Books, Reading, Massachusetts, (1999).
4. See M. Randeria and references therein in *Bose-Einstein condensation*, ed. by A. Griffin, D.W. Snoke and S. Stringari, Cambridge Un. Press, Cambridge (1995).
5. A. G. Leggett, J. Phys. (Paris) **C7**, 19 (1980); M. Houbiers and H. T. C. Stoof, Phys. Rev. A **59**, 1556-1561 (1999); G. Bruun, Y. Castin, R. Dum *et al.*, Eur. Phys. J. D **7**, 433–439 (1999); H. Heiselberg, C. J. Pethick, H. Smith, and L. Viverit, Phys. Rev. Lett. **85**, 2418 (2000).
6. H. T. C. Stoof, M. Houbiers, C. A. Sackett, and R. G. Hulet, Phys. Rev. Lett. **76**, 10 (1996); R. Combescot, Phys. Rev. Lett. **83**, 3766 (1999).
7. H. Feshbach, Ann. Phys. **5**, 357 (1958); E. Tiesinga, B. J. Verhaar, and H. T. C. Stoof, Phys. Rev. A **47**, 4114 (1993).
8. This expression for g is chosen so that a obeys the correct field dependence. For further discussion see E. Timmermans *et al.*, Phys. Rep. **315** 199 (1999).
9. M. Holland, S. J. J. M. F. Kokkelmans, M. L. Chiofalo, and R. Walser, Phys. Rev. Lett. **87**, 120406 (2001).
10. An analagous field-theory is derived for a bosonic model in M. Holland, J. Park, and R. Walser, Phys. Rev. Lett., **86**, 1915 (2001).
11. J. Ranninger and S. Robaszkiewicz, Physica B **53**, 468 (1985); R. Friedberg and T. D. Lee, Phys. Rev. B **40**, 6745 (1989).
12. S. J. J. M. F. Kokkelmans, R. Walser, M. Chiofalo, J. Milstein, and M. Holland, to be published.
13. N. N. Bogoliubov, Nuovo Cimento **7**, 6 (1958); **7**, 794 (1958).
14. John L. Bohn, Phys. Rev. A **61**, 053409 (2000).

HARMONIC POTENTIAL TRAPS FOR EXCITONS IN 3D AND 2D

D.W. SNOKE, S. DENEV, AND V. NEGOITA

University of Pittsburgh Department of Physics and Astronomy, 3941 O'Hara St., Pittsburgh, PA 15260
E-mail: snoke@pitt.edu

L. PFIEFFER

Bell Laboratories, 700 Mountain Ave., Murray Hill, NJ 07974

We review recent experimental work on Bose-Einstein condensation of excitons. It is possible to confine excitons in a harmonic potential exactly analogous to the traps created for alkali gas condensates, in either a three-dimensional or two-dimensional system. We can also image the spatial profile of the excitons in the condensate, which should give a clear signature of condensation. We discuss two semiconductor systems in which this is possible– Cu_2O, in which a three-dimensional trap can be created, and GaAs quantum wells, in which a two-dimensional trap can be created. In Cu_2O, the main limitation on Bose-Einstein condensation of excitons at this time is density-dependent recombination into photons, which limits the total number of excitons. In GaAs-based quantum wells, the main limitation is the mobility of the excitons in the presence of disorder. Advances have been made in overcoming both of these problems.

In the past 15 years, several experiments have attempted to observe Bose-Einstein condensation of excitons in a semiconductor.[1,2,3,4,5,6,7] There are two motivations for this work. The first is the philosophical question of whether Bose condensation can occur for an energy particle. An exciton is a quantum of electronic excitation in a solid. It is a longstanding theoretical prediction[8,9,10,11] that Bose condensation is possible for energy particles such as excitons as long as two conditions are fulfilled: (1) the particles have an effective mass, and (2) the lifetime of the particles is long compared to their interparticle scattering time (thermalization time). Both of these conditions can be satisfied for excitons, and possibly also for phonons.

Excitons normally move freely through semiconductors, with an effective mass equal to the sum of the masses of a conduction-band electron and a valence-band hole. Although excitons are metastable and eventually recombine to photons, with the proper choice of a semiconductor system, excitons can have lifetimes hundreds of times longer than their thermalization time. Because of the light mass of these energy particles (on the order of the free electron mass), Bose condensation can in principle occur at temperatures of tens of Kelvin at typical densities. A Bose condensate of excitons should also be a *weakly interacting gas*, because the excitons are charge neutral, leading to weak exciton-exciton interactions. Excitons also interact weakly with the crystal lattice at low temperature, allowing them to move ballistically over macroscopic distances of tens of microns in some crystals.

The second motivation for this work is that the Bose-condensed state of these energy particles should have unique physical properties. Just as a standard Bose condensate is a superfluid, so also a Bose condensate of excitons should move coherently through a solid. This property could be used for coherent energy transport in optical or optoelectronic devices. For example, one can imagine a solid state laser in which a Bose condensate of excitons is formed, leading to collection of excitation energy from a macroscopic region of the semiconductor, concentrating energy in a small recombination region. Experiments on Bose condensation of excitons are directly related to ultrafast optics experiments on coherent emission from excitons.[13]

Several experiments have been reported using the semiconductors Cu_2O[2,3,4] and GaAs[5,6] in which effects occurred at low temperature and high exciton density which could be associated with Bose condensation. In all of these experiments, however, the interpretation was made difficult by the fact that the excitons were created by a high-power laser focused on the surface of a crystal, which led to a highly nonequilibrium distribution in free expansion. No clear boundary or volume of the excitons was defined.

Our approach has been to create a condensate of excitons in a system completely analogous to the condensates of alkali atoms in optical traps. A harmonic trapping potential for excitons can be created inside a crystal, leading to confinement of the excitons in an energy minimum. An exciton gas in a harmonic potential has a definite equilibrium volume. In this case, a spatial condensation into the ground state in the center of the trap should occur just as with an atomic condensate.[12] We can determine the spatial distribution of the excitons by recording images of the recombination luminescence of the photons, which has intensity proportional to the local exciton density.

1 A Three-dimensional Harmonic Potential for Excitons: Cu_2O

Anything that locally alters the band gap of a semiconductor will put a force on excitons, since a gradient in total energy is a force. A method for collecting excitons in a harmonic potential in a 3D semiconductor crystal has long been known and has been used to cause exciton motion over large distances.[14] This method uses inhomogeneous stress on the surface of the crystal to create a shear stress maximum inside the crystal. Shear stress leads to change of the symmetry of the crystal, causing a splitting of the electronic states, which

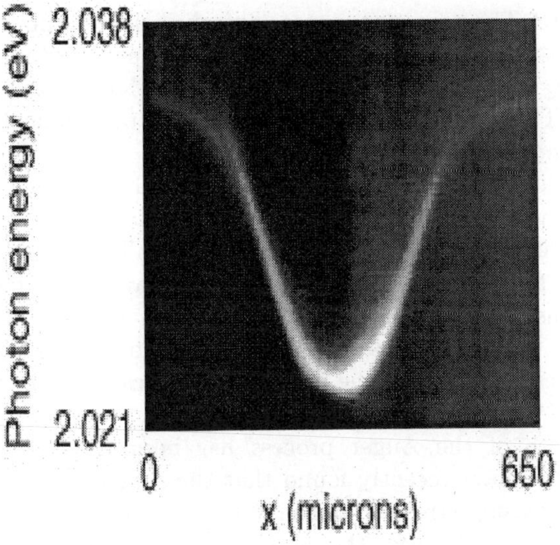

Figure 1. Image of a three-dimensional harmonic potential trap for excitons in Cu_2O.

means that some states will shift to lower energy at high shear stress. A maximum of shear stress therefore corresponds to a exciton energy minimum.

Fig. 1 shows a harmonic oscillator trap for excitons in the the semiconductor Cu_2O created this way. The image gives energy along one axis and spatial position along the other axis. We record images like by using a lens system to project an image of the semiconductor crystal onto the front slit of an imaging spectrometer, which preserves the spatial position of the light in one dimension while dispersing the wavelength of the light in the perpendicular direction. A CCD camera on the back of the spectrometer therefore gives us an image with position along one axis and exciton energy along the other axis. To see the entire potential energy profile, we scan a laser across the crystal and integrate the luminescence from the excitons over time.

The lowest exciton state in this crystal is split into a triplet "orthoexciton" state and a singlet "paraexciton" state. The paraexciton state lies 12 meV below the orthoexciton state, and therefore Bose condensation is expected in this state. The paraexciton state absorbs light very poorly, however, and therefore in a typical experiment, orthoexcitons are created, which then convert down into paraexcitons. Fig. 1 shows the orthoexciton luminescence from the trap. (This image has been recorded in a perpendicular direction from the image shown in a previous paper.[15])

This stress-trapping method was attempted by Trauernicht et al.[14] and abandoned as a method for creating a Bose condensate of excitons because they could not achieve high enough exciton densities. They could not achieve high densities for two reasons. First, they did not have a tunable laser with high instantaneous power. Second, an Auger effect exists in Cu_2O which causes excitons to annihilate at a rate proportional to their density. At high density, this process means that the exciton density can only increase as the square root of the laser power.

The exact rate of the Auger process has been the subject of some controversy.[16,17] We have recently found that the Auger recombination rate increases with increasing stress,[18] and may be very low at zero stress. This means that in creating a trap for the excitons using stress, one also increases the rate of the Auger process which destroys excitons. To avoid this, we can try to create a trap for the excitons using lower stress. The technical problem with this approach is that at low stress, the exciton paraexciton luminescence lies very near to the phonon-assisted orthoexciton luminescence line, which makes observation of the paraexcitons difficult. To avoid this, we can excite the paraexciton ground state directly via two-photon excitation.

In addition to the stress dependence of the Auger process, we have also found that the rate of ortho/para conversion also depends on stress. Surprisingly, as shown in Fig. 2, the rate of conversion becomes *slower* at higher stress.[18] Since in a typical experiment, orthoexcitons are created by the laser, if their rate of conversion to paraexcitons slows at high stress, this will also inhibit Bose condensation. This also implies that we should look for condensation in traps with low stress.

2 A Two-dimensional Harmonic Potential for Excitons: Indirect Excitons in InGaAs

We have also pursued experiments aimed at Bose condensation of excitons in two dimensions. Although Bose condensation is forbidden in two dimensions in translationally-invariant system, in a confining potential, Bose condensation becomes allowed in two dimensions.[19]

A two-dimensional quantum well system is appealing because the total number of excitons required for condensation is much less; also, imaging the

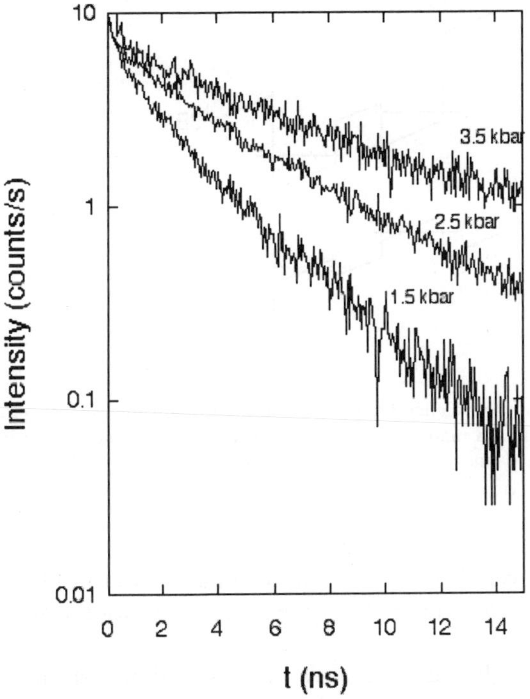

Figure 2. Decay of the orthoexciton luminescence vs. time, for three different crystal stresses.

excitons in a 2D system is very easy. The band structure of GaAs quantum well structures can be tailored to give excitons with long lifetime. Fig. 3 shows a typical band structure when electric field is applied. Two parallel 2D planes are created with a coupling between them. When electric field is applied perpendicular to the planes, the negative electrons become confined in one well, while the positive holes become confined in the adjacent well. Because the rate of exciton recombination into photons is proportional to the probability that the electron and hole of the exciton are at the same point in space, the exciton lifetime depends on the wavefunction overlap of the electron and hole. When the electron and hole are confined in different wells,

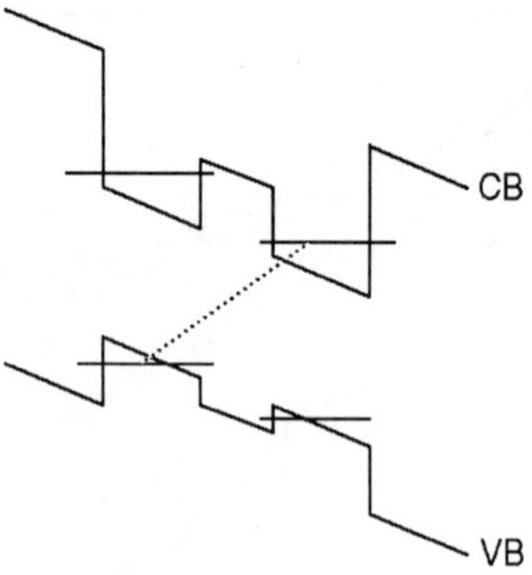

Figure 3. Conduction and valence bands for a double quantum well system in an electric field. Indirect excitons are made from an electron in one well and a hole in the other, as indicated by the dashed line.

the wavefunction overlap is very small, leading to exciton lifetimes of up to 100 ns,[20] which is long compared to typical thermalization times on the order of picoseconds.

In the case of bulk Cu_2O crystals discussed above, shear stress created by inhomogeneous compression of the sample was the main mechanism for the trapping of the excitons. In the case of GaAs-related compounds, this will not work, because the hydrostatic deformation potential is much larger than the shear deformation potential. This means that compression of the sample will raise the total energy of the excitons more than a shear stress will lower it. We can use stress to create an energy minimum, however, by creating a hydrostatic *expansion* in the plane of the quantum wells. We do this by bending the sample slightly with a pin, causing the bottom surface to stretch.[21] The thinner the sample, the deeper and narrower the well will be. Fig. 4 shows a trap created this way in a thin (100 μm) sample with InGaAs quantum wells. The excitons are confined to move in a two-dimensional plane by the barriers, and confined in a macroscopic harmonic potential in the plane.

The limitation on Bose condensation in these 2D quantum wells is not the total density, but the diffusion length of the excitons. In GaAs quantum wells, the mean free path at low density and low temperature is very low, less than a micron, due to localization in random potential minima. These random potential minima arise from disorder in the sample. We can see that disorder gives energy fluctuations in our sample, because this disorder gives a well-known effect on the spectral line width, called "inhomogeneous

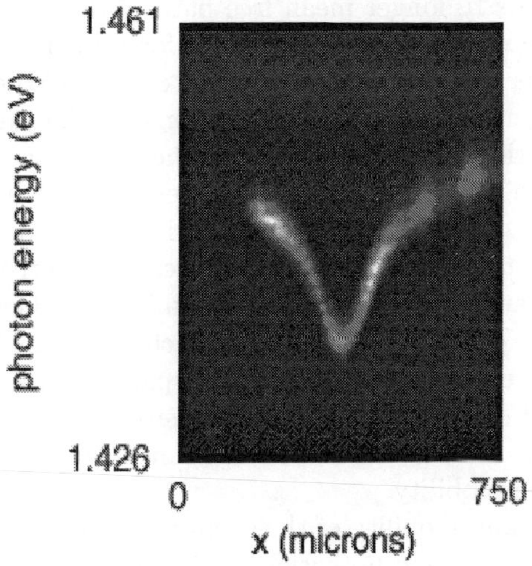

Figure 4. Image of a two-dimensional harmonic potential trap for excitons in $In_{.1}Ga_{.9}As$ quantum wells.

broadening." In the limit of low density and low temperature, the full width at half maximum of the luminescence line in a solid gives the range of energy fluctuations due to disorder in the sample.

This disorder is not due to poor quality of the wells, for example, variation of the width of the wells. Even one monolayer variation in the well width of our wells would give an energy jump of more than 5 meV due to the change in quantum zero-point energy; by comparison, the inhomogeneous broadening in our samples is 3 meV or less. Instead, the random potential fluctuations most likely arise from alloy disorder in the barrier material.

We have reduced this alloy disorder by using InGaAs quantum wells with pure GaAs barriers and buffer layers. Since the electron and hole wavefunctions spill over into the barrier layers, the quality of these layers has a large effect on the disorder felt by the excitons. In undoped samples with pure GaAs buffer and barrier layers, we have observed luminescence with inhomogeneous broadening as low as 0.7 meV. This is remarkable for narrow quantum wells, and also corresponds to longer mean free path of the excitons, up to 10 microns, according to recent measurements. Unfortunately, because the barrier is pure GaAs, it has lower energy, making the overlap of the electron and hole wavefunctions larger, which reduces the lifetime of the excitons to around 2 ns even when electric field is applied across the wells.

In principle, we could look for Bose condensation at higher temperature, so that the excitons have enough energy to hop over the random potential fluctuations. Since the critical temperature scales linearly with exciton density, Bose condensation is theoretically possible at higher temperature and higher density. The problem with this approach is that at higher density, the excitons begin to overlap with each other, so that they are no longer a weakly interacting gas and can undergo a phase transition to a plasma. Therefore it is preferable to stay at low temperature and low exciton density and work to improve the exciton mobility.

As the lifetime and mobility of the excitons is increased, the excitons feel more of the entire harmonic potential. This allows us to look for a macroscopic, quasi-equilibrium Bose condensate exactly analogous to an atomic Bose condensate in a magneto-optical trap.

Acknowledgments

This work has been supported by the National Science Foundation under grant DMR-0102457 and by the Department of Energy under grant DE-FG02-99ER45780. Early contributions to these experiments were made by Karl Eberl of the Max Planck Institute FKF in Stuttgart, Germany.

References

1. D.W. Snoke, J.P. Wolfe, and A. Mysyrowicz, Phys. Rev. Lett. **64**, 2543 (1990).
2. J.L. Lin and J.P. Wolfe, Phys. Rev. Lett. **71**, 1223 (1993).
3. E. Fortin, S. Fafard and A. Mysyrowicz, Phys. Rev. Lett. **70**, 3951 (1993); A. Mysyrowicz, E. Benson, E. Fortin, Phys. Rev. Lett. **77**, 896 (1996).
4. M. Y. Shen, T. Yokouchi, S. Koyama, and T. Goto Phys. Rev. B **56**, 13066 (1997); T. Goto, M. Y. Shen, S. Koyama, and T. Yokouchi, Phys. Rev. B **56**, 4284 (1997).
5. L. V. Butov, A. Zrenner, G. Abstreiter, G. Böhm, and G. Weimann, Phys. Rev. Lett. **73**, 304 (1994).
6. J.C. Kim and J.P. Wolfe, Phys. Rev. B **57**, 9861 (1998).

7. For a general review of experiments on Bose condensation of excitons, see S.A. Moskalenko and D.W. Snoke, *Bose-Einstein Condensation of Excitons and Biexcitons and Coherent Nonlinear Optics with Excitons,* (Cambridge University Press, 2000).
8. S.A. Moskalenko, Fiz. Tverd. Tela **4**, 276 (1962).
9. J.M. Blatt, K.W. Boer, and W. Brandt, Phys. Rev. **126**, 1691 (1962).
10. L.V. Keldysh and A.N. Kozlov, *Zh. Eksp. Teor. Fiz., Pisma.* **5**, 238 (1967); *Zh. Eksp. Teor. Fiz.* **54**, 978 (1968) [*Sov. Phys. JETP* **27**, 521 (1968)].
11. C. Comte and P. Nozières, *J. Phys.* **43**, 1069 (1982); P. Nozières and C. Comte, *J. Phys.* **43**, 1083 (1982).
12. E.g., M.H. Anderson, J.R. Ensher, M.R. Matthews, C.E. Weiman, and E.A. Cornell, Sciencc **269**, 198 (14 July 1995).
13. E.g., E. Runge, R. Zimmermann, Phys. Rev. B **61**, 4786 (2000); T. Karasawa, H. Mino, and M. Yamamoto, J. Lum. **87/89**, 174 (2000).
14. D.P. Trauernicht, A. Mysyrowicz, and J.P. Wolfe, Physical Review B **28**, 3590 (1983); D.P. Trauernicht, J.P. Wolfe, and A. Mysyrowicz, Phys. Rev. B **34**, 2561 (1986).
15. D.W. Snoke and V. Negoita, Phys. Rev. B **61**, 2904 (2000).
16. K. E. O'Hara, J. P. Wolfe, Phys. Rev. B 62, 12909 (2000); K.E. O'Hara, J.R. Gullingsrud, and J.P. Wolfe, Phys. Rev. B **60**, 10872 (1999).
17. G. M. Kavoulakis and A. Mysyrowicz, Phys. Rev. B **61**, 16619 (2000)
18. S. Denev and D.W. Snoke, "Stress Dependence of Exciton Relaxation Processes in Cu_2O," submitted to Physical Review B (preprint: cond-mat/0106412.)
19. See, e.g. P. Nozieres, in *Bose-Einstein Condensation*, A. Griffin, D.W. Snoke and S. Stringari, eds. (Cambridge University Press, 1995).
20. V. Negoita, D.W. Snoke, and K. Eberl, Phys. Rev. B **60**, 2661 (1999).
21. V. Negoita, D.W. Snoke, and K. Eberl, Appl. Phys. Lett. **75**, 2059 (1999).

MEASURING THE FREQUENCY OF LIGHT WITH ULTRA SHORT PULSES

T. W. HÄNSCH, R. HOLZWARTH, M. ZIMMERMANN, AND TH. UDEM

Max-Planck Institut für Quantenoptik, 85748 Garching, Germany
E-mail: Thomas.Udem@mpq.mpg.de

Femtosecond laser frequency comb techniques are vastly simplifying the art of measuring the frequency of light. A single mode-locked femtosecond laser is now sufficient to synthesize hundreds of thousands of evenly spaced spectral lines, spanning much of the visible and near infrared region. The mode frequencies are absolutely known in terms of the pulse repetition rate and the carrier-envelope phase slippage rate, which are both accessible to radiofrequency counters. Such a universal optical frequency comb synthesizer can serve as a clockwork in atomic clocks, based on atoms, ions or molecules oscillating at optical frequencies.

1 Introduction

For more than a century, precise optical spectroscopy of atoms and molecules has played a central role in the discovery of the laws of quantum physics, in the determination of fundamental constants, and in the realization of standards for time, frequency, and length. The advent of highly monochromatic tunable lasers and techniques for nonlinear Doppler-free spectroscopy in the early seventies had a dramatic impact on the field of precision spectroscopy [1,2]. Today, we are able to observe extremely narrow optical resonances in cold atoms or single trapped ions, with resolutions ranging from 10^{-13} to 10^{-15}, so that it might ultimately become possible to measure the line center of such a resonance to a few parts in 10^{18}. Laboratory experiments searching for slow changes of fundamental constants would then reach unprecedented sensitivity. A laser locked to a narrow optical resonance can serve as a highly stable oscillator for an all-optical atomic clock [3,4] that can satisfy the growing demands of optical frequency metrology, fiber optical telecommunication, or navigation. However, until recently there was no reliable optical "clockwork" available that could count optical frequencies of hundreds of THz. Most spectroscopic experiments still rely on a measurement of optical wavelengths rather than frequencies. Unavoidable geometric wavefront distortions have so far made it impossible to exceed an accuracy of a few parts in 10^{10} with a laboratory-sized wavelength interferometer. To measure optical frequencies, only a few harmonic laser frequency chains have been built during the past 25 years which start with a cesium atomic clock and generate higher and higher harmonics in nonlinear diode mixers, crystals, and other nonlinear devices [5,6,7,8]. Phase-locked transfer oscillators are needed after each step, so that such a chain traversing a vast region of the electromagnetic spectrum becomes highly complex, large, and delicate, and requires substantial resources and heroic efforts to build and operate. Most harmonic laser frequency chains are designed to measure just one single optical frequency.

In 1998, our laboratory has introduced a revolutionary new approach that vastly simplifies optical frequency measurements. We could demonstrate that the broad comb of modes of a mode-locked femtosecond laser can be used as a precise ruler in frequency space [9,10]. This work has now culminated in a compact and reliable all-solid-state frequency "chain" which is actually not really a chain any more but requires just a single mode-locked laser [11,12,13,14]. As a universal optical frequency comb synthesizer it provides the long missing simple link between optical and microwave frequencies. For the first time, small scale spectroscopy laboratories have now access to the ability to measure or synthesize any optical frequency with extreme precision. Femtosecond frequency comb techniques have since begun to rapidly gain widespread use, with precision measurements in Cs [9], Ca [15,16], CH_4 [18], H [18], Hg^+ [15,4], I_2 [12,19], Yb^+ [17] and In^+ [20].

The same femtosecond frequency comb techniques are also opening new frontiers in ultrafast physics. Control of the phase evolution of few cycle light pulses, as recently demonstrated [13,21], provides a powerful new tool for the study of highly nonlinear phenomena that should depend on the phase of the carrier wave relative to the pulse envelope, such as above threshold ionization, strong field photoemission, or the generation of soft x-ray attosecond pulses by high harmonic generation.

In the first experiment of its kind, we have applied the frequency comb of a mode-locked femtosecond laser to measure the frequency of the cesium D_1 line [9]. This frequency provides an important link for a new determination of the fine structure constant α. More recently, we have measured the absolute frequency of the hydrogen 1S-2S two-photon resonance in a direct comparison with a cesium atomic fountain clock to within 1.9 parts in 10^{14}, thus realizing one of the most accurate measurement of an optical frequency to date [11,16]. During the past few years, precision spectroscopy of hydrogen has yielded a value for the Rydberg constant that is now one of the most accurately known fundamental constant [22]. Nonetheless, after more than a century of spectroscopic experiments, the hydrogen atom still holds substantial challenges and opportunities for further dramatic advances.

2 Optical Frequency Differences

While it has been extremely difficult in the past to measure an absolute optical frequency, a small frequency difference or gap between two laser frequencies can be measured rather simply by superimposing the two laser beams on a photodetector and monitoring a beat signal. The first experiments of this kind date back to the advent of cw He-Ne-lasers in the early sixties[23]. Modern commercial fast photodiodes and microwave frequency counters make it possible to directly count frequency differences up to the order of 100 GHz. Since the gap between the high frequency endpoint of a traditional harmonic laser frequency chain and an unknown optical frequency to be measured can easily amount to tens or hundreds of THz, there has long been a strong interest in methods for measuring much larger optical frequency differences. Motivated by such problems in precision spectroscopy of atomic hydrogen, we have previously introduced a general, although perhaps not very elegant solution for the measurement of large optical frequency gaps with the invention of the optical frequency interval divider (OFID) which can divide an arbitrarily large frequency difference by a factor of precisely two [24,25]. An OFID receives two input laser frequencies f_1 and f_2. The sum frequency $f_1 + f_2$ and the second harmonic of a third laser $2f_3$ are created in nonlinear crystals. The radio frequency beat signal between them at $2f_3 - (f_1 + f_2)$ is used to phase-lock the third laser at the exact midpoint. Phase-locking of two optical frequencies is achieved electronically by locking the phase of their beat signal to zero or, to reduce $1/f$-noise, to a given offset radio frequency, provided by a local oscillator. Techniques of conventional radio frequency phase-locked loops can be applied. With a divider chain of n cascaded OFIDs, the original frequency gap can be divided by a factor of 2^n. Frequency intervals up to several THz can also be measured with passive optical frequency comb generators [26,27]. These are electro optical modulators that create side bands very efficiently. Beat signals can then be observed with sidebands on different sides of the carrier and frequency gaps on the order of a few THz can be bridged. To measure larger gaps, a chain of OFIDs can be followed by an OFCG.

3 Femtosecond Light Pulses

It has long been recognized that the periodic pulse train of a mode-locked laser can be described in the frequency domain as a comb of equidistant modes, so that such a laser can serve as an active OFCG. More than twenty years ago, the frequency comb of a mode-locked picosecond dye laser has first been used as an optical ruler to measure transition frequencies in sodium [28]. This route

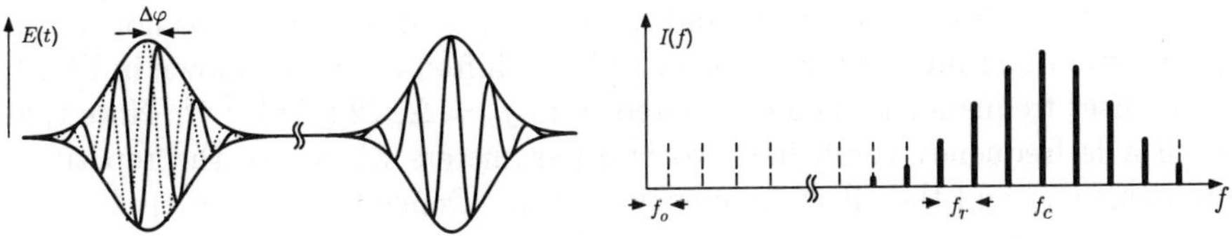

Figure 1. Two consecutive pulses of the pulse train emitted by a mode locked laser and the corresponding spectrum (right). The pulse to pulse phase shift $\Delta\varphi$ results in a frequency offset because the carrier wave at f_c moves with the phase velocity v_p while the envelope moves with the group velocity v_g.

was further pursued in the seventies and eighties [29,30,31], but the attainable bandwidths were never sufficiently large to make it a widespread technique for optical frequency metrology. Broadband femtosecond Ti:sapphire lasers have existed since the beginning of the nineties, but only our recent experiments at Garching have shown conclusively, that such lasers can play a crucial role in this field [10,14]. To understand the mode structure of a fs frequency comb and the techniques applied for its stabilization one can look at the idealized case of a pulse circulating in a laser cavity with length L as a carrier wave at f_c that is subject to strong amplitude modulation described by an envelope function $A(t)$. This function defines the pulse repetition time T, and the pulse repetition frequency $f_r = T^{-1}$ by demanding $A(t-T) = A(t)$ where $T = 2L/v_g$ with cavity mean group velocity v_g. Because of the periodicity of the envelope function the electric field at a given place (e.g. at the output coupler) can be written as

$$E(t) = Re\left(A(t)e^{-2\pi f_c}\right) = Re\left(\sum_q A_q e^{-2\pi(f_c+qf_r)}\right) \tag{1}$$

where A_q are Fourier components of $A(t)$. This equation shows that the resulting spectrum consists of a comb of laser modes that are separated by the pulse repetition frequency. Since f_c is not necessarily an integer multiple of f_r the modes are shifted from being exact harmonics of the pulse repetition frequency by an offset that is chosen to be smaller than f_r:

$$f_n = nf_r + f_o \qquad\qquad n = \text{a large integer} \tag{2}$$

This equation maps two radio frequencies f_r and f_o onto the optical frequencies f_n. While f_r is readily measurable, is not easy to access unless the frequency comb contains more than an optical octave [32]. In the time domain the frequency offset is obvious because the group velocity differs from the phase velocity inside the cavity and therefore the carrier wave does not repeat itself after one round trip but appears phase shifted by $\Delta\varphi$ as shown in Fig. 1. The offset frequency is then calculated from $f_o = \Delta\varphi/2\pi T$ [29,32,33]. Note that such a fs frequency comb has two free parameters which are the repetition frequency f_r and the offset frequency $f_o < f_r$. Depending on the application one or both parameters have to be stabilized.

4 Femtosecond Combs as Frequency Rulers

At the high peak intensities of femtosecond laser pulses nonlinear effects due to the $\chi^{(3)}$ nonlinear susceptibility are considerable even in standard silica fibers. The output spectrum of a femtosecond laser can be broadened significantly via self phase modulation in an optical fiber therefore increasing its useful width even further beyond the time-bandwidth limit of the laser. Now the question arises whether or not this broad frequency comb is equally spaced and can therefore be used as a ruler to measure frequency differences. To test this we have compared the fs comb with an optical frequency interval divider by locking the two input lasers to modes separated by as much as 44 THz. We measured a beat note between the output of the fiber with a nearby mode of the comb and found that the frequency comb is equally spaced even after spectral broadening in a standard single mode fiber at the level of a few parts in 10^{18} [34]. Note that the coherence between the pulses is obviously preserved. Previously we have also shown that the easily accessible repetition rate of such a laser equals the mode spacing within the experimental uncertainty of a few parts in 10^{16} [10].

To phase-lock the pulse repetition rate to a signal provided by a synthesizer one faces the problem of noise multiplication. It is well known that the total noise intensity grows as N^2 when a radio frequency is multiplied by a factor of N [35]. Fortunately the laser cavity acts as a filter and prevents the high frequency noise components from propagating through the frequency comb [32]. For most applications it is desirable to fix one of the modes in frequency space and phase-lock the pulse repetition rate simultaneously. For this purpose it is necessary to control the phase velocity (more precisely the round trip phase delay) of that particular mode and the group velocity (more precisely the round trip group delay) independently. A piezo driven folding mirror changes the cavity length and leaves $\Delta\varphi$ approximately constant as the additional path in air has a negligible dispersion. A mode-locked laser that uses two intracavity prisms to produce the negative group velocity dispersion $(\partial^2\omega/\partial k^2)$ necessary for Kerr-lens mode-locking provides us with a means for independently controlling the pulse repetition rate. We use a second piezotransducer to slightly tilt the mirror at the dispersive end of the cavity about a vertical pivot that ideally corresponds to the mode f_n [32]. We thus introduce an additional phase shift $\Delta\Phi$ proportional to the frequency distance from f_n, which displaces the pulse in time and thus changes the round trip group delay. In the frequency domain one could argue that the length of the cavity stays constant for the mode while higher (lower) frequency modes experience a longer (shorter) cavity (or vice versa, depending on the sign of $\Delta\Phi$). In the case where only dispersion compensation mirrors are used to produce the negative group velocity dispersion one can modulate the pump power or manipulate the Kerr lens by slightly tilting the pump beam [14]. Although the two controls (i.e. cavity length and pump power) are not independent they affect the round trip group delay T and the round trip phase delay differently and this is which allows us to control both, f_o and f_r.

5 Absolute Optical Frequencies

For the absolute measurement of optical frequencies one has to determine frequencies of several 100 THz in terms of the definition of the SI second represented by the cesium ground state hyperfine splitting of 9.2 GHz. Extending our principle of determining large frequency differences to the intervals between harmonics or subharmonics of an optical frequency leads naturally to the absolute measurement of optical frequencies. In the most simple case this is the interval between an optical frequency f and its second harmonic $2f$ as illustrated in Fig. 2. But of course other intervals can be used as well. Such an optical frequency synthesizer directly relates a radio frequency to an optical frequency f simply by multiplying the known mode spacing f_r (i.e. a radio frequency) with the number of modes between the harmonics (e.g. between f and $2f$).

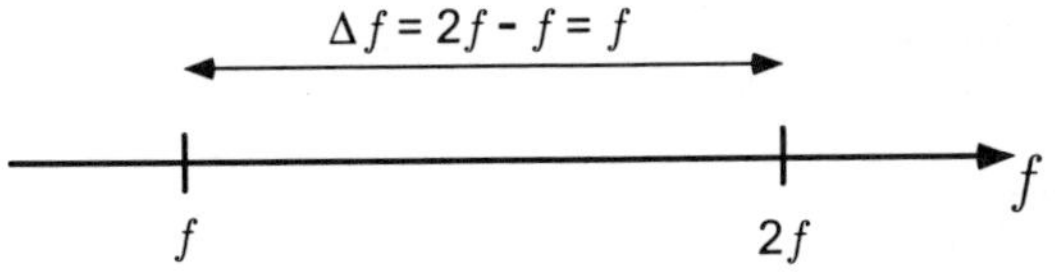

Figure 2. The principle of absolute optical frequency measurements. The interval Δf between f and $2f$ is just equal to the frequency f itself.

6 Frequency Combs Spanning more than an Octave

The first absolute measurement of an optical frequency with a fs frequency comb [11] has inspired further rapid advances in the art of frequency metrology. In collaboration with P. St. Russell, J. Knight and W. Wadsworth from the University of Bath (UK) we have used novel microstructured photonic crystal fibers (PCF) [36] to achieve further spectral broadening of femtosecond frequency combs. In the fiber used here light is guided in a pure silica core with a diameter of approximately 1.5 μm by surrounding it with an array of air holes. The remarkable dispersion characteristics attainable with the large effective index step (including zero group velocity dispersion well below 800 nm) and the high peak intensities associated with the short pulses and the small core size, enables one to observe a range of unusual nonlinear optical effects [37], including very effective spectral broadening to more than an optical octave even with the moderate output power from the laser oscillator. Similar experiments have been reported by S. Cundiff, J. Hall and coworkers in Boulder using a fiber fabricated at Bell Laboratories [12,13].

With an octave wide spectrum we can directly access the interval between an optical frequency f and its second harmonic $2f$. As shown in Fig. 3 such an optical frequency synthesizer [11,12,13,14] allows the direct comparison of radio and optical frequencies without the need of any optical frequency interval dividers or further nonlinear steps. Our $f : 2f$ optical frequency synthesizer is based on a 25 fs Ti:sapphire high repetition rate ring laser (GigaOptics, model GigaJet). While the ring design makes it almost immune to feedback from the fiber, the high repetition rate increases the available power per mode. A PZT-monted folding mirror is used to lock f_r and the pump intensity is used to f_o to a precise reference frequency provided by an atomic clock or GPS

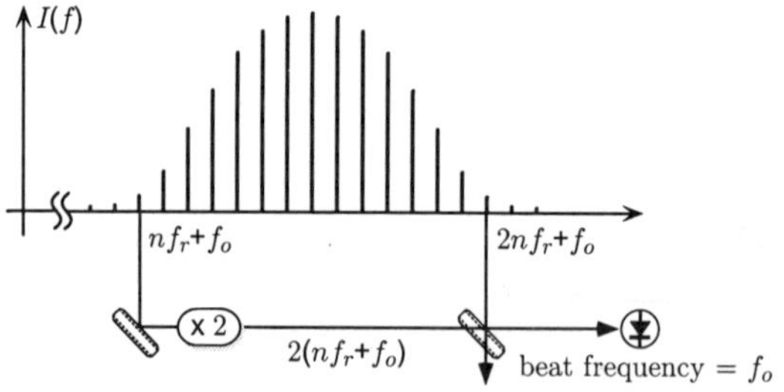

Figure 3. To obtain the frequency offset f_o the "red" portion of an octave spanning comb is frequency doubled and a beat note with "blue" wing is observed.

receiver [14]. The set-up requires only 1 square meter on our optical table with the potential for further miniaturization. At the same time it supplies us with a reference frequency grid across much of the visible and infrared spectrum with comb lines that distinguished with a wavemeter. This makes it an ideal laboratory tool for precision spectroscopy and a compact solid state system for all optical clocks [4].

7 Validation of the $f : 2f$ Optical Synthesizer

To check the integrity of the broad frequency comb and evaluate the overall performance of the $f : 2f$ optical synthesizer we have compared it with a similar set-up that is based on bridging the frequency gap between two optical frequencies with the ratio $3.5f$ and $4f$ [11]. We produced two signals around 353.5 THz (848 nm) derived from the same 10 MHz quartz oscillator. After averaging all data we obtained a mean deviation from the expected beat frequency of 71 ± 179 mHz. This corresponds to a relative uncertainty of 5.1×10^{-16} [14]. An even more accurate testing was recently performed by the NIST group [38]. No systematic effect was visible in these experiments.

Other important applications of this frequency domain technique in the time domain where the carrier offset slippage frequency is an important parameter and needs to be controlled for the next generation of ultrafast experiments. In collaboration with F. Krausz (Vienna technical university) we have applied fs comb techniques to control the phase evolution of ultra-short pulses lasting for only a few optical cycles [21]. Future applications of precise optical frequency measurements also include the search for variations in the fundamental constants and the test of CPT invariance with anti-hydrogen now underway at CERN. We believe that the development of accurate optical frequency synthesis marks only the beginning of an exciting new period of ultra-precise physics. Finally we would like to thank our collaborators, without their help the work presented here would not have been possible.

References

1. *Nonlinear Spectroscopy, Proceedings of the International School of Physics "Enrico Fermi"*, ed. N Bloembergen (North Holland Publ. Co., 1977).
2. *Frontiers in Laser Spectroscopy, Proceeedings of the International School of Physics "Enrico Fermi"*, ed. T W Hänsch and M Inguscio (North Holland Publ. Co., 1995).
3. S A Diddams *et al.*, *Science* **293**, 825 (2001).

4. J C Bergquist *et al.*, in this volume.
5. Th Udem *et al.*, *Phys. Rev. Lett.* **79**, 2646 (1997).
6. H Schnatz *et al.*, *Phys. Rev. Lett.* **76**, 18 (1996).
7. C Schwob *et al.*, *Phys. Rev. Lett.* **82**, 4960 (1999).
8. J E Bernard *et al.*, *Phys. Rev. Lett.* **82**, 3238 (1999).
9. Th Udem *et al.*, *Phys. Rev. Lett.* **82**, 3568 (1999).
10. Th Udem *et al.*, *Opt. Lett.* **24**, 881 (1999).
11. J Reichert *et al.*, *Phys. Rev. Lett.* **84**, 3232 (2000).
12. S A Diddams *et al.*, *Phys. Rev. Lett.* **84**, 5102 (2000).
13. D Jones *et al.*, *Science* **288**, 635 (2000).
14. R Holzwarth *et al.*, *Phys. Rev. Lett.* **85**, 2264 (2000).
15. Th Udem *et al.*, *Phys. Rev. Lett.* **86**, 4996 (2001).
16. J Stenger *et al.*, *Phys. Rev.* A **53**, 021802(R) (2001).
17. J Stenger *et al.*, physics/0103040.
18. M Niering *et al.*, *Phys. Rev.* A **84**, 5456 (2000).
19. A Yu Nevsky *et al.*, *Opt. Commun.* **192**, 263 (2001).
20. J v Zanthier *et al.*, *Opt. Lett.* **25**, 1729 (2000).
21. A Apolonski *et al.*, *Phys. Rev. Lett.* **85**, 740 (2000).
22. P J Mohr *et al.*, *Rev. Mod. Phys.* **72**, 351 (2000).
23. A Javan *Fundamental and Applied Laser Physics, Proceedings of the Esfahan Symposium*, ed. M. S. Feld, A. Javan, and N. A. Kurnit (John Wiley & Sons, New York 1973).
24. T W Hänsch *Proceedings of The Hydrogen Atom*, ed. T W Hänsch (Springer, Berlin 1989).
25. H R Telle *et al.*, *Opt. Lett.* **15**, 532 (1990).
26. M Kourogi *et al.*, *IEEE J. Quantum Electron.* **31**, 2120 (1995).
27. L Brothers *et al.*, *Opt. Lett.* **19**, 245 (1994).
28. J N Eckstein *et al.*, *Phys. Rev. Lett.* **19**, 245 (1994).
29. A I Ferguson *et al.*, *Appl. Phys.* **18**, 257 (1979).
30. Y V Baklanov *et al.*, *Appl. Phys.* **12**, 97 (1977).
31. D M Kane *et al.*, *Appl. Phys.* **39**, 171 (1986).
32. J Reichert *et al.*, *Opt. Commun.* **172**, 59 (1999).
33. D J Wineland *et al. Proceedings of The Hydrogen Atom*, ed. T W Hänsch (Springer, Berlin 1989).
34. R Holzwarth PhD thesis, Ludwig-Maximilians-Universität, Munich, Germany (2000).
35. F L Walls *et al.*, *IEEE Trans. Instrum. Meas.* **24**, 210 (1975).
36. J C Knight *et al.*, *Opt. Lett.* **21**, 1547 (1996).
37. W J Wadsworth *et al.*, *Electron. Lett.* **36**, 53 (2000).
38. S A Diddams *et al.*, submitted to *Opt. Lett.*

COHERENT OPTICAL FREQUENCY SYNTHESIS AND DISTRIBUTION

JUN YE, JOHN L. HALL, JOHN JOST, LONG-SHENG MA, AND JIN-LONG PENG

JILA, National Institute of Standards and Technology and University of Colorado, Department of Physics, University of Colorado, Boulder, CO 80309-0440 USA

E-mail: ye@jila.colorado.edu

We demonstrate a simple optical clock based on an optical transition of iodine molecules, providing a frequency stability superior to most rf sources. Combined with a femtosecond-laser-based optical comb to provide the phase coherent clock mechanism linking the optical and microwave spectra, we derive an rf clock signal of comparable stability over an extended period. In fact, the stability (5×10^{-14} at 1 s) of the cw laser locked on the iodine transition is transferred to every comb component throughout the optical octave bandwidth (from 532 nm to 1064 nm) with a precision of 3.5×10^{-15}. Stability characterization of the optical clock is below 3×10^{-13} at 1 s (currently limited by the microwave sources). The long-term stability of this simple optical standard is demonstrated to be better than 4.6×10^{-13} over a year. To realize a genuine optical frequency synthesizer, another widely tunable single-frequency cw laser has been employed to randomly access the stabilized optical comb and lock to a desired comb component. The goal is to generate on demand a highly stable single-frequency optical signal in any part of the visible spectrum with a useful output power.

The recent revolution in the physical science brought by the beautiful merger of cw-based ultra-precision laser work and ultrafast lasers and associated nonlinear optics has enabled profound progress in both areas. Optical frequency measurements have been reduced to a simple task even while the highest level of measurement precision has been achieved [1-5]. Control of the carrier-envelope phase is now possible [6,7]. Pulse trains from independent mode-locked lasers have been synchronized below the 10 fs level and their carrier waves phase locked, leading to coherent pulse synthesis [8,9]. Arbitrary and yet phase coherent synthesis of optical spectra, either in terms of selecting desired discrete components in the frequency spectrum or by way of specifying preferred pulse shape and duration in the time domain, now appears possible.

To complement the rapid development of high performance optical frequency standards, it is important to establish an optical comb with excellent phase coherence among its individual components. The phase stability needs to exceed that of the optical standards. With this capability, we will be able to transfer the stability of a single optical oscillator to the entire comb set over its vast bandwidth, and also derive clock signals in the microwave/rf domain without any stability compromise. Optical standards based on single ions and cold atoms promising potential stability around 1×10^{-16} at 1 s [10] and potential accuracy at 1×10^{-18} [11] may very well become future national standards, but such systems would require elaborate designs. On the other hand, excellent candidates in cell-based optical frequency standards do exist, such as the one presented in this paper, that would offer compact, simple, and less expensive system configurations, albeit at the cost of performance degradation by perhaps two decades. Along with optical combs, a competent laboratory would be able to realize a network of microwave and optical frequencies at a level of stability and reproducibility that surpasses the properties of basically all normal commercially available frequency sources, but with a reasonable cost. Easy access to the resolution and stability offered by the optical standards would greatly facilitate application of frequency metrology both to precision experiments for fundamental physics and to practical devices.

To reach that goal, it is important to understand and implement an optimized control scheme of the optical comb that avoids the limitation in phase coherence between the two ends of comb spectrum. In our previous work, an entangled control scheme could achieve only *frequency*-locking across the comb spectrum, with residual frequency noise exceeding 100 Hz at 1-s [12]. In this contribution we demonstrate that an orthogonal control of the (100 MHz fs-laser based) optical comb can lead to Hz-level ($\leq 3.5 \times 10^{-15}$) *phase*-tracking stability across the entire optical octave. Furthermore, the orthogonalization procedure permits independent control of both degrees of freedom associated with the optical comb, leading to a clock work mechanism using only one comb parameter. Recent work [13] uses a two-parameter-control scheme to transfer the stability of a cold ion based optical standard to the comb lines at the 2×10^{-16} level. Clearly, with a mature technical solution to the "gearbox problem" at hand, all future progresses in optical domain and rf domain standards can be utilized in both spectra.

The octave-spanning optical comb and the associated control scheme are shown in Figure 1. A Kerr-lens mode-locked (KLM) femtosecond (fs) laser [14] generates a repetitive (~ 10 ns repetition interval) pulse train, with a corresponding rigorous periodicity in the spectral domain. To permit the coverage of an entire optical octave, the bandwidth of the comb emitted from the laser is further broadened by launching the pulse train into a microstructure fiber [15]. There are two degrees of freedom associated with the comb frequencies. The interval between adjacent frequencies in this periodically spaced "comb" is directly defined by the laser pulse repetition-rate. The other degree of freedom is the rate of slipping of the carrier-envelope phase of these short pulses. The generation of ultrashort pulses requires that the group velocity (v_g) dispersion inside the laser cavity is minimized across the pulse's frequency spectrum. This criterion is not directly related to the frequency comb spacing, since the individual mode frequencies correspond to eigenmodes of the phase-velocity (v_p) of the light, and in general we have $v_g \neq v_p$. With these considerations, each optical comb frequency is therefore effectively given by $f_n = nf_{rep} + f_{ceo}$. Here n represents the integer (~ 3 × 10^6) harmonic number of the optical comb line relative to the repetition rate, f_{rep} ; and f_{ceo} is the comb offset frequency from the exact harmonics of f_{rep} [16].

The two variables of the comb can be expressed as: $f_{rep} = v_g/l_c$ and $f_{ceo} = v_0(1-v_g/v_p)$, where v_0 is the laser carrier frequency and l_c the cavity length. If one is interested only in a stable f_{rep} , for example to generate a clock signal in microwave, then control of l_c seems to be a natural choice and is sufficient. Only when the entire comb spectrum needs to be stabilized do we then need another control freedom, either the pump laser power to influence both v_g and v_p, or a swivel mirror reflecting the dispersed spectrum inside the laser cavity to control v_g [1]. Therefore some degree of signal mixing/feedforward is attractive. We would use a fast servo loop acting on l_c to stabilize f_{rep} , with use of the second control loop to influence mainly f_{ceo}. The inevitable variation in v_g caused by the second loop is then compensated by a properly scaled and opposite-sign feedforward signal to the first loop.

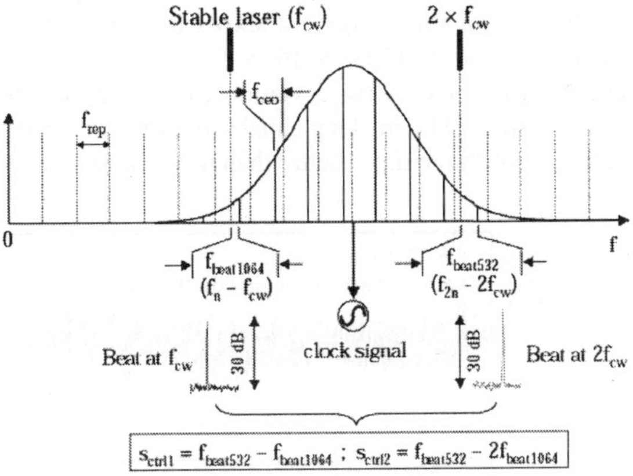

Figure 1. The schematics of the octave-bandwidth optical comb and the stabilized cw laser used for control of the comb. The beat notes of the cw laser and its second harmonic against the corresponding comb components located respectively at the two comb ends are shown. The two beat signals serves as the two observed variables that are used to derive the orthogonal control signals for the comb.

The remaining question is how to have two experimental observables to recover the information relating to f_{rep} and f_{ceo}. In our case we use two beat signals between a cw stabilized laser and its second harmonic against two respective comb components near the two ends of the comb spectrum. The cw reference laser is a Nd:YAG laser (frequency f_{cw}) with its second harmonic ($2f_{cw}$) locked on an iodine transition near 532 nm: this system offers the stability of 5 × 10^{-14} at 1-s [17]. Both beat signals,

$f_{beat1064} = n \times f_{rep} + f_{ceo} - f_{cw}$ and $f_{beat532} = 2n \times f_{rep} + f_{ceo} - 2f_{cw}$, are recovered with about 30 dB signal-to-noise ratio at 100 kHz bandwidth, as shown in Fig. 1. These beat signals are regenerated electronically using the rf tracking-oscillator/filter approach, then mixed in the following way to produce control signals related to f_{rep} and f_{ceo}, namely $s_{ctrl1} = f_{beat532} - f_{beat1064} = n \times f_{rep} - f_{cw}$ and $s_{ctrl2} = f_{beat532} - 2f_{beat1064} = -f_{ceo}$. Therefore the frequency/phase variations arising in both f_{rep} and f_{ceo} are now directly manifested in the two control variables s_{ctrl1} and s_{ctrl2} and are linked to the optical frequency standard f_{cw}. These two signals can then drive the two servo transducers mentioned above to close the feedback loops.

To demonstrate the effectiveness of our control scheme, we first show the stabilization of f_{rep} to the optical standard. Essentially we need to use only the information of s_{ctrl1} to control l_c and thus f_{rep}. This approach magnifies the noise of f_{rep} relative to the optical standard by a factor ~ 3×10^6. In doing so, we can leave the variable f_{ceo} free-running since it has been effectively taken out of the control equation. In practice, we use l_c to control the phase of s_{ctrl1} to that of another stable oscillator in the rf domain (which translates the optical frequency by a small offset with no degradation of stability). Figure 2 shows the time record of the frequency differences between f_{cw} and $2.813988 \times 10^6 \times f_{rep}$, with a standard deviation of 0.8 Hz at a 1-s counter gate time. Allan deviation calculated from this time record is shown in the bottom trace. The tracking capability of the comb system, at a level of 10^{-15} or better, is more than ten times better than the current optical standard itself.

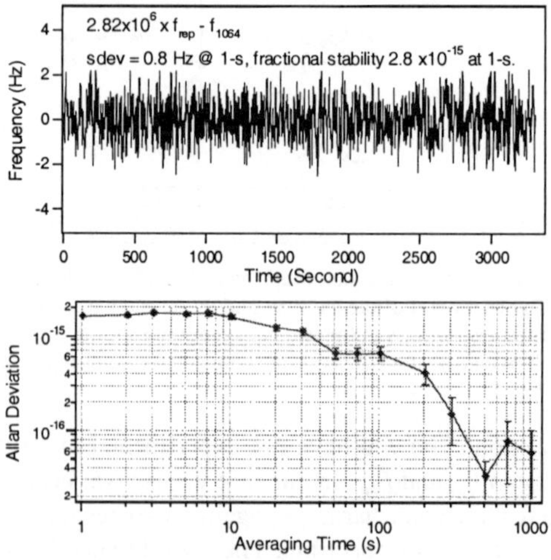

Figure 2. Tracking stability of the comb repetition frequency to the cw reference laser. (a) Time record of the frequency difference between the cw reference laser and the 2.82 millionth harmonic of frep. (b) The associated Allan deviation calculated from the time record.

With the excellent tracking property of the comb system, we expect the stability of the derived clock signal of f_{rep} to be basically that of the optical standard, namely 5×10^{-14} at 1-s. Such an optically derived clock would give its natural time stamps at the $1/f_{rep}$ interval and/or its integer multiples. To characterize the system, a reality check would be to compare the optical clock signal against other well-established microwave/rf frequency standards. The international time standard, Cs clock, should certainly be one of the references; however, the short term stability of a Cs atomic clock is only ~ 5×10^{-12} at 1-s. For improved short term characterization of the fs comb clock, we also use a hydrogen maser signal transmitted over a 2-km fiber, and another in-house highly stable crystal oscillators (short term stability better than 5×10^{-13} at 1-s), which is slowly slaved to the Cs reference for correcting the frequency offset and drift [18]. Figure 3 summarizes the comparison results of the optical clock against all three rf references. The upper graph shows part of the time record of the beat signal between an 8 GHz synthesized frequency from the crystal oscillator against the 80th harmonic of f_{rep} (~ 100 MHz). We use the combination of high harmonic orders and heterodyne beat to help circumvent the resolution limit of frequency counters. The standard deviation of the beat frequency at 1-s averaging time is 0.0033 Hz. The resultant Allan deviation is shown as the curve in triangles in the bottom graph of the figure. Use of a more stable hydrogen maser signal further reduces the Allan deviation of the beat, to be just below 3×10^{-13} at 1-s (shown with open circles). The beat between the optical and the Cs clock is represented by the curve in diamonds. For comparison, we also display the Allan deviation associated with the Cs atomic clock ("worst case" specification) in circles and the Allan deviation of the iodine stabilized laser in squares. The data of the optical standard itself was obtained from heterodyne experiments between two similar laser stabilization systems. We note that the superior stability of our optical clock is currently not yet revealed by the microwave-clock based tests. A microwave source with a better short term stability can be substantially more expensive, even more than our optical system. Use of two optical clocks would of course be the ultimate choice to perform thorough cross-checks of these new devices. Similar work is being pursued in other labs [13].

So far we have made an optical comb that has a well-defined frequency spacing, but the absolute frequencies are uncertain since f_{ceo} is left floating. An attractive approach to stabilize the entire comb spectrum is to transfer the stability of a single optical standard to the whole set of the comb components throughout the optical bandwidth. To accomplish this task, we need the information carried by s_{ctrl2} to exert servo action on the comb by the second control parameter, in our case, the swivel mirror. When this second loop is activated, the impact on the first loop where f_{rep} is being stabilized through l_c is small. This is partly due to the fact that the dependence of f_{rep} and f_{ceo} on their respective control variables is to a large degree already well separated. The other part of the reason is that fluctuations of f_{ceo} develop on a slower time scale compared with that of f_{rep} and therefore a correspondingly slower servo loop is sufficient for stabilization of f_{ceo} . Nevertheless we take part of the second servo signal and after appropriate signal conditioning we feedforward this information to the first servo loop. The resulting loop performance is improved by about a factor of two. (See Figure 4)

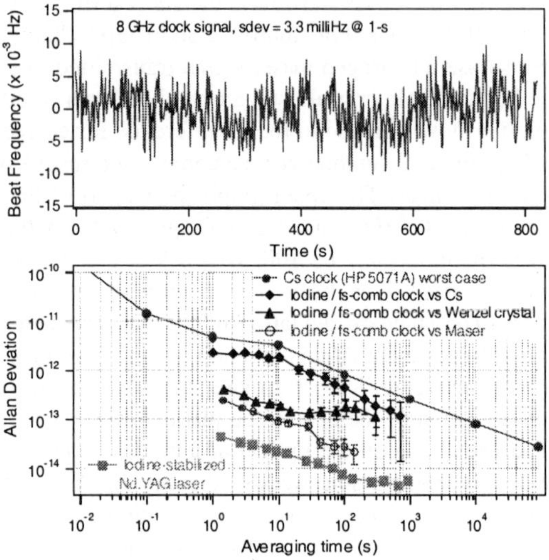

Figure 3. Characterization of the clock signal derived from the iodine stabilized laser. The upper graph shows part of the time record of the beat signal between an 8 GHz synthesized frequency of the crystal oscillator referenced to a Cs clock against the 80th harmonic of f_{rep} of the comb. The lower graph shows the relevant Allan deviations: squares for iodine stabilized laser; circles for the upper stability limit of the Cs atomic clock; triangles, open circles, and diamonds for the beat between the optical clock and the crystal oscillator, the maser, and the Cs clock, respectively.

We use the two original optical beats, namely $f_{beat1064}$ and $f_{beat532}$ that are responsible for generating the control observables but are otherwise outside the servo loops, to characterize the performance of the orthogonal control of the comb. Figure 4 shows the counting record of the two beat frequencies of $f_{beat1064}$ and $f_{beat532}$. Both signals are shown with their mean values removed but indicated in the figure. Again the counter gate time is 1 s and the standard deviations of the two beat signals are 1.7 Hz for $f_{beat532}$ and 1.5 Hz for $f_{beat1064}$. This result indicates that every comb component over the entire optical octave bandwidth is following the cw laser standard at a level of 3.5×10^{-15}, again a factor of about ten times better than the current optical standard itself. The future implication of this work is very clear: With an appropriately chosen optical standard, we can establish an optical frequency grid with lines repeating every 100 MHz over an octave optical bandwidth and with every line stable at the one Hz level.

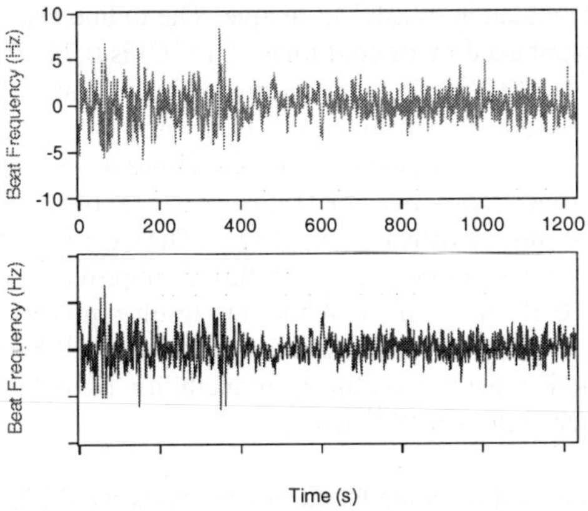

Figure 4. Orthogonal control of the entire optical comb, showing Hz-level stability for both beat signals of the cw laser against a comb component at 1064 nm (bottom trace) and the second harmonic of the cw laser against its corresponding comb line at 532 nm (upper trace). Better orthogonalization in the control loops leads to reduced noise after 400 s.

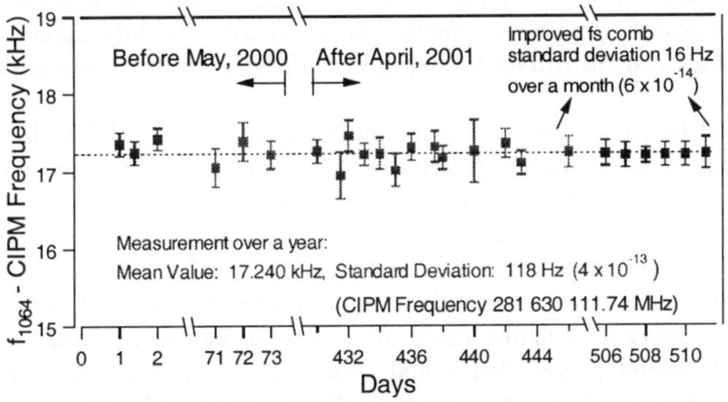

Figure 5. Long term reproducibility of intercomparison between the Cs clock and the iodine-stabilized laser. The measurement period covers more than one year with the stability level at the 4×10^{-13}. This represents the upper limit of the long-term reproducibility of our current optical standard as well as the reproducibility of the comb-based frequency transfer mechanism.

The long-term stability/reproducibility of the iodine-stabilized laser is characterized by comparison against the Cs clock over a period of more than one year. In this long-term comparison, basically we measured repeatedly the stabilized laser frequency using the Cs-referenced optical comb. During the measurement period, we changed a number of parameters associated with the comb and its generation, including f_{rep} , laser power, spectrum and pulse width, and the nonlinear fiber lengths, etc. The reference Nd:YAG laser is stabilized on the R(56) 32-0 a_{10} iodine transition via a modulation transfer technique. The iodine cell is 1.2 m long and its vapor pressure maintained by its cold finger (-15 ^{0}C) is 0.787 Pa. The pump (probe) laser power is ~ 1.0 (0.25) mW with collimated beam diameters of ~ 1.9 mm. The result of this rf – optical frequency intercomparison is shown in Figure 5 and is consistent with our previous measurement [2]. However, we are now able to show the measurement uncertainty over the entire year is about 126 Hz, or ~ 4×10^{-13}. Long-term drift is not statistically significant. During the last month of the data record, after the optical comb system was further improved, the standard deviation was reduced to 16 Hz (6×10^{-14}). While this result represents the lowest level of measurement uncertainty associated with any compact, cell-based optical frequency standards, we notice the long-term reproducibility of cold atom based optical standards has reached the 4×10^{-14} level [5].

A portable version of this iodine based optical clock would have a great impact in field applications: we will be able to make precision measurements in length and time with a single device. The frequency stability in both microwave and optical domains is exceptional, surpassing basically all common sources save only national-standards-scale massive installations. The stability (5×10^{-14} at 1 s) and reproducibility (4×10^{-13}) of the cw laser locked on the iodine transition can be further improved, possibly by another factor of ten. We fully expect such simple optical clock systems will be developed by various interested laboratories.

An optical frequency grid with stable lines repeating every 100 MHz over a large optical bandwidth is useful for a number of applications. However, often times we desire a single-frequency optical-"delta"-function that can be tuned to any preferred frequency position on demand. Realization of such an optical frequency synthesizer (analogous to its radio-frequency counterpart) will add a tremendously useful tool for modern optics-based experiments. One can foresee an array of diode lasers, each covering a successive tuning range of ~ 10 – 20 nanometers and emitting some reasonably useful power, that would collectively cover most part of the visible spectrum. Each diode laser frequency will be controlled by the stabilized optical comb, and therefore related to the absolute time/frequency standard, while the setting of the optical frequency will be done through computer control to any desired value. For the first step, we have constructed an electronic control system that allows a widely tunable diode laser to tune through an targeted spectral region at a 10 MHz step while maintaining reference to the stabilized optical comb. A self-adaptive searching algorithm first tunes the laser to within the specified wavelength region with the aid of a LambdaMeter wavelength measurement device. The heterodyne beat between the diode laser and the comb lines is then detected and properly processed. An auxiliary rf source provides a tunable frequency offset for the optical beat. Frequency-tuning of the diode laser is accomplished by a diode-laser-servo-control that locks the optical beat to the tunable radio frequency. Once the laser frequency tuning exceeds 100 MHz (going through one comb spacing), we reset the radio frequency offset back to the original value to start the process over again. The laser frequency can thus be tuned smoothly in an "inch-worm" fashion through the comb structure. In experiment we verify this tuning process by using independent optical resonances to monitor the diode laser frequency. With improved servo bandwidth, the stability of the iodine-stabilized Nd:YAG laser will be faithfully transferred to another cw laser lying hundreds of THz away.

Acknowledgements

We thank T. Fortier, R. Shelton, S. Cundiff, H. Kapteyn, S. Diddams, L. Hollberg, J. Bergquist, J. Kitchin and J. Levine for useful discussions. The work at JILA is supported by NASA, NIST, NSF and the Research Corporation. JY and JLH are staff members of the Quantum Physics Division of NIST Boulder.

References

1. Th. Udem *et al.*, Phys. Rev. Lett. **82**, 3568 (1999).
2. S. A. Diddams *et al.*, Phys. Rev. Lett. **84**, 5102 (2000).
3. M. Niering *et al.*, Phys. Rev. Lett. **84**, 5496 (2000).
4. J. Ye *et al.*, Phys. Rev. Lett. **85**, 3797 (2000).
5. Th. Udem *et al.*, Phys. Rev. Lett. **86**, 4996 (2001).
6. D. J. Jones *et al.*, Science **288**, 635 (2000).
7. A. Apolonski *et al.*, Phys. Rev. Lett. **85**, 740 (2000).
8. L.-S. Ma *et al.*, Phys. Rev. A **64**, Rapid Comm. 021802(R) (2001).
9. R. K. Shelton *et al.*, Science **293**, 1286 (2001).
10. L. Hollberg *et al.*, IEEE J. Quant. Electron., in press (2001).
11. R. J. Rafac *et al.*, Phys. Rev. Lett. **85**, 2462 (2000).
12. J. Ye *et al.*, Opt. Lett. **25**, 1675 (2000).
13. S. A. Diddams *et al.*, Science **293**, 826 (2001).
14. M. T. Asaki *et al.*, Opt. Lett. **18**, 977 (1993).
15. J. Ranka *et al.*, Opt. Lett. **25**, 25 (2000).
16. H. Telle *et al.*, Appl. Phys. B **69**, 327 (1999).
17. J. Ye *et al.*, IEEE Trans. Instrum. & Meas. **48**, 544 (1999); J. L. Hall *et al.* ibid, 583 (1999).
18. Cs clock: HP 5071 A; Crystal oscillator: Wenzel 5/10 Blue Top Ultra Low Noise Oscillator; Hydrogen maser: ST-22, Clock-13, NIST, Boulder. We thank our NIST colleagues for the fiber-based delivery of the maser signal to JILA. Mentioning of commercial products is for technical communications only.

A SINGLE ^{199}HG^{+} ION OPTICAL CLOCK*

J.C. Bergquist, S.A. Diddams, C.W. Oates, E.A. Curtis,
L. Hollberg, R.E. Drullinger, W.M. Itano, and D.J. Wineland
National Institute of Standards and Technology, 325 Broadway,
Boulder, Colorado 80305

Th. Udem
Max Planck Institute für Quantenoptik, Hans-Kopfermann Strasse 1,
85748 Garching, Germany

Telephone:1 303 497 5459
Fax:1 303 497 7375
e-mail: berky@boulder.nist.gov

1. Introduction

Optical clocks based on laser-cooled atoms and ions potentially have superior stability and accuracy over present-day time standards, but until recently, no practical device was fast enough to count optical cycles in order to generate time. Now that obstacle has been circumvented using pulsed lasers to span an octave from the infrared to the ultraviolet with a grid of equally spaced, phase-coherent frequencies [1-8]. Here we report on work at NIST toward the realization of a highly stable and accurate optical timepiece [9] that uses a femtosecond laser to phase-coherently divide the frequency of a well stabilized laser that is locked to a narrow transition of a single, laser-cooled ^{199}Hg^{+} ion [10,11].

2. Atomic clock basics

All clocks consist of two major components: a device that generates periodic events, and a means for counting, accumulating, and displaying these events. For example, the swing of a pendulum can provide the periodic events that are counted, accumulated, and displayed by a set of gears driving a pair of clock hands. Atomic clocks add a third component: the resonance of a well-isolated atomic transition, which is used to control the oscillator frequency. If the frequency of the oscillator is made to match the transition frequency between two nondegenerate (and largely unperturbed) atomic states, then the clock will have improved long-term stability and accuracy. For an atomic clock based on a microwave transition, high-speed electronics count and accumulate a defined number of cycles of the reference oscillator to mark a unit of time. The same basic concepts apply for an atomic clock based on an optical transition at a much higher frequency. However, if all other factors are equal, the stability of the optical standard will be proportionally higher. This is the principal advantage of an optical clock over a microwave clock. The stability can be quantitatively expressed by the Allan deviation $\sigma_y(\tau)$ which provides a convenient measure of the fractional frequency instability of a clock as a function of averaging time τ [12]. For an oscillator locked to an atomic transition of frequency f_0 and linewidth Δf,

$$\sigma_y(\tau) \approx \left\langle \frac{\Delta f_{rms}}{f_0} \right\rangle_\tau \approx \frac{\Delta f}{\pi f_0} \sqrt{\frac{T}{N\tau}} \tag{1}$$

where Δf_{rms} is the measured frequency fluctuation, N is the number of atoms, and T is the cycle time (the time required to make a single determination of the line center) with $\tau > T$. This expression assumes that other noise sources are reduced to a sufficiently small level such that quantum-mechanical atomic projection noise is the dominant stability limit [13,14]. In this case, $\sigma_y(\tau)$ decreases as the square root of the averaging time for all clocks, so a ten-fold decrease in the short-term Allan deviation leads to a hundred-fold reduction in averaging time τ to reach a given instability. This point is particularly important if we ultimately hope to reach a fractional frequency uncertainty of 10^{-18}, which is anticipated as the limit for many optical clocks [15]. In this case, $\sigma_y(\tau) < 1\times10^{-15}\tau^{-1/2}$ is clearly desirable to avoid inordinately long averaging times.

3. The optical frequency standard

Our optical frequency standard [10] is based on a single, nearly motionless and unperturbed $^{199}Hg^+$ ion that is confined in a cryogenic, spherical Paul trap. The ion is cooled, state-prepared and detected by light that is scattered on the strongly allowed $^2S_{1/2}$ - $^2P_{1/2}$ transition at 194 nm. The $^2S_{1/2}$ ($F = 0$, $m_F = 0$) - $^2D_{5/2}$ ($F = 2$, $m_F = 0$) electric-quadrupole allowed transition at 282 nm (τ ($D_{5/2}$) $\approx$ 0.09 s) provides the reference for the optical standard (Fig. 1). The natural linewidth of the S-D resonance is about 2 Hz at 1.064 PHz, and recently a Fourier-transform-limited linewidth of only 6.7 Hz (Q = 1.5×10^{14}) was observed [10]. We lock the frequency-doubled output of a narrowband and stable 563 nm dye laser [16] to the narrow S-D resonance by the method of electron shelving, whereby each transition to the metastable D-state is detected by the suppression of the scattering of many 194 nm photons on the strongly allowed S-P transition [17,18]. The short-term (1-10 s) fractional frequency instability of the probe laser is $\leq 5 \times 10^{-16}$, which matches well to the best stability predicted by the shot-noise-limited detection of the S-D transition of a single mercury ion. The fractional frequency instability of the single ion mercury standard is expected to be $\leq 1 \times 10^{-15}\ \tau^{-1/2}$ with a fractional frequency uncertainty that approaches 10^{-18} [10]. However, in our present realization of the $^{199}Hg^+$ optical standard, the stability obtained from locking to the ion was degraded

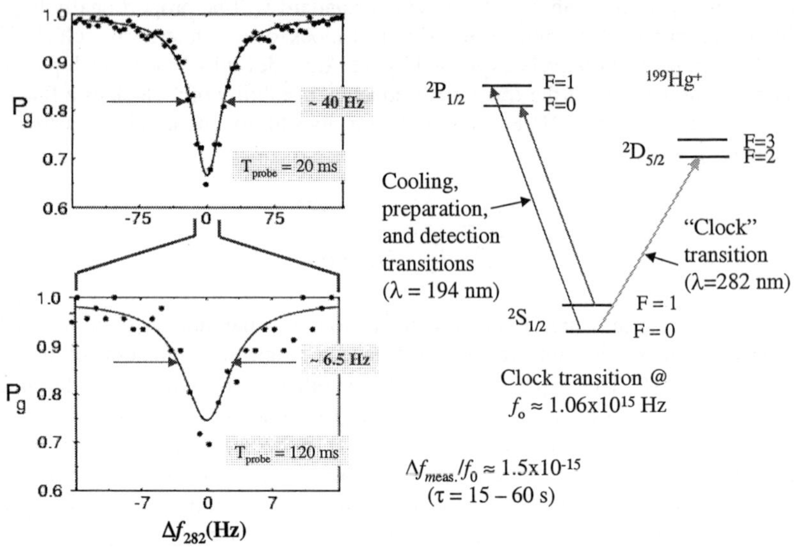

Figure 1. Quantum-jump absorption spectra of the $^2S_{1/2}$(F=0) • $^2D_{5/2}$(F=2) •m_F = 0 electric-quadrupole transition and a partial energy level diagram of $^{199}Hg^+$ with the relevant transitions indicated. •f_{282} is the frequency detuning of the 282 nm probe laser, and P_g is the probability of finding the atom in the ground state. The upper spectrum is obtained with a Rabi excitation pulse 20 ms long (averaged over 292 sweeps) and the lower spectrum corresponds to an excitation pulse 120 ms long (averaged over 46 sweeps). The linewidths are consistent with the Fourier-transform limit of the respective pulse times [10].

primarily because the signal contrast was less than 100 % (see the spectra in Fig. 1) and there was substantial dead time between probes of the S-D resonance. Hence, $\sigma_y(\tau) \leq 2 \times 10^{-15}$ for averaging times up to ~ 30 s (where the frequency stability of the laser is dominantly controlled by the cavity), at which point $\sigma_y(\tau)$ began to average down as $\tau^{-1/2}$ (as the control of long term frequency fluctuations is transferred to the ion) [9,10].

4. The frequency divider

Figure 2 shows the optical clock in more detail, consisting of the optical frequency standard and the femtosecond-laser-based optical clockwork [9]. The Hg^+ standard provides high accuracy and stability, but to realize a countable clock output, we must phase-coherently convert the optical signal to a lower frequency. The

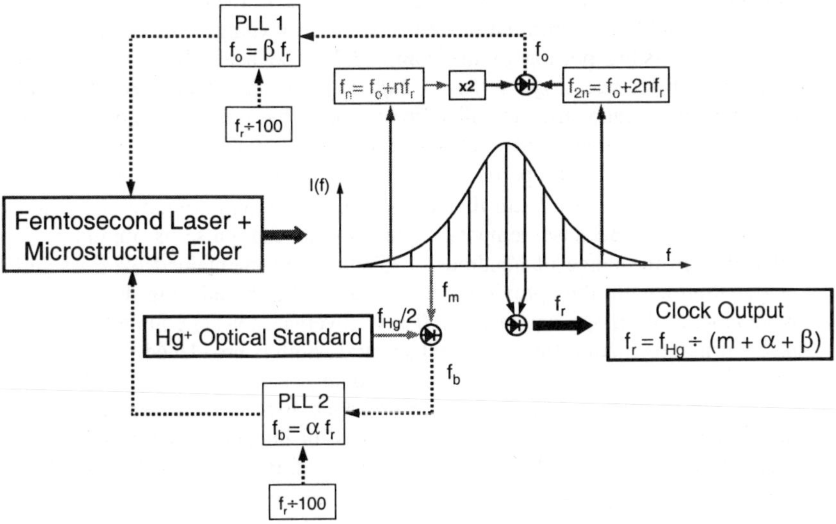

Figure 2. Schematic of the self-referenced all-optical atomic clock. Solid lines represent optical beams, and dotted lines represent electrical paths. The femtosecond laser, having repetition rate f_r, combined with the spectral broadening microstructure fiber produces an octave-spanning comb of frequencies in the visible/near infrared spectrum, represented by the array of vertical lines in the center of the figure. As depicted above this comb, the low frequency portion of the comb is frequency-doubled and heterodyned against the high frequency portion, yielding the offset frequency f_o that is common to all modes of the comb. Additionally, an individual element of the comb is heterodyned with the optical standard laser oscillator ($f_{Hg}/2$ = 532 THz) whose harmonic is locked to the clock transition frequency of the single $^{199}Hg^+$ ion. This yields the beat frequency f_b. Two phase-locked loops (PLL) control f_o and f_b with the result that the spacing (f_r) of the frequency comb is phase-locked to the Hg^+ optical standard. Thus f_r is the countable microwave output of the clock, which is readily detected by illuminating a diode with the broadband spectrum from the frequency comb. See the text for further details.

clockwork that divides the 1.064 PHz optical frequency to a countable microwave frequency f_r is based on a femtosecond laser and a novel microstructure optical fiber. The Ti:sapphire femtosecond ring laser emits a train of pulses (~25 fs duration) at the nominal repetition rate of f_r = 1 GHz [19]. The frequency-domain spectrum of the pulse train is a uniform comb of phase-coherent continuous waves separated by f_r. The frequency of the nth mode of this comb is $f_n = nf_r+f_o$ [20,21], where f_o is a frequency offset common to all modes that results from the difference between the group- and the phase-velocity inside the laser cavity [22]. If the frequency comb of the laser covers an entire octave, then f_o can be measured by frequency doubling an infrared mode (n) and heterodyning it with an existing mode ($2n$) in the visible portion of the comb [3,7]. The heterodyne signal yields the frequency difference $2(nf_r + f_o) - (2nf_r + f_o) = f_o$. Only recently with the development of microstructure silica fibers [23,24] has the required octave-spanning spectrum been attained with high repetition rate, low-power, femtosecond lasers. The unique dispersion properties of the microstructure fiber provide guidance in a single spatial mode (~1.7 μm diameter) with zero group velocity dispersion near 800 nm [23]. Because temporal spreading of the pulse is minimized, peak intensities in the range of hundreds of GW/cm^2 are maintained over a significant propagation length, thus providing enhanced spectral broadening due to self-phase modulation. With approximately 200 mW (average power) coupled into a 15 cm piece of microstructure fiber, the total spectral width is broadened from ~15 THz to ~300 THz (spanning from ~520 nm to ~1170 nm).

In addition to f_o, a second heterodyne beat f_b is measured between an individual comb element $f_m = mf_r + f_o$ (m is an integer) and the 532 THz local oscillator of the Hg^+ standard. As shown in Fig. 2, two phase-lock loops (PLL) are employed to control f_o and f_b, thereby fixing the clock output f_r. PLL-1 forces $f_o = \beta f_r$ by controlling the pump power of the femtosecond laser [7]. Similarly, PLL-2 changes the cavity length of the femtosecond laser with a piezo-mounted mirror, such that $f_b = \alpha f_r$. The constants α and β are integer ratios implemented with frequency synthesizers that utilize $f_r/100$ as a reference. In this manner, the frequencies of both PLL's are phase-coherently linked to f_r such that all oscillators employed in the clock are referenced to the 532 THz laser oscillator itself. When f_o and f_b are phase-locked, every element of the femtosecond comb, as well as their frequency separation f_r, is phase-coherent with the laser locked to the Hg^+ standard [25]. With no other frequency references as an input, we realize all aspects of a high accuracy, high-stability, optical atomic clock: a stable laser that is locked to a narrow atomic reference, and whose frequency is phase-coherently divided to give a pulsed microwave output that can be recorded with a counter [9].

5. High-stability clock output

With both PLL's closed, the ~1 GHz microwave output has the value of $f_r = f_{Hg}/(m + \alpha + \beta)$. If we choose the signs of beats f_o and f_r such that $\alpha = -\beta$, then f_r would be an exact sub-harmonic of $f_{Hg}/2$. The stability of the 532 THz laser should be transferred to each element of the femtosecond comb, in addition to f_r. We obtain f_r from the bandpass-filtered photocurrent generated with ~5 mW of the broadened comb light incident on a p-i-n photodiode. We have measured the instability of f_r by subtracting it from the output of a synthesizer that is referenced to a hydrogen maser for which $\sigma_y(1\text{s}) \approx 2.5 \times 10^{-13}$. The stability of this difference frequency is then analyzed with both a high-resolution counter and a dual-mixer time measurement system [26]. Both results are consistent with the resolutions of the respective measurements and the maser stability, demonstrating that the 1 s stability of f_r is at least as good as that of the hydrogen maser [9].

Before we can conclusively state that a microwave signal with stability matching that of the optical standard can be obtained from the optical clock, f_r needs to be compared to an oscillator with stability significantly better than the hydrogen maser. This reference oscillator could be either the microwave output of a second optical clock, or the high stability output of a cryogenic microwave oscillator [27]. Lacking these, we can still verify the stability of the comb in the optical domain, and thereby infer the expected stability of f_r, by comparing one element of the optical comb to a second optical standard with high stability. For example, we can detect, filter, and count the heterodyne beat signal between a single element of the comb at 456 THz and a frequency-stabilized diode laser locked to the $^1S_0$ - $^3P_1$ intercombination transition of a laser-cooled ensemble of calcium atoms [28,29]. The Allan deviation of the heterodyne signal between the Hg^+ stabilized comb and the Ca-stabilized optical standard is shown as the triangles in Fig. 3. For $\tau < 10$ s the Allan deviation averages down roughly as $9 \times 10^{-15}\ \tau^{-1/2}$, which is higher than that of the Hg^+ standard (recall that the Hg^+ standard stability for $\tau < 10$ s is primarily controlled by the cavity (see Sec. 3)), but consistent with that expected of the Ca standard in its present configuration. It is also consistent with the monotonic increase in the 1 s instability of the heterodyne beat frequency as the stability of the Ca standard is degraded by using shorter probe times and lower resolution signals [9]. However, for $\tau > 10$ s, frequency and phase fluctuations introduced by the 180 m long optical fiber that transmits the 532 THz light to the femtosecond system begin to degrade the stability observed between the Ca standard and the Hg-referenced comb. We have measured the fiber-induced noise by double-passing the light through the optical fiber and find that the average fractional frequency fluctuations are between 2 to 4×10^{-15} for $1\ \text{s} < \tau < 10$ s. For $\tau > 10$ s, the fiber noise averages down as $1/\tau$. Finally, for $\tau > 30$ s the instability of the Hg^+ standard in its present configuration contributes to the measured instability at approximately the same level as the Ca standard. Nonetheless, the measured fractional frequency instability decreases with averaging to $\sim 1.5 \times 10^{-15}$ at 100 s.

More recently, we have implemented active cancellation [30,31] of this fiber noise and have further improved the signal-to-noise in the Ca standard. Data taken under these conditions reveal a fractional frequency instability of 7×10^{-15} at $\tau = 1$ s. These results are plotted as the square data points in Fig. 3 [9]. In this case we cannot place great significance in the stability for $\tau > 1$ s for two reasons. First, the Allan deviation for averaging times $\tau > 1$ s is calculated from the juxtaposition of 1 s averages. Such data analysis is known to result in errors for certain noise processes [32]. Second, for this specific data the 532 THz laser oscillator was not locked to the $^{199}Hg^+$ ion, and therefore it was necessary to subtract out the smooth and predictable drift (~1 Hz/s) of the Fabry-Perot cavity to which this laser is stabilized. However, neither of these affect the measured 1 s Allan deviation, for which we find an upper limit of 7×10^{-15} for the optical comb. Again, this 1 s

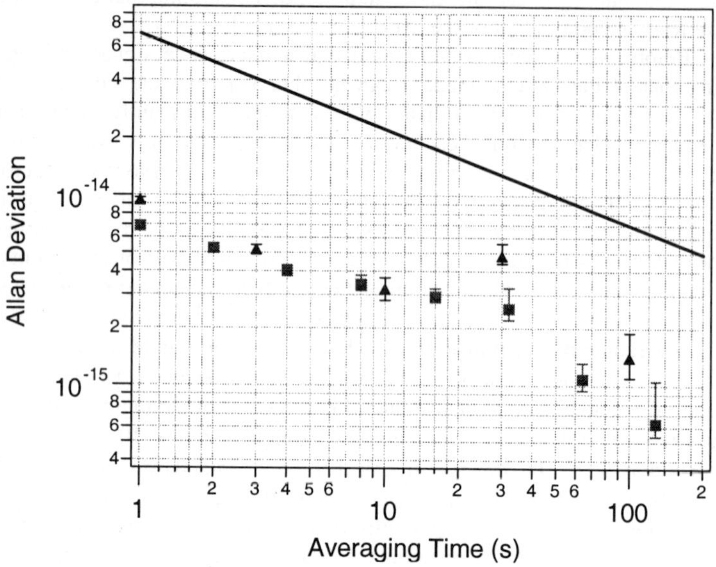

Figure 3. Measured stability of the heterodyne signal between one element of the femtosecond comb and the Ca optical standard at 456 THz (657 nm). The femtosecond comb is phase locked to the 532 THz laser oscillator. The triangles are the stability data without cancellation of the additive fiber noise. The squares are the measured stability with active cancellation of the fiber noise and improved stability in the Ca standard. These results are about an order of magnitude better than the best stability reported with a cesium microwave standard, which is designated by the solid line [34].

instability is consistent with that of the Ca standard in its improved configuration. Similar stability in the ~1 GHz clock output remains to be verified.

Finally, when f_r and f_0 are detected and counted with respect to the frequency of the hydrogen maser (which acts as a transfer standard to the NIST realization of the SI second [33]), an absolute measurement of the $^{199}Hg^+$ clock transition can be made [11], f_{Hg} = 1 064 721 609 899 143 (10) Hz. The statistical uncertainty of our measurements is about +/- 2 Hz, limited in part by the fractional frequency instability of the maser at our measurement times and in part by the accuracy determination of the cesium standard. The systematic uncertainty of +/- 10 Hz assigned to f_{Hg} is based on theoretical arguments in lieu of a full experimental evaluation. A second $^{199}Hg^+$ standard has been constructed toward making a full evaluation.

6. Conclusion

In conclusion, we have constructed an optical clock based on the 1.064 PHz (282 nm) electric-quadrupole transition in a laser-cooled, single $^{199}Hg^{+}$ ion. The optical frequency is phase-coherently divided to provide a microwave output using a mode-locked femtosecond laser and a microstructured optical fiber. The short-term (1 s) instability of the optical output of the clock is measured against an independent optical standard to be $\leq 7 \times 10^{-15}$. This optically-referenced femtosecond comb provides a countable output at 1 GHz, which ultimately could be used as a higher accuracy reference for time scales, synthesis of frequencies from the RF to the UV, comparison to other atomic standards, and tests of fundamental properties of nature.

Acknowledgements

This report is based in part on references 9 and 10. The authors are grateful to A. Bartels (GigaOptics GmbH) for his valuable assistance with the femtosecond laser. We are also indebted to R. Windeler (Lucent Technologies) for providing the microstructure optical fiber. We further acknowledge many illuminating discussions with J. Hall, S. Cundiff, J. Ye, and F. Walls. This work was supported in part by the Office of Naval Research.

7. References

1. Th. Udem, *et al., Phys. Rev.Lett.* **82**, 3568 (1999).
2. S. A. Diddams *et al., Phys. Rev. Lett.* **84**, 5102 (2000).
3. D. J. Jones *et al., Science* **228**, 635 (2000).
4. M. Niering *et al., Phys. Rev. Lett.* **84**, 5496 (2000).
5. R. Holzwarth *et al., Phys. Rev. Lett.* **85**, 2264 (2000).
6. J. Reichert *et al., Phys. Rev. Lett.* **84**, 3232 (2000).
7. R. Holzwarth *et al., Phys. Rev. Lett.* **85**, 2264 (2000).
8. J. Stenger *et al., Phys. Rev. A* **63**, 021802R (2001).
9. S.A. Diddams, *et al., Science* **293,** 825 (2001).
10. R. Rafac, *et al., Phys. Rev. Lett.* **85**, 2462 (2000).
11. Th. Udem, *et al., Phys. Rev. Lett.* **86**, 4996 (2001).
12. J. A. Barnes *et al, IEEE Trans. Inst. & Meas.* **20**, 204 (1971).
13. W. M. Itano *et al., Phys Rev. A* **47**, 3554 (1993).
14. G. Santarelli *et al., Phys. Rev. Lett.* **82**, 4619 (1999).
15. A. A. Madej and J. E. Bernard, in *Frequency Measurement and Control*, A. N. Luiten, ed. (Springer-Verlag, Berlin, 2001), pp. 153-194.
16. B. C. Young, F. C. Cruz, W. M. Itano, J. C. Bergquist, *Phys. Rev. Lett.* **82**, 3799 (1999).

17. H. Dehmelt, *Bull. Am. Phys. Soc.* **20**, 60 (1975).
18. J.C. Bergquist *et al., Phys. Rev. A* **36**, 428 (1987).
19. A. Bartels, T. Dekorsy, and H. Kurz, *Opt. Lett.* **24**, 996 (1999).
20. A. I. Ferguson, J. N. Eckstein, T. W. Hänsch, *Appl. Phys.* **18**, 257 (1979).
21. J. Reichert, *et al., Opt. Comm.* **172**, 59 (1999).
22. D.J. Wineland *et al.*, in *The Hydrogen Atom*, G.F. Bassani, M Inguscio and T.W. Hänsch, eds. (Springer-Verlag, Heidelberg, 1989) pp. 123-133.
23. J. K. Ranka, R. S. Windeler, A. J. Stentz, *Opt. Lett.* **25**, 25 (2000).
24. W. J. Wadsworth *et al., Electron. Lett.* **36**, 53 (2000).
25. We count the phase-locked beats f_o and f_b with a high resolution counter and verify that they fluctuate about their nominal respective phase-locked values < 100 mHz in 1 s. This implies that the stability of the 532 THz laser oscillator is transferred to the femtosecond comb with a relative uncertainty $< 2 \times 10^{-16}$ in 1 s.
26. S. Stein *et al.*, in *Proceedings of the 1982 IEEE International Frequency Control Symposium*, (IEEE, Piscataway, NJ, 1982), pp. 24-30.
27. S. Chang, A. G. Mann, A. N. Luiten, *Electron. Lett.* **36**, 480 (2000).
28. C. W. Oates, F. Bondu, R. W. Fox, L. Hollberg, *J. Phys. D*, **7**, 449 (1999).
29. C. W. Oates, E. A. Curtis, L. Hollberg, *Opt. Lett.* **25**, 1603 (2000).
30. B. C. Young *et al.* in *Laser Spectroscopy XIV*, R. Blatt, J. Eschner, D. Leibfried, F. Schmidt-Kaler, eds. (World Scientific, Singapore, 1999), pp. 61-70.
31. L.-S. Ma, P. Jungner, J. Ye, J. L. Hall, *Opt. Lett.* **19**, 1777 (1994).
32. P. Lesage, IEEE Trans. Inst. & Meas. **32**, 204 (1983).
33. S. R. Jefferts, *et al.*, in *Proceedings of the 2000 IEEE International Frequency Control Symposium*, Kansas City, 7-9 June, 2000 (IEEE, Piscataway, NJ, 2000), pp. 714-717.
34. P. Laurent *et al.*, in *Laser Spectroscopy XIV*, R. Blatt, J Eschner, D. Liebfried, F. Schmidt-Kaler, Eds. (World Scientific, Singapore, 1999), pp. 41-50.

ATOMIC CLOCKS AND COLD ATOM SCATTERING

CHAD FERTIG, RONALD LEGERE,* J. IRFON REES, AND KURT GIBBLE

The Pennsylvania State University, 104 Davey, University Park, PA 16802, kgibble@phys.psu.edu

SERVAAS KOKKELMANS† AND BOUDEWIJN J. VERHAAR

Eindhoven Technological University, Eindhoven, The Netherlands, B.J.Verhaar@phys.tue.nl

Abstract: We demonstrate a Rb fountain clock which has a small cold collision shift that is cancelled by detuning the microwave cavity. To further enhance the stability and accuracy, we are currently developing a new state detection technique that directly measures population differences and are juggling atoms in the fountain. Fountain clocks also enable a novel quantum atom-optics precision scattering measurement.

1 Introduction

Laser cooling has dramatically shifted the relative importance of various systematic errors of fountain clocks relative to beam clocks since there is a very large frequency shift due to the collisions of cold atoms. This requires operating with a small atomic density so that optimizing the performance of fountain clocks will demand shot-noise limited detection of a large number of atoms, and possibly spin squeezed atomic states.

We demonstrate a prototype of a laser-cooled ^{87}Rb fountain clock and measure the frequency shift due to cold collisions. The shift is fractionally 30 times smaller than that in a laser-cooled Cs clock, allowing a Rb clock to use higher densities for the same clock accuracy. We observe a density dependent pulling by the microwave cavity and use it to cancel the collision shift.

To reach the atom shot noise limit, we are developing a new atomic state detection scheme that directly measures the population difference of the two clock states using FM absorption spectroscopy. We also demonstrate a juggling Rb fountain clock that we will use to study the collisions of juggled balls of atoms. Juggling can significantly improve the clock's short-term stability without requiring greater signal-to-noise or a larger cold collision frequency shift. Finally, a juggling fountain clocks also enables a novel quantum atom-optics precision scattering measurement.

---

* Present address: MIT Lincoln Laboratory, 244 Wood Street, Lexington MA 02420.

† Present address: JILA, University of Colorado Boulder, Box 440, Boulder, CO 80303.

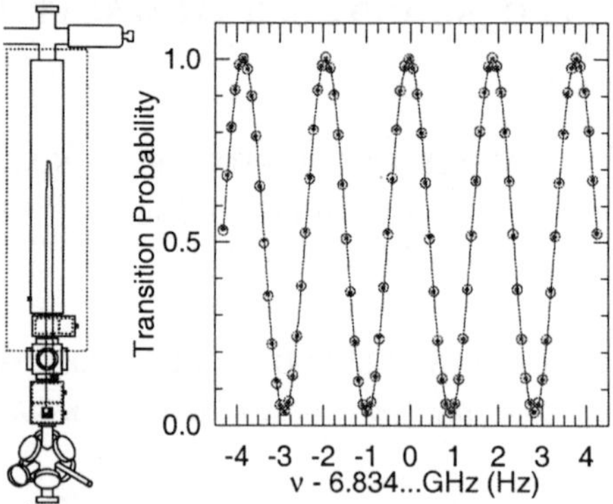

Figure 1. Schematic of ^{87}Rb fountain clock and Ramsey fringes at 6.834 GHz. The large circles are the data and the small are a fit to the data. The linewidth is 0.95 Hz and the S/N=200.

2 Laser-cooled ^{87}Rb fountain clock

The most serious systematic error in laser-cooled clocks is the frequency shift due to cold collisions.[1] Tiesinga *et al.* first calculated this shift[2] and Gibble and Chu[3] measured the shift for laser-cooled Cs clocks to be $\delta\nu/\nu = -1.7\times10^{-12}$ at a density of 10^9 cm^{-3}. Due to Cs_2 molecular resonances, the frequency shift cross section has nearly the maximal value of $\lambda_{dB}{}^2/2\pi$, where λ_{dB} is the de Broglie wavelength. This large cross section led us to examine clocks based on other atoms, for which the cold collision shift might be smaller.[4,5]

A schematic of our ^{87}Rb fountain clock is shown in Figure 1. Using 0.9W of light from a Ti:Sapphire laser delivered by an optical fiber, atoms are collected from the room temperature Rb vapor in the vapor-cell MOT. The atoms are launched upwards and cooled to 1.8 μK in the moving frame. The atoms then pass through 2 microwave cavities that are normally used to prepare half of the atoms in the $5S_{½}$ $|F=1,m=0\rangle$ state.[6] Here, to achieve high density so we can best measure the small collision shift, the selection cavities are not used. Instead, we optically pump the atoms into $|1,-1\rangle$ just below the clock cavity. Microwaves from a horn then transfer atoms from $|1,-1\rangle$ state to the $|2,0\rangle$ state, and then a variable number of atoms from $|2,0\rangle$ to $|1,0\rangle$. Any atoms remaining in F=2 are cleared with light tuned to the $2\rightarrow3'$ transition. In this way, we can vary the density preparing a maximum of 70% of the atoms in $|1,0\rangle$ at a temperature of 5 μK, with fewer than 1% in $|1,\pm1\rangle$. To prepare the atoms in $|2,0\rangle$, we add another microwave pulse to transfer the atoms in $|1,0\rangle$ to $|2,0\rangle$ and then repump any atoms in F=1 with light tuned to $1\rightarrow2'$.

The state prepared atoms enter the magnetic shielding and experience two 6.8 GHz microwave pulses during their up and down passages through the rectangular TE_{102} clock cavity. We detect the transition probability using a laser tuned to $2\rightarrow3'$

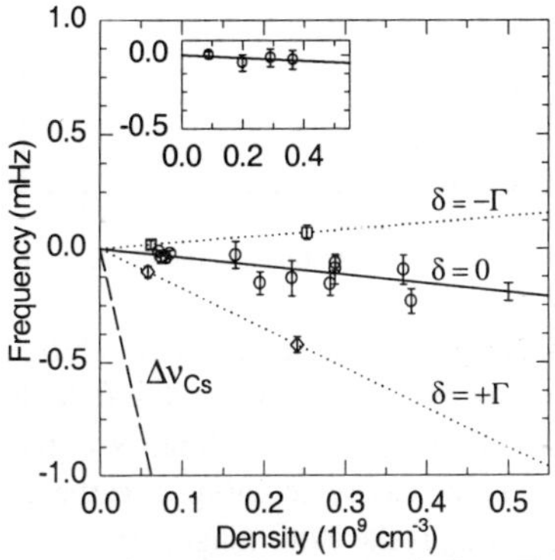

Figure 2. Measured ^{87}Rb cold collision shift (circles and solid line). The shift is -0.38(8) mHz for a density of $1.0(6)\times10^9$ cm^{-3}. The dashed line is the measured shift for Cs.[3] The dotted lines (diamonds and squares) show the density dependence for clock cavity detunings of $\delta=\pm\Gamma$. The inset shows a cancelled collision shift for $\delta = -30$ kHz and atoms prepared in $|1,0\rangle$.

producing Ramsey fringes as in Figure 1. To normalize the fringes, we also detect the total number of atoms by repumping the population in $|1,0\rangle$ using a laser beam tuned $1\rightarrow2'$, followed by a second detection laser pulse. In Figure 1, the interrogation time is T=0.526 s which gives a linewidth of $\Delta\nu$=0.950 Hz. With our detection signal-to-noise S/N=500 on a single launch, the atomic transition frequency of 6,834,682,610.9 Hz[7] can be determined in 1 s with a precision of $\delta\nu/\nu=\Delta\nu/\pi\nu S/N=9\times10^{-14}$. However, the short-term instability of our local oscillator limits the S/N to 200 or $\delta\nu/\nu = 2.1\times10^{-13}$ for 1s of averaging.

During the precession time above the clock cavity, collisions between cold atoms shift the phase of their coherence producing a frequency shift of the clock. To measure this frequency shift, we vary the atomic density on successive fountain launches and look for a relative shift of the Ramsey fringes. In Figure 2 we show a series of measurements of the frequency as a function of density(circles) for atoms prepared in $|1,0\rangle$ and in $|2,0\rangle$. The extrapolated shift for a density of $1.0(6)\times10^9$ cm^{-3} is −0.38(8) mHz.[8,9] We also show the shift for Cs which is fractionally 30 times larger.[3] The measured shift agrees with that calculated in Ref. [5] and also a recent reanalysis of the ^{87}Rb interactions.[10,11]

The measured frequency differences in Figure 2 have a precision of $\pm2\times10^{-15}$. At the 10^{-15} level, there are several potential error sources. The only source that is explicitly density dependent is the pulling of the transition frequency by the coupling of the atoms to the microwave cavity. In NMR, the effect is known as radiation damping where the field radiated by the magnetization of the sample builds up in the microwave cavity and causes the Bloch vector to decay.[12] In hydrogen masers, it is called cavity pulling and is used to cancel the collisional frequency shift.[13] Here, more apparent than the decay of the atomic coherence is a small phase shift. When the cavity is detuned from the atomic transition

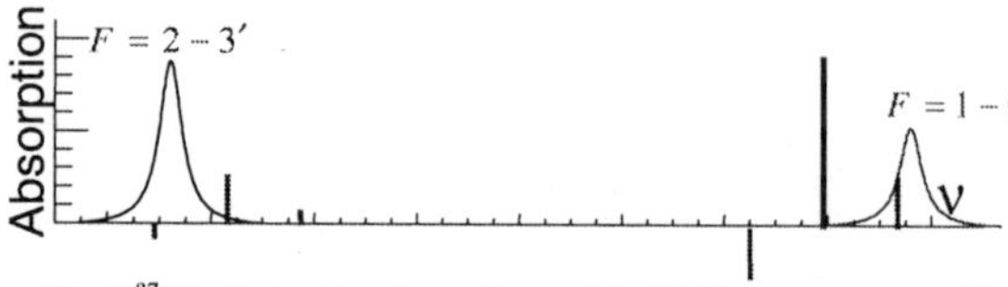

Figure 3. Schematic of ^{87}Rb clock state absorption and the FM laser spectrum. The frequency of the FM is near 50 MHz.

frequency, the field radiated by the atoms is phase shifted relative to the applied field. The Bloch vector precesses about the total field leading to a phase error proportional to $\mu_0 \hbar \mu_B{}^2 N\omega\delta/(\delta^2+\Gamma^2)V_{cav}$.[14] Here, N is the number of atoms, ω is the transition frequency, δ is the cavity detuning, Γ is the cavity HWHM, V_{cav} is the effective cavity volume, and μ_B is the Bohr magneton.

In Figure 2 we also show the measured density dependent frequency shift when we detune the clock cavity by $\pm\Gamma$ for atoms prepared in $|1,0\rangle$ (diamonds and squares). The cavity detuning can significantly influence the density dependence. An intentional detuning of the cavity cancels the cold collision shift as shown in the inset. This spin-exchange tuning[13] has advantages for insuring immunity to long-term variations in the number of trapped atoms. Moreover, density extrapolations can be more accurate since the extrapolation does not depend on accurate density ratios.[4]

We tune the clock cavity by $\delta=\pm\Gamma$ by changing its temperature by $\mp$2.5K. To measure the frequency response of the cavity, we detect the AC Zeeman shift of the clock states due to a strong microwave sideband that is injected into the clock cavity.[6] The loaded Q of our TE_{102} copper clock cavity is 13,000 and its resonance is measured with an accuracy of 5 kHz.

3 Direct detection of population differences

Because the cold collision frequency shift in ^{87}Rb fountain clocks is fractionally at least 30 times smaller than that for Cs,[8,9] ^{87}Rb clocks can operate at higher densities and with consequently smaller projection noise. Nonetheless, the ^{87}Rb cold collision frequency shift cross-section is sufficiently large to suggest using as few atoms as possible. To retain sufficient short-term stability to reach the clock's potential accuracy, the detection of the transition probability in the clock should be at least nearly shot-noise limited. Accuracies exceeding 10^{-16} will likely require a short-term stability approaching 10^{-14} per launch. This stability of 10^{-14} per launch requires a detection signal-to-noise ratio S/N of nearly 5,000:1 for ^{87}Rb.

The detections schemes currently used in Cs fountain clocks detect the atomic population in F=4 and then either the population in F=3 or the total number of atoms.[8, 15-17] In this way, the transition probability can be normalized to the total number of atoms launched, which is typically stable to not better than 1%. The temporal separation of these two detection pulses aliases instabilities in the detection system, such as the detection laser frequency noise, into the detected signal, limiting the S/N. To date, the best achieved is S/N = 2300:1.[15]

We have demonstrated a new state detection scheme that directly detects a population difference to avoid this difficult source of noise. We use FM absorption spectroscopy using 2 lasers tuned near the $5S_{1/2}$ F=2 → $5P_{3/2}$ F′=3′ and the 1→0′ transitions as shown in Figure 3. Because the lasers tuned near 2→3′ & 1→0′ are on opposite sides of the lines, the partial absorption of each sideband by the atoms creates AM that is phase shifted by 180 degrees as shown in Figure 4 (a). Thus, when the AM's produced by the populations in each clock state have equal magnitudes, the laser light impinging upon the detection photodiode has no AM and the output of the demodulation mixer is zero. Because the detection of both populations is simultaneous, phase locking the lasers probing the two transitions can eliminate the sensitivity to laser frequency noise. McGuirk *et al.* have demonstrated another realization of this idea by stopping the atoms in two spatially resolved regions and detecting the same state with beams derived from the same laser.[18]

The most difficult aspect of our scheme is efficiently detecting the F=1 population. While the 1→0′ transition is a closed system, the atoms can quickly optically pump into a non-absorbing m_F state. By applying a magnetic field perpendicular to the linear polarization of the laser, the atomic absorption is much higher (Figure 4(b)). In Figure 3, we adjust the laser probing 2→3′ to be much less intense than the laser probing 1→0′. Our work to date has almost entirely focused on reducing the ratio of the background detection noise, when there are no atoms, relative to the size of the 1→0′ signal when atoms are present.

The most troublesome background noise is residual AM from the EOM phase modulator. We choose the modulation frequency to be not close to piezo-acoustic resonances in the EOM, which can cause a very large and spurious AM. Spatial inhomogeneities in EOM also lead to AM noise. We control the inhomogeneity by launching the laser beam through a single-mode optical fiber after it passes through

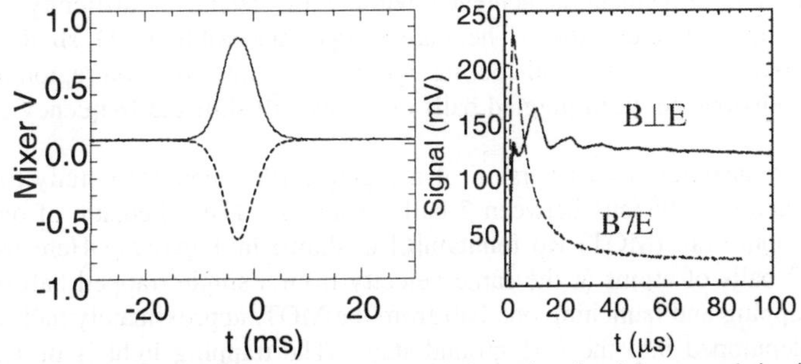

Figure 4. a) Demodulated time-of-flight FM absorption signal from population in the F=1 state (solid) and F=2 state (dashed). With equal populations, the demodulating mixer output is 0. b) Absorption for the 1→0′ transition as a function of time for B perpendicular to the laser polarization (solid) and parallel (dashed).

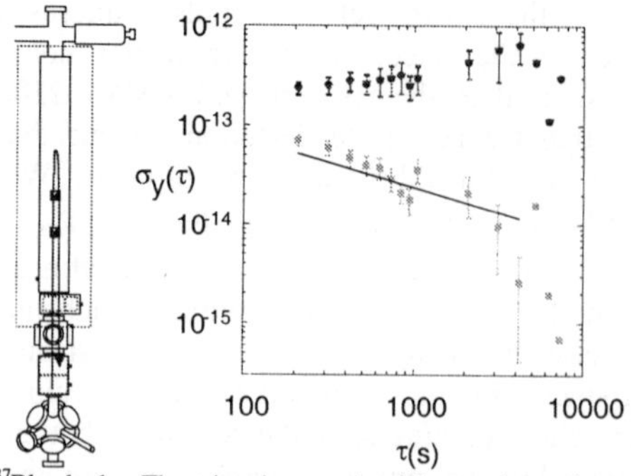

Figure 5. Juggling ^{87}Rb clock. The triangles are the Allan variance for the ^{87}Rb clock versus a quartz oscillator showing the expected deviations of the oscillator. Considering each juggled ball as an individual clock, the squares show the relative stability of these two "clocks" without the oscillator instability.

the EOM. We then servo the AM by detecting the residual AM after the fiber and feeding back with a DC voltage applied across the EOM.[19] Currently this leads to a noise level of 0.26 mV.

4 A juggling Rb clock

One can achieve higher stabilities and eliminate the dead time (reducing the requirements for the local reference oscillator) by *juggling* atoms in the fountain[20] by launching at a rate faster than the inverse of interrogation time. With a S/N = 4600 on a single launch, the frequency uncertainty would be $\delta\nu/\nu = \Delta\nu/(\pi\ \nu\ S/N) = 1\times10^{-14}$. If the cycle time is 1 s, then the fractional instability of the clock after 1s of averaging is $\sigma_y(\tau{=}1s) = 1\times10^{-14}\ \tau^{-1/2}$. By launching balls of atoms at a rate of 25 s^{-1}, the dead time is eliminated and gives a short-term instability of $\sigma_y(\tau) = 2\times10^{-15}\ \tau^{-1/2}$. This improvement is achieved without the technical difficulty of higher S/N or higher atomic densities. There are 2 important problems: 1) shutters must be used to block the trapping and cooling light from the interrogation region of the clock; 2) collisions between juggled balls of atoms will shift the frequency of the clock.

We have demonstrated a first version of a juggling Rb fountain to study the frequency shift due to collisions between 2 balls of atoms. The schematic of our single magneto-optic trap (MOT) Rb fountain[6] is shown in Figure 5. Here we rapidly launch 2 balls of atoms at the same velocity from a single trapped ball of atoms. After trapping and launching one ball from the MOT, approximately half of the atoms are depumped into the F=1 ground state. The trapping light is turned back on, recapturing the atoms remaining in F=2. A carefully masked repumping beam keeps these atoms trapped in F=2 until they are launched, 10 to 30 ms later. This second ball is also depumped after launch and then the F=1 m_F=0 atoms in

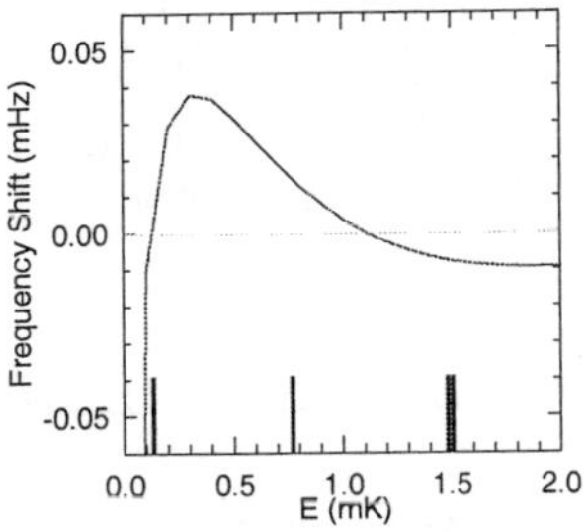

Figure 6. Juggling frequency shift for s-waves. The juggling shift is zero for time delays of 22ms (0.12mK) and 66 ms (1.1mK). The energy spectrum shown corresponds to juggling pattern for alternate launch delays of 22 and 55 ms.

each ball are transferred to the F=2 $m_F=0$ state with microwaves. The atoms remaining in F=1 are then cleared with a laser beam tuned to the $1 \rightarrow 0'$ transition.

In Figure 5 we also show the Allan variance for this juggling clock versus the microwaves supplied to the clock that are generated from a quartz oscillator. The triangles show the stability of a "clock" based on each of the two juggled balls of atoms as compared to the quartz oscillator, in agreement with the previously measured stability of the oscillator. We also compare the relative stability of the "clock" based on the first juggled ball against the "clock" based on the second juggled ball, which removes the long-term drift of the quartz oscillator.

In Figure 6 we show a calculation of the s-wave juggling frequency shift for ^{87}Rb. The first Ramsauer-Townsend[21] frequency shift null occurs at 0.12 mK, corresponding to a time delay of 22 ms. Unfortunately, the peak of the s-wave juggling shift is very nearly at a time delay of 44 ms. Therefore, for a juggling rate of 1/(22 ms), the energy for collisions between every other ball would be near the peak of the juggling shift. To cancel the juggling shift, one can launch with a more sophisticated pattern. A pattern that cancels the s-wave shift launches balls with alternate delays of 22 and 55 ms. The energy spectrum for this pattern is also shown in Figure 6. Each ball collides with 1 ball with a 22 ms delay for which there is no shift, with 1 ball at 55 ms delay, for which the shift is positive, and with 2 balls at 77 ms, where the shift is negative and half as large as the shift at 55 ms, so that there is no net frequency shift. It is likely that a pattern similar to this will allow a high juggling rate and cancel the total juggling collision shift.

5 Direct measurement of scattering phase shifts

We are also currently building a juggling atomic fountain to perform a new type of scattering experiment that accurately measures atomic scattering phase shifts using atomic clock techniques. The first juggled ball will be prepared in one of the clock states and the microwaves excite and probe the atoms during successive passages through the cavity. The second ball is prepared in an internal energy eigenstate.

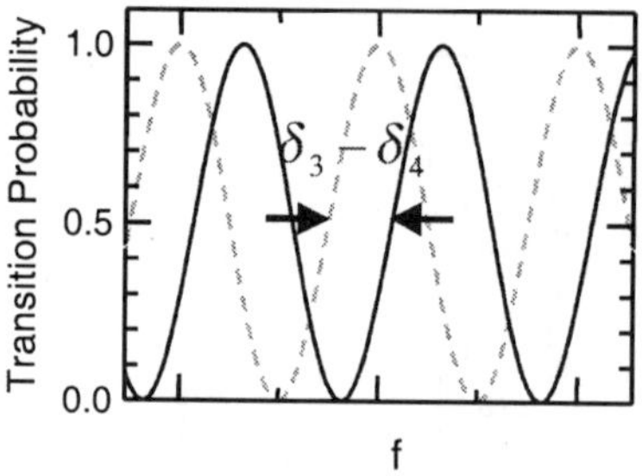

Figure 7. The phase shift (solid) of the Ramsey fringes when detecting only the scattered part of each atom in a juggling atomic clock, is $\delta_3 - \delta_4$, where δ_3 & δ_4 are the s-wave phase shifts of the two clock states in the juggling collision. For reference, the Ramsey fringe when the entire atom is detected is also shown (dashed).

When the two balls collide, the atoms in the first ball are in a coherent superposition of the two clock states and each of these states scatters off of the second ball, experiencing an s-wave phase shift. In a clock, the unscattered as well as the scattered components of each atom are detected, resulting in a density dependent frequency shift.[22,23] Here, however, we exclude the unscattered part of each (clock) atom in the first ball and detect only the scattered fraction after it returns through the microwave cavity. Because the scattering differentially shifts the phases of the two internal states of the clock atoms, the microwaves convert this phase difference into a population difference on the return passage. By detecting the populations, the difference in the s-wave scattering phase shifts can be directly measured as a phase shift of a Ramsey fringe pattern as shown in Figure 7. This phase shift does not depend on the atomic density (in the single collision limit); the number of scattered atoms, the amplitude of the Ramsey fringe, is proportional to the density.

This technique will enable extremely precise measurements of the differences of scattering phase shifts for a variety of spin states in a range of well-defined collision energies and can dramatically enhance our understanding of low energy atomic scattering. In other areas of physics, density independent scattering phase shifts are also directly measured. These include correlation measurements,[24] ratios of cross sections and generally any measurement of the angular dependence of low energy scattering. Here, cold-atom clock techniques allow us to perform a phase measurement as a frequency measurement; in essence, this enables the high potential accuracy.

6 Acknowledgments

We acknowledge financial support from the NASA Microgravity program, the NSF, a NIST Precision Measurement Grant, and the Stichting voor Fundamenteel Onderzoek der Materie. C.F. acknowledges support from a NASA GSRP fellowship. We also acknowledge stimulating discussions with Mark Kasevich on atom shot-noise limited detection.

References

1. See K. Gibble and S. Chu, Metrologia **29**, 201 (1993).
2. E. Tiesinga, B. J. Verhaar, H. T. C. Stoof, and D. van Bragt, Phys. Rev. A **45**, 2671 (1992).
3. K. Gibble and S. Chu, Phys. Rev. Lett. **70**, 1771 (1993).
4. K.Gibble and B. J. Verhaar, Phys. Rev. A **52**, 3370 (1995).
5. S. J. J. M. F. Kokkelmans, B. J. Verhaar, K. Gibble, and D. J. Heinzen, Phys. Rev. A. **56**, 4389 (1997).
6. C. Fertig and K. Gibble, IEEE Trans. Instr. Meas. **48**, 520 (1999).
7. S. Bize, Y. Sortais, M. S. Santos, C. Mandache, A. Clairon, and C. Salomon, Europhys. Lett, **45**, 558 (1999).
8. C. Fertig and K. Gibble, Phys. Rev. Lett. **85**, 1622 (2000).
9. Y. Sortais, S. Bize, C. Nicolas, A. Clairon, C. Salomon, and C. Williams, Phys. Rev. Lett. **85**, 3117 (2000).
10. S. Kokkelmans and B. J. Verhaar, private communication (2000).
11. Carl Williams, private communication (2000).
12. S. Bloom, J. Appl. Phys. **28**, 800 (1957).
13. S. B. Crampton, Phys. Rev. **158**, 57 (1967).
14. M. E. Hayden and W. N. Hardy, J. Low Temp. Phys. **99**, 787 (1995).
15. G. Santarelli, Ph. Laurent, P. Lemonde, A. Clairon, A. G. Mann, S. Chang, A. N. Luiten, and C. Salomon, Phys. Rev. Lett. **82**, 4619 (1999).
16. S.R. Jefferts, D.M. Meekhof, J. Shirley, and T.E. Parker, submitted to Metrologia (2000).
17. S. Weyers, A. Bauch, U. Hubner, R. Schroeder, and Ch. Tamm, IEEE Tran. Ultrason. Ferr. Freq. Contr. **47**, 432 (2000).
18 J. M. McGuirk, G. T. Foster, J. B. Fixler, and M. A. Kasevich, Optics Lett. **26**, 364-366 (2001).
19. N. C. Wang and J. L. Hall, J. Opt. Soc. Am. B **2**, 1527 (1985).
20. R. Legere and K. Gibble, Phys. Rev. Lett. **81**, 5780 (1998).
21. Ramsauer, Ann. der Physik **72**, 345 (1923); Townsend and Bailey, Phil. Mag. **43**, 593 (1922).
22. E. Tiesinga, B. J. Verhaar, H. T. C. Stoof, and D. van Bragt, Phys. Rev. A 45, 2671 (1992);
23. K. Gibble and S. Chu, Phys. Rev. Lett. 70, 1771 (1993).
24. See D.K.McDaniels et al., Phys. Lett. **1**, 295 (1962).

CONTINUOUS COHERENT LYMAN-α EXCITATION OF ATOMIC HYDROGEN

K. S. E. EIKEMA,* A. PAHL, B. SCHATZ, J. WALZ,† AND T. W. HÄNSCH

Max-Planck-Institut für Quantenoptik,
Hans-Kopfermann-Str. 1, 85748 Garching, Germany
E-mail:jcw@mpq.mpg.de

The *1S–2P* transition in atomic hydrogen has been observed for the first time with almost natural line width. The required narrow bandwidth Lyman-α radiation is produced by continuous four-wave-mixing in mercury which yields up to 20 nW of optical power. This unique source was employed to excite hydrogen in a strongly collimated and liquid nitrogen cooled atomic beam. We observed a line width of 119(2) MHz, which is several times narrower than could be obtained before. These results show that optical cooling and detection of antihydrogen with continuous Lyman-α radiation has excellent prospects.

1 Introduction

The *1S–2P* transition is the first and strongest dipole allowed transition from the ground state of atomic hydrogen. Yet, to drive this transition has proven to be rather difficult because of the lack of suitable narrow bandwidth laser sources at 121.56 nm (Lyman-α). The major appeal of the *1S–2P* transition is its use for laser cooling of hydrogen in the ground state, and for sensitive optical detection. Both properties are especially important for experiments with antihydrogen. Measuring transition frequencies between atomic states and comparing them between hydrogen and antihydrogen would open a new field of precise CPT tests [1,2]. It would possibly even make an experimental observation of the effect of gravity on antimatter feasible [3,4]. Production of antihydrogen is, however, notoriously difficult. A few high velocity antihydrogen atoms have been produced recently [5,6], but atoms cold enough to be confined magnetically have not been observed yet. Standard methods such as sympathetic and evaporative cooling, or *1S–2S* precision spectroscopy, are severely hampered by the low numbers of antihydrogen atoms that are expected and the rapid annihilation on contact with normal matter. However, both cooling and detection can be performed optically and efficiently by driving the *1S–2P* transition. Furthermore, the sensitivity of *1S–2S* spectroscopy could be enhanced by employing both transitions in a shelving scheme [7]. Here

*PRESENT ADDRESS: VRIJE UNIVERSITEIT, AMSTERDAM
†PRESENT ADDRESS: CERN, GENEVA, SWITZERLAND

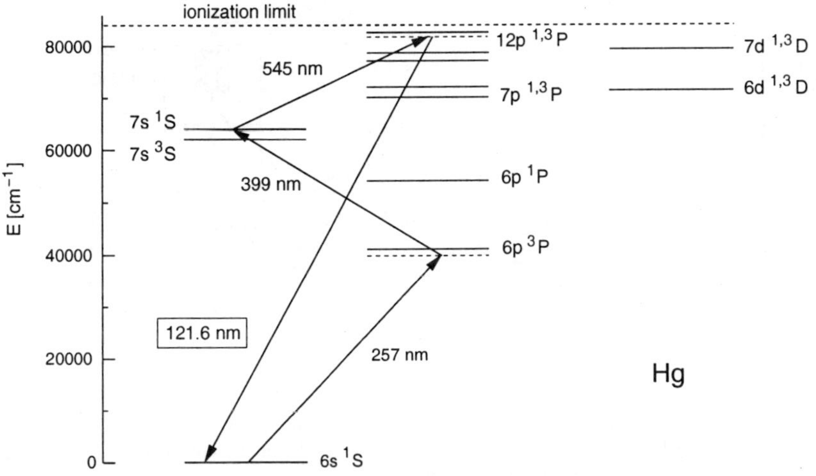

Figure 1. Four-wave-mixing scheme in mercury for continuous wave Lyman-α.

we report on the first near-natural line width atomic hydrogen excitation of the *1S–2P* transition with narrow bandwidth continuous Lyman-α [8,9]. The results of this demonstration with hydrogen are a direct measure of what can be achieved with antihydrogen as the internal energy structure is expected to be the same to high precision.

2 Continuous wave Lyman-α generation in mercury

Laser cooling and spectroscopy with *pulsed* Lyman-α has been demonstrated for magnetically trapped hydrogen [10] before. However, the use of a *continuous* source of Lyman-α is strongly preferred as it allows for more efficient laser cooling without saturation and has a more favorable duty cycle of 100%. The bandwidth of cw sources is superior which allows for a higher selectivity of sub-levels in a magnetic trap. This should enable low-loss laser cooling close to 3 mK (determined mainly by the Doppler limit).

Our source of Lyman-α radiation is based on continuous four-wave-mixing (FWM) in natural mercury vapor which produces the sum frequency of three incident laser beams [8,9] (see Fig. 1). We employ fundamental beams at wavelengths of 257 nm and 399 nm for an exact two-photon resonance with the *7s* $^1S_0$ state of mercury (mass 202 isotope, 30% abundance). The third incident beam is at a wavelength of 545 nm, to reach the photon energy of Lyman-

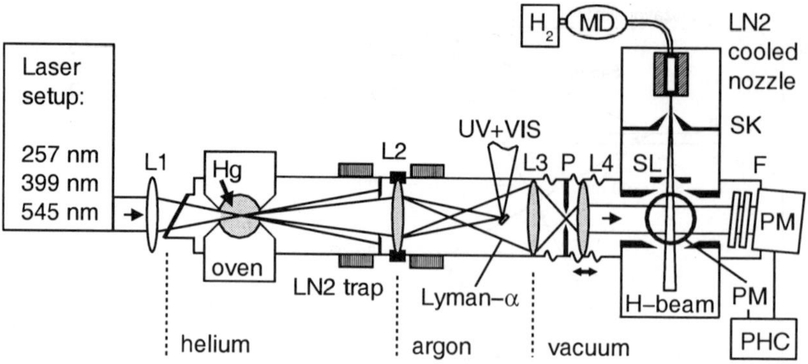

Figure 2. Schematic of the setup for continuous Lyman-α generation and *1S–2P* atomic hydrogen spectroscopy. In this figure VIS+UV stands for the fundamental laser beams, L1 for a quartz lens (while L2..4 are MgF_2 lenses), SL for slit, F for 2 Lyman-α filters, PM for solar-blind photomultiplier, MD for microwave dissociator, PHC for photon counting, SK for skimmer, P for pinhole, and LN2 for liquid nitrogen.

α, and originates from a Rhodamine 110 dye laser. Both ultraviolet (UV) wavelengths are produced by frequency doubling in four-mirror enhancement resonators. Up to 850 mW of light at 257 nm is generated this way in a BBO crystal from a 2.2 W single-mode argon-ion laser at 514.5 nm. A similar resonator containing LBO as nonlinear crystal is used to frequency double a single-mode frequency-stabilized Ti:Sapphire laser. From 2.05 W at 798 nm, up to 920 mW of 399 nm light has been produced. Both crystals are heated to 50 °C, and flushed with dust-free oxygen to protect the surfaces from degradation. VUV radiation is generated by strongly focusing the three fundamental beams with a single fused silica lens (*L1*, $f = +15$ cm) in a vapor zone of mercury (Fig. 2). All beams have the same parallel polarization.

The astigmatism of the beams produced by frequency doubling is compensated by cylindrical lenses. Telescopes with spatial filtering are employed to ensure equal focusing properties of plano-convex lens L1. The average Rayleigh length is $\approx$0.8 mm, which is 2 – 3 times larger than the diffraction limit. This is due to the use of a single lens, non-Gaussian intensity profiles, and limitations of the astigmatism compensation optics.

The interaction zone with mercury vapor is 15 mm long and is produced in a special mercury oven which produces a near rectangular density profile. The mercury pressure is inferred from temperature measurements and can be set up to 40 mbar (220 °C). A buffer gas of $\approx$70 mbar helium in the oven ensures that the surrounding optics remain free of mercury.

To measure the VUV yield and to perform *1S–2P* spectroscopy, the intensity of the fundamental beams has to be suppressed by many orders of magnitude. For this purpose we use the chromatic aberration of a MgF_2 lens (*L2*). VUV light focuses after lens *L2* at an almost two times shorter distance than the fundamental beams. In this manner most of the VUV can pass a tiny (3 mm) mirror placed in the focus of the fundamental wavelengths. Although the mirror casts a shadow (≈20% intensity loss) in the VUV beam, the collinear arrangement is very convenient for the alignment of the apparatus. Lens *L2* has been one of the most critical objects in the setup. Without special measures taken, the high power 257 nm beam induces a rapid buildup of contaminants on the surface of the lens. As some parts of the setup cannot be baked, we assume that the deposits constitute of organic molecules which can absorb virtually all VUV radiation. This effect strongly limited the sustainable VUV power that could be produced before [8]. We now solved this problem by using liquid nitrogen traps around the lens. Furthermore, the lens itself is mounted with indium seals and heated to 65 °C. In addition argon buffer gas is used in the middle section to reduce diffusion of contaminant molecules. As a final measure we exploit the fact that the VUV beam has a ≈1.7 times smaller divergence than the fundamental beams. Only the central part of the beam containing the VUV is therefore led through and fills the lens of 20 mm effective diameter, so that the UV intensity is kept as low as possible. These combined steps make it now possible to operate the source for many weeks at full power without any degradation of the VUV output.

To measure the VUV yield a solar-blind photomultiplier (Hamamatsu R1459) was placed directly after the first MgF_2 lens and the tiny mirror, with 3 narrow bandwidth Lyman-α filters (Acton Research) in front of it. In Fig. 3 a typical VUV yield curve is shown. Corrected for the detection efficiency of 2.4×10^{-4} we achieved a maximum yield at Lyman-α of 20 nW (1.2×10^{10} photons/s) in 22 mbar Hg. This yield is nearly constant over a Hg pressure-range of 20 to 40 mbar. The fundamental laser power in the interaction zone was 570 mW 257 nm, 500 mW 399 nm, and 1.2 W at 545 nm. The VUV yield is strongly wavelength dependent as can be seen in Fig. 3, due to the level structure of mercury and phasematching effects. As a result up to 10 times more VUV at 122.1 nm can be produced (200 nW), which is the highest yield reported so far by several orders of magnitude compared to continuous FWM in other metal vapors at these short wavelengths.

Possible saturation of the VUV yield has been investigated for the three fundamental wavelengths by varying the individual intensities over roughly 2 orders of magnitude. No clear sign of saturation was seen within the accuracy of the measurement. To compare these results with theory [11] we calculated

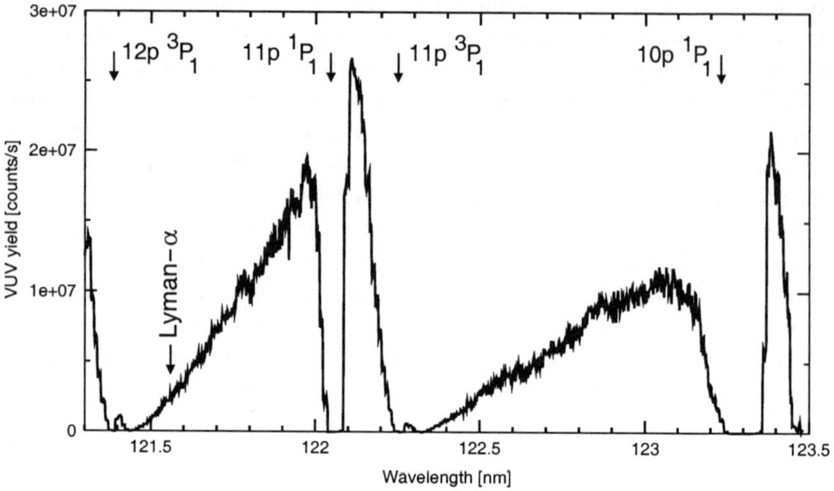

Figure 3. VUV yield of cw four-wave-mixing in mercury vapor near Lyman-α.

the Lyman-α power for the maximum mercury pressure (40 mbar) and laser power (see above) we can attain in the current oven. A maximum of $\approx$30 nW is predicted for focusing to a Rayleigh-length of 0.6 mm. The experimental yield compares rather well with this calculation, especially given the less than ideal focusing properties of the fundamental beams at present. The highest VUV yield is expected for even tighter focusing and higher mercury density to maintain phase-matching. However, this is a difficult combination to achieve, and at high density the pressure broadening of the $7s\ ^1S_0$ state will reduce the crucial two-photon resonant enhancement. Saturation due to ground state depletion can be estimated based on the maximum rates per atom in the center of a focus with 0.8 mm Rayleigh-length (assuming Gaussian beams). The calculated one-photon absorption rate is 1.1×10^3 s^{-1} (pressure broadened [12] at 40 mbar Hg), and the two-photon absorption rate is 3.6×10^3 s^{-1} (Doppler broadened, hyperpolarizability from [11]). Both rates are much lower than that of the relaxation processes such as decay of excited states (10–100 ns), or diffusion of new atoms into the focal region (timescale $<1\ \mu$s). Therefore no saturation is expected, even for considerably higher fundamental laser power. The also measured 556 nm power dependence of the VUV production peak at 122.1 nm, however, does show signs of saturation. Photomultiplier saturation was avoided in this case by using an oxygen-argon mixture in the second section, which reduced the VUV power by an order of magnitude. This leaves three-photon absorption as an explanation. However, this process is too weak to be of importance, so that further investigations are required to explain the observed power dependence at 122.1 nm.

3 Hydrogen 1S–2P spectroscopy

Application of the Lyman-α beam for *1S–2P* spectroscopy requires strong suppression of the fundamental UV wavelengths. For spectroscopy the Lyman-α beam is therefore focused first through a 0.5 mm pinhole with lens *L3* to reduce Lyman-α and UV stray light (see Fig. 2. This pinhole typically transmits 90% of Lyman-α, and reduces UV light by at least 2 orders of magnitude. A third MgF_2 lens (*L4*) collimates the Lyman-α beam after which it crosses a beam of atomic hydrogen. Blackened guiding tubes reduce stray light and leave a gap of 16 mm wide for the hydrogen beam and fluorescence detection. Due to the three lenses and the pinhole, 5–10% of the generated Lyman-α reaches the interaction zone ($\approx$1–2 nW).

A beam of hydrogen is produced in a differentially pumped chamber by microwave-dissociation of H_2 at 0.5 mbar. The atomic hydrogen is channeled through teflon tubing and emanates from a 0.5 mm diameter nozzle. This nozzle is cooled with liquid nitrogen, and constructed with an inner surface of 0.5 mm thick teflon to reduce recombination to molecular hydrogen. A skimmer of 1 mm diameter orifice and at 20 mm distance from the nozzle is used for differential pumping. Both the source and interaction chamber are pumped by two cascaded turbo-pumps.

Accurate collimation of both the Lyman-α and atomic beam is required to measure the *1S–2P* with a near-natural line width of $\approx$100 MHz. Hydrogen at 1K already results in a Doppler broadening of 1 GHz. Near-natural line width can only be accomplished for an effective transversal temperature below a few mK. To achieve this goal the hydrogen beam is strongly collimated by a 0.3 mm narrow slit just before the interaction zone. Given the distance to the nozzle of $\approx$280 mm, this amounts to an effective collimation of $\approx$1:400. The Lyman-α beam divergence is minimized by adjusting collimation lens *L4* until the recorded *1S–2P* resonance line shows no further narrowing.

The inset in Fig. 4 shows the excitation scheme in hydrogen. Almost all of the measurements have been performed on the (strongest) $1\,^2S_{1/2}$–$2\,^2P_{3/2}$ transition that can be used for laser cooling. The ground state hyperfine splitting of 1420.4 MHz can be resolved easily. The hyperfine structure of the *2P* states is smaller than the natural line width. However, as indicated in Fig. 4, from $F = 0$ only the $F' = 1$, and from $F = 1$ mainly the $F' = 2$ component of the *2P* state is excited. As a result the expected double peak separation

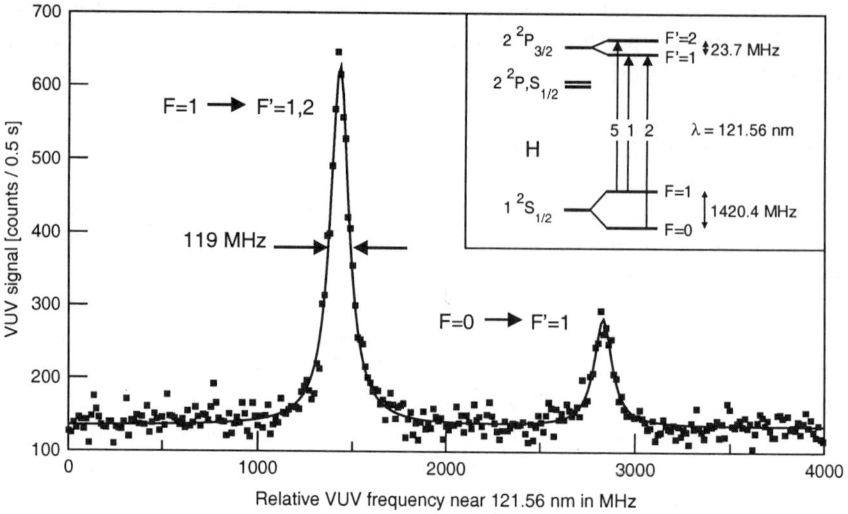

Figure 4. Recording of the *1s* $^2S_{1/2}$–*2p* $^2P_{3/2}$ transition in a beam of atomic hydrogen with continuous coherent Lyman-α (gate time 0.5 s per data point). Inset: the excitation scheme (not drawn to scale). The numbers in the arrows indicate the relative transition intensity, the natural width of the *2P* is 99.7 MHz.

for excitation to the *2* $^2P_{3/2}$ is 1400.5 MHz, 20 MHz smaller than the ground state hyperfine splitting.

A *1S–2P* excitation spectrum is obtained by photon counting the resulting fluorescence with a solar-blind photomultiplier (Hamamatsu R1459, 23 mm diameter). It is placed 22 mm above the interaction region, perpendicular to the Lyman-α and atomic beam. Another solar-blind photomultiplier is placed directly in the Lyman-α beam to measure the generated VUV intensity. That photomultiplier has two narrow-bandwidth filters (each with Lyman-α transmission $\approx$17%, 257 nm blocking >1:200) in front to suppress spurious UV signal.

Fig. 4 shows a typical experimental *1S–2P* spectrum taken out of 24 recordings. A full spectrum takes about 10 minutes to record. On resonance a maximum signal count rate of 1.2 kHz has been observed. Fitting the measured lines with a Lorentzian results in widths of 119(2) MHz and 120(3) MHz for excitation from the $F = 1$ and $F = 0$ hyperfine states respectively. A 2 MHz larger line width is expected for the $F = 1$ transition compared to the $F = 0$ transition, which is smaller than the experimental accuracy. The expected double peak structure separation of 1400.5 MHz is reproduced to high accuracy with a measured value of 1396(6) MHz. These results clearly demonstrate the great potential for continuous Lyman-α laser cooling and spectroscopy. It should be noted that no special measures were taken yet to stabilize the Lyman-α frequency. Stability and scan range was checked by recording the markers of an etalon (free spectral range 750 MHz) in the 545 nm beam, relative to the *1S–2P* resonances. Analysis of the 24 recordings showed that drifts up to 12 MHz occurred over the period between the two peaks in the *1S–2P* spectrum. On average, however, correction for drift had only limited influence on the measured transition width (3 MHz) or splitting (1 MHz).

The bandwidth of the Lyman-α light hardly contributes to the measured *1S–2P* linewidth, as it is expected to be well below 10 MHz. This estimation is based on the bandwidths of the fundamental beams which are each less than a few MHz. Time-of-flight and magnetic field ($B<40\mu T$) broadening is negligible, so that the Doppler effect is mainly responsible for excess transition width. Gaussian-like broadening is therefore expected, dominated by the velocity spread of the hydrogen atoms in combination with a deviation from perfect perpendicular alignment and residual Lyman-α divergence. Convolution of a 100 MHz Lorentzian with a 55 MHz Gaussian gives a good agreement with the measured width and shape. As a rough measure, an atom-flux corrected velocity of 1400 m/s with a spread of 1250 m/s can be assumed for our liquid-nitrogen cooled hydrogen source [13]. A deviation of just 5 mrad from perpendicular alignment is then enough to account for the measured line width.

4 Conclusions and Outlook

Our continuous source of Lyman-α has almost reached the average power of the pulsed source [10] which was previously used for laser cooling atomic hydrogen, but with an almost 2 orders of magnitude narrower bandwidth. Therefore the continuous source is very promising to perform laser cooling and first spectroscopy of antihydrogen in the ground state. Laser cooling of magnetically trapped antihydrogen [14] is, however, not the same as in hydrogen. For hydrogen use has been made of collisions to re-thermalize the cooling in one dimension to all degrees of freedom [10]. In this manner only a single beam of Lyman-α is required, and lossy optics can be avoided. Due to the low number of antihydrogen atoms that are expected to be produced, this method will be very inefficient as virtually no collisions will occur. One probably has to resort to either multiple Lyman-α beams at the cost of strong reduction of optical power, or rely on a coupling between the different degrees of freedom in a Ioffe trap to cool in all dimensions. Either way, laser cooling antihydrogen will take much more time and effort than cooling hydrogen.

Further improvement of the Lyman-α source is therefore still very desirable as it directly translates into faster laser cooling and more sensitive detection of antihydrogen. One order of magnitude in power can be gained when natural mercury is replaced with isotopically pure mercury. The high cost has prohibited the realization of this idea so far. Another option is to increase the fundamental laser power and improve beam quality. This could be accomplished by changing to a solid-state laser system, and possibly by an enhancement cavity for 545 nm (see e.g. [15]). As saturation does not seem to be limiting yet, further headroom for at least two orders of magnitude more Lyman-α radiation clearly exists.

Acknowledgments

This work is supported by the MPG and BMBF.

References

1. See e.g. J. Eades and F. J. Hartmann, Rev. Mod. Phys. **71**, 373 (1999).
2. R. Bluhm, V. A. Kostelecký, and N. Russell, Phys. Rev. Lett. **82**, 2254 (1999).
3. J. S. Bell in *Fundamental Symmetries*, edited by P. Bloch, P. Pavlopoulos, and R. Klapisch (Plenum, New York, 1987), p. 1.
4. G. Gabrielse, Hyperfine Interact. **44**, 349 (1988).
5. G. Baur *et al.*, Phys. Lett. B **368**, 251 (1996).
6. G. Blanford *et al.*, Phys. Rev. Lett. **80**, 3037 (1998).
7. T. W. Hänsch and C. Zimmermann, Hyperfine Interact. **76**, 47 (1993).
8. K. S. E. Eikema, J. Walz, and T. W. Hänsch, Phys. Rev. Lett. **83**, 3828 (1999).
9. K. S. E. Eikema, J. Walz, and T. W. Hänsch, Phys. Rev. Lett. **86**, 5679 (2001).
10. I. D. Setija *et al.*, Phys. Rev. Lett. **70**, 2257 (1993).
11. A. V. Smith and W. J. Alford, J. Opt. Soc. Am. **4**, 1765 (1987).
12. R. E. Drullinger, M. M. Hessel, and E. W. Smith, J. Chem. Phys. **66**, 5656 (1977).
13. A. Huber *et al.*, Phys. Rev. **A 59**, 1844 (1999).
14. E. L. Surkov, J. T. M. Walraven, and G. V. Shlyapnikov, Phys. Rev. **A 49**, 4778 (1994).
15. J. Nolting and R. Wallenstein, Opt. Commun. **79**, 437 (1990).

A MEASUREMENT OF THE FINE STRUCTURE CONSTANT

Joel M. Hensley, Andreas Wicht, Edina Sarajlic and Steven Chu
Physics Department, Stanford University, Stanford, CA 94305

Using an atom interferometer method, we measure the recoil velocity of cesium due to the coherent scattering of a photon. This measurement is used to obtain a preliminary value of $\hbar/M_{Cs}$ and the fine structure constant, α, with an uncertainty $\Delta\alpha/\alpha = 7.3 \times 10^{-9}$.

1. Introduction

The fine structure constant, α appears in many measurements of fundamental physical constants.[1,2] A summary of the most accurate determinations of α, along with our preliminary value from this work, is shown in Fig. 1.

The fine structure constant α can be written as

$$\alpha^2 = \left(\frac{2R_\infty}{c}\right)\left(\frac{m_p}{m_e}\right)\left(\frac{M_{Cs}}{m_p}\right)\left(\frac{h}{M_{Cs}}\right). \tag{1}$$

The Rydberg constant R_∞,[1] and the mass ratios, M_{Cs}/m_p,[3] and m_p/m_e,[1] have been determined with accuracies of 0.008, 0.20, and 2.1 ppb respectively. Thus, a measurement of h/M_{Cs} of comparable precision in conjunction with the other measured quantities, yields an improved value of α.

A value of h/M_{Cs} can be obtained by measuring the photon recoil velocity due to the recoil of an accurately known photon momentum.[4] An atom in the ground state $|g\rangle$ and moving with momentum $\mathbf{v}$ will recoil with velocity $v_{rec} = \hbar k/M$ if excited into state $|e\rangle$. Energy conservation demands

$$\omega - \omega_{eg} = \mathbf{k}\cdot\mathbf{v} + \hbar k^2/2M, \tag{2}$$

where $\hbar\omega_{eg}$ is the energy separating the states $|g\rangle$ and $|e\rangle$. A counter-propagating photon of frequency ω' can induce a transition back to the ground state, where ω and ω' are shifted by an amount

$$\omega - \omega' = (\mathbf{k} + \mathbf{k}')\cdot\mathbf{v} + (\hbar/2M)(\mathbf{k} + \mathbf{k}')^2. \tag{4}$$

The $\omega_{rec} = 2\pi f_{rec} \equiv \hbar k^2/M$ is defined as the recoil frequency and $v_{rec} \equiv \hbar k/M$ is the recoil velocity.

Since the frequency of the light used to induce the transition has been accurately measured,[5] ω_{rec} can be determined if the initial velocity of the atoms is known. The recoil shift appears as a spectral doublet in saturation absorption spectroscopy, which selects atoms with zero velocity with respect to counter-propagating laser beams.[6]

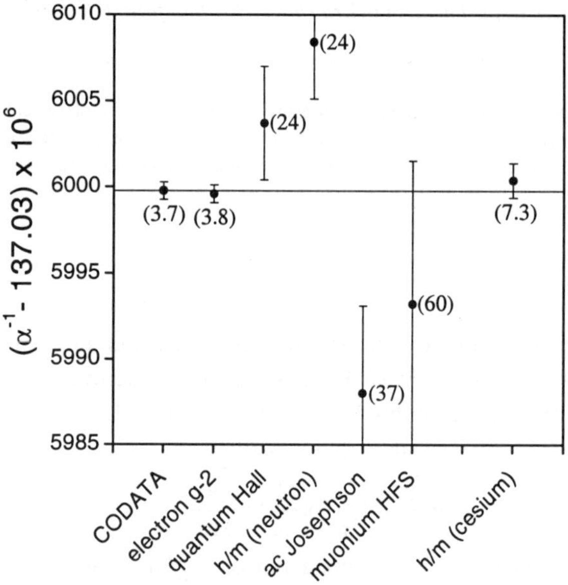

Fig. 1 Determinations of the most accurate values of the fine structure constant α (from Table XV of Ref. 1), the CODATA value, and our preliminary value.

2. Experimental method

The velocity dependent term in Eq. 4 can also be eliminated by replacing the 2 excitation pulses with 2 pairs of π/2-pulses as shown in Fig. 2.[7,8,9] This so-called Ramsey-Bordé interferometer was first used as an extension of the Ramsey method of separated oscillatory fields into the optical domain.[10] Since the sequence of 4 π/2-pulses creates two interferometers, the photon recoil measurement is transformed into a differential measurement of phase differences between two atom interferometers. This configuration has two advantages over a measurement based on two π-pulses: (1) The recoil shift is independent of the atom's initial velocity and *all* shifts that are position independent because of the differential nature of the measurement. (2) The frequency resolution is determined by the time between each pair of π/2-pulses while the duration of the π/2-pulses determines the spectral width of the pulses. Thus, by using fairly short π/2 pulses, a relatively large number of atoms in the atomic fountain can be addressed without sacrificing frequency resolution.

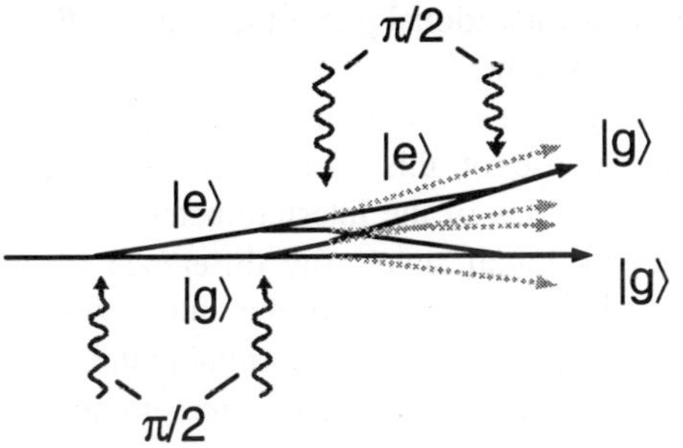

Fig. 2 A space-time trajectory of a Ramsey-Bordé interferometer beginning with an atom initially at rest. Each arrow indicates the direction of the k-vector of the $\pi/2$-pulses. The dotted lines show atomic paths that do not contribute to the interference signal.

Using this interferometer configuration as a basic starting point, we improve the resolution of the measurement in several ways. (*i*) The precision of this interferometer is increased by inserting 30 π-pulses between the two sets of $\pi/2$-pulses so $\omega_{rec} \rightarrow (30{+}1)\omega_{rec}$. (*ii*) The measurement time is increased by using an atomic fountain,[11] where the time between $\pi/2$-pulses is 0.12 seconds. (*iii*) The transitions are between different ground states of the atom. A two-photon Raman transition with counter-propagating beams is used to impart photon recoils with $\mathbf{k}_L \rightarrow \mathbf{k}_{eff} = \mathbf{k}_1 - \mathbf{k}_2$, where $\mathbf{k}_1$, $\mathbf{k}_2$ are the wavevectors of the counter-propagating photons. (*iv*) Our current atom interferometer uses an adiabatic method of transferring momenta to the atoms first introduced by Gaubatz, *et al.*[12] By generating the time delayed pulses, we were able to tailor the shape of the pulses to construct an atom interferometer where the coherent transfer is ~94% efficient.[8]

We calculate the phase shift of each path of the atom interferometer using Feynman's path integral formulation of quantum mechanics.[13] In this formulation, the phase shift is divided into two contributions. The first contribution is due to the free evolution of the wavefunction $\Delta\Phi_{free} = S_{cl}/\hbar$, given by

$$S_{Cl}/\hbar = \frac{1}{\hbar}\int dt\left\{\frac{1}{2}mv^2 - m[g+\delta g(z)]z(t) - \hbar\omega_i\right\}, \tag{5}$$

where the integral is over the classical trajectory. The Lagrangian includes terms due to the kinetic energy, the gravitational energy including a gravity gradient,[14] and the internal energy state of the atom. The second contribution to the phase shift is due to the interaction of the atoms with the optical field. In the short pulse limit, the transitions $|g\rangle \rightarrow |e\rangle$ and $|e\rangle \rightarrow |g\rangle$ add an additional phase term $\exp[+i(k_{eff}z - \omega_L t + \phi_L)]$ and $\exp[-i(k_{eff}z - \omega_L t + \phi_L)]$, respectively, where k_{eff} is the effective k-vector of the Raman pulse, z is the position of the atomic wavefunction at time t, ω_L is the optical frequency, and ϕ_L any additional phase factor of the light. The terms k_{eff}, z, ω_L, and ϕ_L are all functions of time.[13]

The phase of each atom interferometer is measured relative to a microwave signal generator. If the phase of this local oscillator tracks exactly the phase difference between the two arms of the interferometer, the atom originally in state $|g_1\rangle$ is returned back to its initial state. Any difference in phase accrued by the microwave oscillator and the two arms of the interferometer produces Ramsey oscillations between states $|g_1\rangle$ and $|g_2\rangle$. By determining the peak of the central fringe of the interferometer, we measure the ω_{rec}. The measurement is made less sensitive to experimental imperfections by measuring the phase difference between the two atom interferometers, $\Phi_1 - \Phi_2$. Systematic effects that are common to both interferometers are then subtracted out. A second pair of interferometers Φ_3, Φ_4 is created by reversing the directions of all the momentum impulses to create a pair of "inverted" interferometers. The photon recoil frequency ω_{rec} is determined from the phase difference

$$\Delta\Phi = [(\Phi_1 - \Phi_2) + (\Phi_3 - \Phi_4)] = -8(N+1)(\omega_{rec} - \omega_{fixed})T - 4k_{eff}|\langle\delta g\rangle|\,(T+T')T, \qquad (6)$$

where $|\langle\delta g\rangle|$ is the average difference in the gravitational potential for the two trajectories, and T' is time between the 2nd and 3rd $\pi/2$-pulses. The quantity, ω_{fixed} is the value assumed by the local oscillator. We set $\omega_{fixed} = 2\pi \times 15\,006.278\,875$ based on values of physical constants known in March, 1998. The phase shift we measure then gives us the difference from this value and the new value. In calculating the gravity gradient correction, we evaluate $|\langle\delta g\rangle|$ by using Eq. 5 to calculate the action $S/\hbar$.

3. Systematic effects

We checked for systematic effects by varying a large number of experimental parameters. In the limited amount of space, we only describe a few the effects that were considered to give the reader a sense of how we searched for systematic effects.

3.1 Beam alignment and wavefront curvature

We can align counter-propagating beams to within $\pm$ 15 μrad. The optical alignment was observed to drift for about an hour after a re-alignment to roughly 1 ppb from the ideal alignment. For this reason we choose to correct all our data by 1 ppb with a $\pm$ 0.4 ppb estimated uncertainty in this correction.

Another systematic effect is the Gouy phase shift of a laser beam in the vicinity of the focal point.[15] For ω_0 = 1 cm, this phase shift results in a 0.8 ppb correction in h/M if the beam waist is centered with respect to the atom interferometers. The positions of the 2m focal length lens was moved between 2 to 4 centimeters away from the collimating position and change in phase shift with the de-collimation was measured to be 1.4 ± 3.9 ppb per 1 cm of change in the position of the lens. The lens can be set to ±0.2 mm uncertainty with the use of a precision optical flat.

3.2 AC Stark shifts

Unequal AC Stark shifts in the two arms of the interferometer can shift the interference fringes. The AC Stark shifts of concern are of the form $\Omega^2/4\Delta$, where Ω is the Rabi frequency and Δ is the detuning of the light from resonance. No unaccounted change in the measured recoil velocity was observed as the laser frequencies were varied. The detuning studies establish an upper limit of 0.1 ppb uncertainty due to AC Stark shifts.

The optical wave fronts used in the experiment are set relative to an actively stabilized vibration-isolated platform suspended just above the vacuum chamber.[16] This beat note provides the feedback signal used to compensate for the vibrations of the optical mounts, air currents, etc., that introduce phase noise into the Raman beams. The vibration isolation system used in this experiment requires a "tracer beam" that follows the same optical path as the Raman beams. To verify that there was no effect, we increased the intensity and duration that the tracer beam by a combined factor of 6700. The photon recoil value remained the same to within 10 ± 25 ppb.

3.3 Coriolis forces

Phase shifts may arise because the atom interferometer may have some spatial area due to misalignment of an atom trajectory with respect to the direction of the momentum impulses induced by the Raman pulses. This area will cause a Sagnac phase shift $\Delta\phi = 2(M/\hbar)\mathbf{\Omega}\cdot\mathbf{A}$, where $\mathbf{\Omega}$ is the angular velocity of the earth and $\mathbf{A}$ is the enclosed area. To verify the insensitivity to rotations, data were taken with intentionally misaligned launch and Raman beam directions. To further amplify any gyroscope effect, the entire experimental apparatus was rocked back and forth sinusoidially with maximum angular velocity ~19 times earth's rotation. With the systematic effect magnified in this way, we established an upper limit ± 1.0 ppb uncertainty due to Coriolis forces.

3.4 Missed photon kicks

The experiment demands that all the atoms receive exactly N additional photon recoils. If an atom were to miss a Doppler-sensitive π-pulse, it would be in the bright state at the beginning of the next pulse. The next π-pulse would then induce incoherent, single-photon transitions resulting in no net (averaged) phase shift in the interferometer. The most serious concern is that some of the atoms may experience a Doppler-free transition induced by two *co*-propagating beams, where one of the beams is due to a reflection from an optical surface. We minimize back reflections by tilting all the optics in the near vicinity of the vacuum can. We also avoid illuminating the atoms with π-pulses close to the apogee of their trajectory where they would be most sensitive to Doppler free transitions.

An atom that misses one momentum impulse will be drastically shifted in phase by $\Delta\phi = (\hbar k^2_{eff}/M_{Cs})T$. If the fraction of atoms that miss a pulse is small, we can choose the time T so that $\Delta\phi$ is modulo 2π. With this choice of T, the fringe pattern with a small number of missed recoils will be the same as the fringe pattern with no missed recoils.

As a check on our ability to eliminate the chance of missed photon kicks, 30 π-pulses were added near the apogee of the atomic trajectory so that the laser would have a 100× higher probability of exciting a Doppler-free transition as compared to normal operating conditions. We then scanned the time interval T over a range where $\Delta\phi$ changed by 2π. The phase shift observed had an amplitude of 17 ± 16 ppb, which is consistent with no phase shift. Because of the enhancement factor of 100, we place a 0.16 ppb uncertainty on this error.

3.5 Phase errors due to $\pi/2$ pulses

If there is a phase error introduced by the $\pi/2$-pulses that is independent of the time T between $\pi/2$-pulses, then the phase difference between a pair of atom interferometers (neglecting the gravity gradient term) will vary as

$$\Delta\Phi = \Phi_1 - \Phi_2 = -4\pi\,(N+1)(\omega_{rec} - \omega_{fixed})T - \phi_{err}\,. \qquad (7)$$

Thus, if the phase shift $\Delta\Phi$ is measured as a function T, a linear dependence in phase shift is expected. The slope of the line, $-4\pi(N+1)(\omega_{rec} - \omega_{fixed})$, yields the recoil velocity and the intercept at T=0 measures the phase shift error.

Fig. 3 shows the phase difference $\Phi_1 - \Phi_2$ plotted for data taken under two conditions. Curve (a) represents data taken while two independent acousto-optic modulators (AOMs) were used to shape the Raman pulses. The AOMs and the rf attenuators have been shown to add phase shifts to the laser light. The offset, ϕ_{err}, was reduced by adding an additional AOM that turned both laser beams off simultaneously. With this change, the value of ϕ_{err} became consistent with zero, as shown in curve (b).

The determination ω_{rec} from the slope of Eq. 7 is one of our most powerful controls over potential systematic effects. The analysis eliminates a large number of systematic shifts such as those introduced by non-ideal RF electronics, the AOMs used to tailor the pulses, various phase-locked loops affected by the switching of

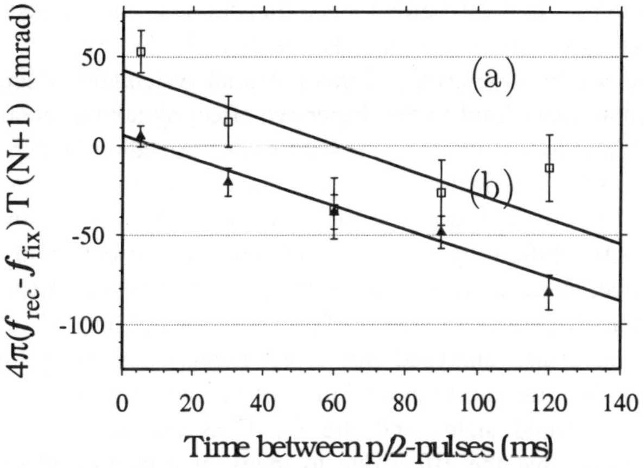

Fig. 3. The interferometer phase difference $\Delta\Phi = \Phi_1 - \Phi_1$ versus the time T between the $\pi/2$ pulses. Pulse shaping AOMs are controlled by microwave attenuators that introduce attenuation dependent phase shifts. Unfortunately, even with the common acousto-optic modulator (AOM) installed, long term drifts in ϕ_{err} were still observed.

frequencies, pulsing MOT coils that might affect the vibration isolation platform, etc. Our value of the photon recoil frequency is based on this data analysis method.

3.6 Index of refraction effects

Cs atoms in the vacuum chamber will produce a non-unity index of refraction for the light. This will change $k_{eff} = k + k'$, which will affect the recoil measurement by (a) changing the momentum imparted to the atoms by the light, and (b) by changing the phase difference $k_{eff}\Delta z$, where Δz is the separation between the two interferometers. The shift in the recoil measurement due to room temperature Cs atoms has been shown to be less than 0.06 ppb by varying the pressure of Cs in the vacuum chamber.

The contribution from the cloud of cold atoms is more complicated. In addition to the usual dispersive features defined by the 5 MHz wide absorption of the other D_1-lines, there are much sharper dispersive features due to electromagnetically induced transparency effects.[17] The width of the induced transparency dip is defined by the linewidth of the adiabatic π-pulses and is ~ 200 kHz. Transitions between the $|F=4, m_F=0\rangle$ and the $|F=3, m_F=0\rangle$ states are made with light tuned exactly to the $|F'=4, m_F=+1\rangle$ excited state, so there is shift due to the index of refraction. However, the adiabatic pulses are only ~94% efficient, and other Zeeman ground states become populated. The population of these other states will contribute to a non-zero effect since transitions to the excited states are tuned 50 and 100 kHz from resonance in a bias field of 72 mG.

After we became aware of this potential systematic effect, we determined an experimental upper limit to the dispersive effect by taking recoil data, switching between high and low atom densities with each succeeding launch. The number of atoms was reduced by a factor of 4 from normal operating conditions by collecting atoms in the MOT before launching for less time. The recoil value taken with the reduced density shifted by +7.3 ± 10.5 ppb, consistent with no effect. The dispersive effect is linear with atomic density, so this measurement places an upper limit to the density dependent effects of (7.3 ± 10.5)(4/3) ppb.

We have also analyzed this effect numerically using the optical Bloch equations that describe the time evolution of all the Zeeman sub-levels of the $S_{1/2}$ F=3 and F=4 ground states and the $P_{1/2}$ F′=3 excited states of cesium. The calculation shows that the effect due to index of refraction changes is less than 1ppb. In this paper, we include the experimentally determined uncertainty in our preliminary value of α, pending further verification of our calculations. However, because we expect that there will be no correction larger than 1 ppb to our final result, so we have not included a 9.7 ppb correction to our value of f_{rec}.

4. Summary of the major systematic effects

A selected list of the systematic effects that were considered is given in Table 1. A more complete description is given elsewhere.[18] In the cases where the experimental test is consistent with zero and there are good reasons why there should not be a systematic effect when the parameter is varied, the uncertainty is kept, but no correction is applied to the value of ω_{rec}. The significant uncertainties (greater than 0.4 ppb) listed in Table 1 are added in quadrature.

Table 1. Systematic error budget

Systematic effect	**Experimental limit** (ppb)	**Theoretical limit** (ppb)	**Correction to h/M** (ppb)
Optical beams			
Gouy phase shift		- 0.89 ± 0.04	+ 0.89 ± 0.4
Wavefront curvature	+ 0.035 ± 0.039	< 0.04	
Relative angle	- 1.0 ± 0.4		+1.0 ± 0.4
Polarization	± (1.5 ± 2.0)		±2.0
Magnetic and Electric fields			
Linear term	-1.0 ± 2.0		0 ± 2.0
Quadratic term	+0.15 ± 0.1		- 0.15 ± 0.1

ac Stark effect from tracer laser beam	< 0.004	< 0.008	
Other			
Gravity gradient			- 9.7 ± 1.0
Dispersion due to cold atoms	-9.7 ± 14	in progress	0 ± 14
Sagnac effect from launch misalignment	< 0.3		
Sagnac effect from beam misalignment	< 1.0		0 ± 1.0

5. The value of h/M_{Cs} and α

The preliminary value of the photon recoil frequency is derived from data that measures the slope of the recoil shift versus T. The weighted average of 31 data sets is $f_{rec} = f_{fix}$ [1 - (124.98 ± 4.88) × 10^{-9}]. The chi-square is $\chi^2/(N-2) = 1.5$. This value of f_{rec} is adjusted by the values listed in right hand column of Table 1. In frequency units,

$$f_{rec} = 15\,006.276\,88\ (23)\ (7)\ (22), \tag{8}$$

where the quantities in parentheses are the total (15 ppb), the statistical (4.9 ppb), and systematic (14 ppb) uncertainties. Without the density dependent uncertainty, the systematic error in h/M reduces to 3.2 ppb. All the uncertainties are reduced by a factor of two when used to calculate α.

Combined with the other measurements needed to determine α,[1,3,4] we arrive at a preliminary value for α^{-1} of

$$\alpha^{-1} = 137.036\,000\,3\ (10)\ \ (7.3\text{ ppb}). \tag{9}$$

We are currently undergoing a final review of our data analysis and numerical simulations of the atom interferometer using the optical Bloch equations. If the initial results are verified, we should be able to reduce uncertainty to α to 3.1 ppb. The h/M_{Cs} contribution to this uncertainty is α to 2.9 ppb.

This work was supported, in part by grants from the National Science Foundation, the Air Force Office of Scientific Research, the Department of the Navy and the National Reconnaissance Office. A.W. acknowledges support from the Alexander von Humboldt Foundation.

REFERENCES

[1] P. J. Mohr and B. N. Taylor, CODATA recommended values of the fundamental constants: 1998, *Rev. Mod. Phys.* **72**, 351-495 (2000).

[2] T. Kinoshita, The fine structure constant, *Rep. Prog. Phys.* **59** 1459-1492 (1996).

[3] M. P. Bradley, *et al.*, Penning trap measurements of the masses Cs-133, Rb-87, Rb-85, and Na-23 with uncertainties ≤ 0.2 ppb, *Phys. Rev. Lett.* **83** (1999) 4510-4513.

[4] T. Udem, J. Reichert, R. Holzwarth, T. W. Hänsch, Absolute optical frequency measurement of the cesium D-1 line with a mode-locked laser, *Phys. Rev. Lett.* **82,** 3568-3571 (1999).

[5] T. Udem, J. Reichert, R. Holzwarth, T. W. Hänsch, Absolute optical frequency measurement of the cesium D-1 line with a mode-locked laser, *Phys. Rev. Lett.* **82,** 3568-3571 (1999).

[6] J.L. Hall, Ch. J. Bordé, and K. Uehara, Direct optical resolution of the recoil effect using saturated absorption spectroscopy. *Phys. Rev. Lett.* **37**, 1339-1342 (1976).

[7] D.S. Weiss, B.C. Young, and S. Chu, *Phys. Rev. Lett.* **70**, 2706-2709 (1992); *Appl. Phys. B* **59,** 217-256 (1994); M. Weitz, B.C. Young, and S. Chu, *Phys. Rev. A* **50**, 2438-2566 (1994); *Phys. Rev. Lett.* **73,** 2563-2566 (1994).

[8] M. Weitz, B.C. Young, and S. Chu, *Phys. Rev. A* **50**, 2438-2566 (1994); *Phys. Rev. Lett.* **73,** 2563-2566 (1994).

[9] B.C. Young, unpublished Ph.D. thesis, (1997); S. Chu, Les Houches Lectures in Physics, Session LXII, eds. R. Kaiser, C. Westbrook, and F. David, (Springer, Berlin, 2001) pp 317-370.

[10] Ch. J. Bordé Atomic interferometry with internal state labelling, *Phys. Lett. A* **140**, 10-12 (1989).

[11] M.A. Kasevich, E. Riis, S. Chu, and R.G. DeVoe, Rf Spectroscopy in an Atomic Fountain, *Phys. Rev. Lett.* **63**, 612 (1989).

[12] U. Gaubatz, *et al.*, *Chem. Phys. Lett.* **149**, 463 (1988).

[13] M. Kasevich and S. Chu, Atomic interferometry using stimulated Raman transitions, *Phys. Rev. Lett.* **67**, 181-183 (1991). Also see P. Storey and C. Cohen-Tannoudji, *J. Phys. II France* **4**, 1999-2027 (1994) for a good tutorial of this calculation.

[14] We are indebted to M. Kasevich for pointing out that this systematic effect does not subtract away in our experiment. Note that the final end points of the atomic trajectories have to be taken as the detection point as discussed in A. Peters, K.Y. Chung and S. Chu, High precision gravity measurements using atom interferometry, *Metrol.* **38**, 25-61(2001).

[15] H. Kogelnik and T. Li, *Appl. Opt.* **5**, 1550 (1966); A.I. Siegman, *Lasers*, University Science Books, Mill Valley, CA, 1986.

[16] J.M. Hensley, A. Peters and S. Chu, Active low frequency vertical vibration isolation. *Revs. of Sci. Inst.* **70**, 2735-2741 (1999).

[17] K. J. Boler, A. Imamogulu, and S. E. Harris, Observation of electro-magnetically induced transparency. *Phys. Rev. Lett.* **66**, 2593-2596 (1991).

[18] A. Wicht, J. M. Hensley, E. Sarajlic and S. Chu, Proc. of the 6th Symp. on Freq. Stds., ed. P. Gill (World Scientific, Singapore, 2002).

TOWARDS GRAVITATIONAL WAVE ASTRONOMY – FROM EARTH AND FROM SPACE

K. DANZMANN AND A. RÜDIGER

Max-Planck-Institut für Gravitationsphysik, Albert-Einstein-Institut, and Universität Hannover, Callinstr. 38, 30167 Hannover, Germany
E-mail: kvd@mpq.mpg.de

The existence of gravitational waves is the most prominent of Einstein's predictions that has not yet been directly verified. The space project LISA is approved by ESA as a cornerstone mission in the field of "Fundamental Physics", and is currently the object of a joint ESA/NASA study aimed at launch in 2011. This space project shares its goal and principle of operation with the ground-based interferometers currently under construction: the detection and measurement of gravitational waves by laser interferometry. Ground and space detection differ in their frequency ranges, and thus the detectable sources. At low frequencies, ground-based detection is limited by seismic noise, and yet more fundamentally by 'gravity gradient noise', thus covering the range from a few Hz to a few kHz. On five sites worldwide, detectors of armlengths from 0.3 to 4 km are nearing completion. They will progressively be put in operation between 2001 and 2003. Future enhanced versions are being planned, with scientific data not expected until 2008, i.e. near the launch of the space project LISA. It is only in space that detection of signals below, say, 1 Hz is possible, opening a wide window to a different class of interesting sources of gravitational waves. The project LISA consists of three spacecraft in heliocentric orbits, forming a triangle of 5 million km sides. A technology demonstrator, designed to test vital LISA technologies, is to be launched, aboard a SMART-2 mission, in 2006.

1 Introduction

At this ICOLS, devoted to advanced technologies in optics and lasers, it may be surprising to have a presentation on waves that are not electromagnetic: gravitational waves. But it will turn out that, in order to detect and measure these, we will require the most advanced technologies in optics, lasers, and interferometry.

Efforts to observe these gravitational waves with ground-based interferometers have gone into their final phase of commissioning, and the international collaboration on placing a huge interferometer into an interplanetary orbit is close to reaching final approval.

We will briefly discuss the characteristics of large GW detectors being built right now. In this talk we will learn how the detectors on ground and in space differ, where aims and technologies overlap, and what can scientifically be gained from the complementarity of these researches.

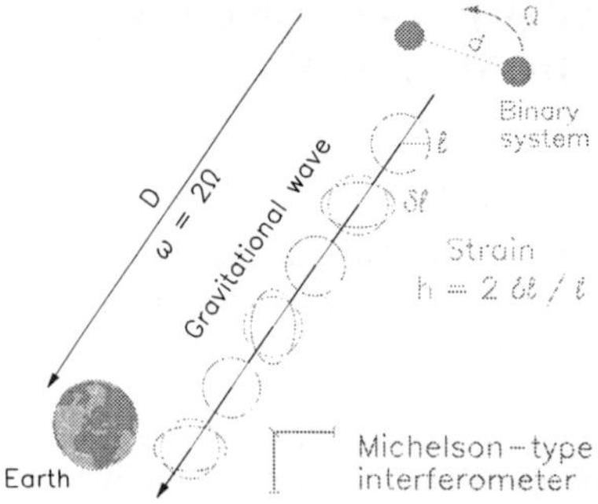

Figure 1. Generation and propagation of a gravitational wave emitted by a binary system.

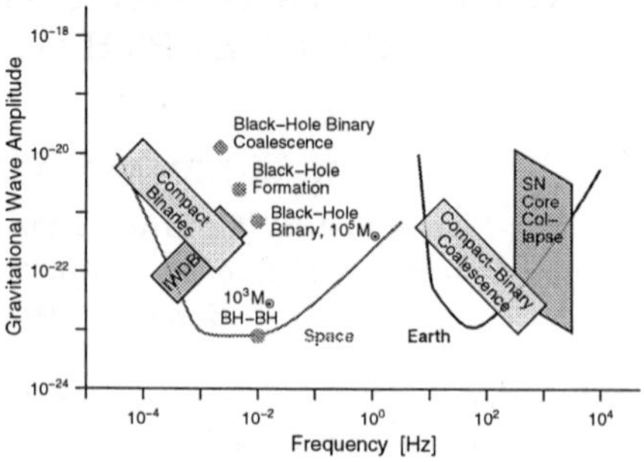

Figure 2. Some sources of gravitational waves, with sensitivities of *Earth* and *Space* detectors.

2 Gravitational waves

Gravitational waves of measurable strengths are emitted only when large cosmic masses undergo strong accelerations, for instance – as shown schematically in Figure 1 – in the orbits of a (close) binary system. The effect of such a gravitational wave is an apparent strain in space, transverse to the direction of propagation, that makes distances ℓ between test bodies shrink and expand by small amounts $\delta\ell$, at twice the orbital frequency: $\omega = 2\,\Omega$. The strength of the gravitational wave, its "amplitude", is generally expressed by $h = 2\,\delta\ell/\ell$. An interferometer of the Michelson type, typically consisting of two orthogonal arms, is an ideal instrument to register such differential strains in space.

"So where's the problem ?" one might ask. It lies in the magnitude, or rather: the smallness, of the effect.

2.1 Strength of gravitational waves

From a supernova, or the in-spiral of a neutron star binary out at the Virgo cluster (a cluster of about 2000 galaxies, 10 Mpc away), we could expect a strain of something like $h \approx 10^{-22}$. So all we have to do is to measure – in a Michelson interferometer of kilometer dimensions – path changes in the order of 10^{-19} m. *Hopeless ?* The sensitivities obtained with prototypes of ground-based interferometers bear evidence that it is within reach.

And yet, despite the smallness of the interaction, gravitational waves are by no means a "weak" phenomenon. On the contrary, they are linked with cosmic events of high energy transfers. Two examples show this clearly. The *binary system* containing the Hulse-Taylor pulsar PSR 1913+16, much publicised through the 1993 Nobel prize, loses its orbital energy primarily through the emission of gravitational radiation; no other loss mechanism comes anywhere near it. A *supernova*, on the other hand, in its final milliseconds of collapse, emits more power than the (visible) luminescence of all the stars of the universe combined. Although most of this is in neutrinos, an appreciable part is also emitted in gravitational waves, from the rebound of the core.

2.2 Complementarity of ground and space observation

Figure 2 shows some typical sources of gravitational radiation. They range in frequency over a vast spectrum, from the kHz region of supernovae and final mergers of compact binary stars down to mHz events due to formation and coalescence of supermassive black holes. Indicated are sources in two clearly separated regimes: events in the range from, say, 5 Hz to several kHz (and only these will be detectable with terrestrial antennas), and a low-frequency regime, 10^{-5} to 1 Hz, accessible only with a space project such as LISA. In the following sections we will see how the sensitivity profiles of the detectors come about. No detector covering the whole spectrum shown could be devised.

Clearly, one would not want to miss the information of either of these two (rather disjoint) frequency regions. The upper band ("Earth"), with supernovae and compact binary coalescence, can give us information about relativistic effects and equations of state of highly condensed matter, in highly relativistic environments.[1] Binary inspiral is an event type than can be calculated to high post-newtonian order, as shown, e.g., by Buonanno and Damour.[2] This will allow tracing the signal, possibly even by a single detector, until the final merger, a much less predictable phase. The ensuing phase of a ring-down of the combined core does again lend itself to an approximate calculation, and thus to an experimental verification. Chances for detection are reasonably good, but not by wide margins.

The events to be detected by LISA, on the other hand, may have extremely high signal-to-noise ratios, and failure to find them would shatter the very foundations of our present understanding of the universe. The strongest signals will come from events involving (super-)massive black holes, their formation as well when galaxies with their BH cores collide. But also the (quasi-continuous) signals from neutron-star and black-hole binaries are among the events to be detected ('Compact Binaries' in Figure 2). Interacting white dwarf binaries inside our galaxy ('IWDB' in Figure 2) may turn out to be so numerous that they cannot all be resolved as individual events. Catastrophic events such as the Gamma-ray bursts are not yet well enough understood to estimate their emission of gravitational waves, but there is a potential of great usefulness of GW detectors mainly at low frequencies. The combined observation with electromagnetic and gravitational waves could lead to a deeper understanding of the violent cosmic events in the far reaches of the universe[3].

3 Ground-based interferometers

The underlying concept of all our ground-based detectors is the Michelson interferometer (see schematic in Figure 3), in which an incoming laser beam is divided into two beams travelling along different (perpendicular) arms. On their return, these two beams are recombined, and their interference (measured with a photodiode PD) will depend on the difference in the gravitational wave effects that the two beams have experienced.

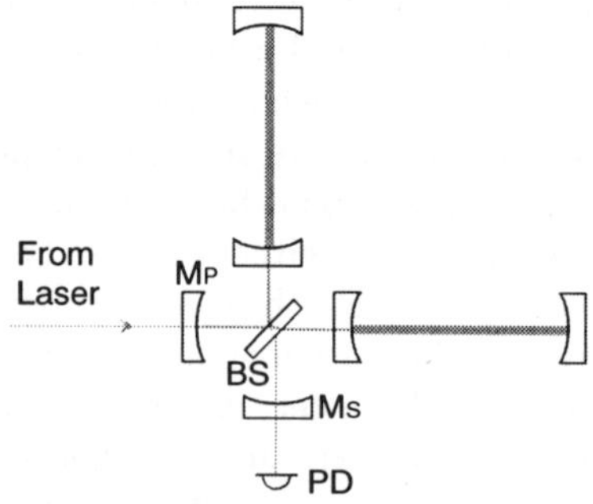

Figure 3. Advanced Michelson interferometer with Fabry-Perots in the arms and extra mirrors M_P, M_S for power and signal recycling.

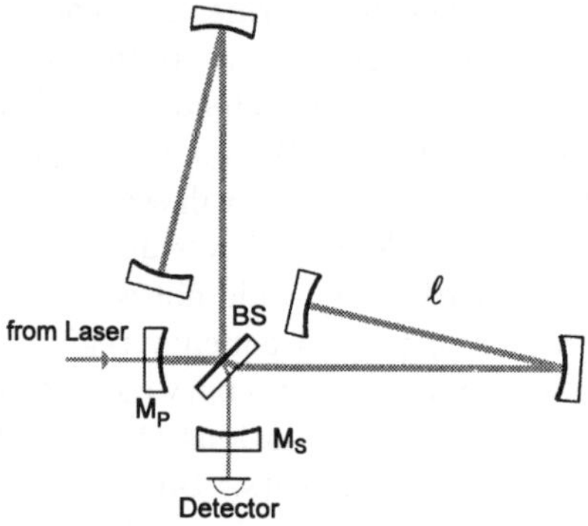

Figure 4. The DL4 configuration with dual recycling to be used in GEO 600

The changes δL in optical path become the larger the longer the optical paths L are made, optimally about half the wavelength of the gravitational wave: e.g. to a seemingly unrealistic 150 km for a 1 kHz signal. Schemes were devised to make the optical path L significantly longer than the geometrical arm length ℓ, which is limited on Earth to only a few km. One way is to use 'optical delay lines' in the arms, the beam bouncing back and forth between two concave mirrors (a modest version of this is shown in Figure 4). The other scheme is to use Fabry-Perot cavities (Figure 3), again with the aim of increasing the interaction time of the light beam with the gravitational wave. For GW frequencies beyond the inverse of the storage time τ, the response of the interferometer will, however, roll off with frequency, as $1/f\tau$.

3.1 The detector prototypes

After pioneering work by Rai Weiss at MIT (1972), other groups at Munich/Garching, at Glasgow, then Caltech, Paris/Orsay, Pisa, and later in Japan and Australia, also entered the scene. Their prototypes range from a few meters up to 30, 40, and even 100 m. An alternative detection scheme, a Sagnac configuration, is being investigated at Stanford.

It is a fortunate feature that on our way to the large-scale detectors we were able to go through generations of ever-improving prototypes. Even though some of these prototypes had reached the sensitivities of cryogenic resonant-mass antennas, they were never meant to be used as detectors, but rather as test-beds for verifying new schemes and configurations devised to overcome otherwise limiting noise effects.

The "phase noise" reduction achieved in these prototypes approaches that required in full-fledged terrestrial interferometers, and it is by many orders of magnitude better than required (at low frequencies) for a space mission.

3.2 The large-scale projects

Table 1 gives an impression of the wide international scope of the interferometer efforts, listed according to size of detector. All of the large-scale projects will use low-noise Nd:YAG lasers ($\lambda = 1.064\,\mu$m), pumped with laser diodes for high overall efficiency. A wealth of experience has accumulated on highly stable and efficient lasers, and also the space mission will profit from that.

The largest is the US project named LIGO.[4] It comprises *two* facilities at two widely separated sites. Both will house a 4 km interferometer, Hanford an additional 2 km one. At both sites construction has long been completed, installation of the optics in the vacuum enclosures is done, and first locking of the interferometers has been achieved. These three interferometers are designed for coincidence operation, allowing autonomous measurements.

Next in size (3 km) is the French-Italian project VIRGO [5], being built near Pisa, Italy. An elaborate seismic isolation system, with seven-stage pendulums (see Section 4.2), will allow measurement down to GW frequencies of 10 Hz or even below, but still no overlap with the space interferometer LISA.

For the detector of the British-German collaboration, GEO 600 [6], with a de-scoped length of 600 m, construction and installation of most of the optics in the vacuum system are finished, power recycling with a single arm was successfully tested, and with the beam splitter now installed, locking of the power-recycled Michelson is under way. GEO 600 will employ advanced optical techniques such as "signal recycling" to make up for the shorter arms.

In Japan, on a site at the National Astronomical Observatory in Tokyo,

Table 1. Prototypes and projects of ground-based GW detectors

Country:	**USA**	**FRA ITA**	**GER GBR**	**JPN**
Institute:	MIT, Caltech	CNRS INFN	AEI Glasgow	NAO, U-Tokyo, ICRR

Large Interferometric Detectors: the current generation

Project name:	**LIGO**	**VIRGO**	**GEO 600**	**TAMA 300**
Arm length ℓ:	4 km / 2 km 4 km	3 km	600 m	300 m
Site (State)	Hanford (WA) Livingston (LA)	Pisa ITA	Hannover GER	Mitaka JPN

Large Interferometric Detectors: the future generation

Planning (start):	1995	1999	1998
Arm length ℓ:	4 km 4 km	3 km	3 km
Site (State)	Hanford (WA) Livingston (LA)	EUROPE	Kamioka JPN
Project name:	**Advanced LIGO**	**EURO**	**LCGT**
special features:	active isolation, suspension, RSE	high seismic rejection; cryogenic, diffractive optics, tunable	cryogenic, underground

construction, vacuum system, and optics installation of the detector called TAMA 300 [7] with 300 m armlength are completed, and several data runs of the Michelson, as yet without power recycling, have been successful, with encouragingly long in-lock duty cycles. TAMA is, just as LIGO and VIRGO, equipped with standard Fabry-Perot cavities in the arms.

Australia (not included in table) also had to cut back from earlier plans of a 3 km detector. Currently a 80 m prototype detector is being built to investigate new interferometry configurations. The design and the site will allow later extension to 3 km.

As it happens, these projects are scheduled such that first scientific operation can be expected around the years 2002/03. It is fortunate that the various projects are rather well in synchronism. For the received signal to be meaningful, coincident recordings from at least two detectors at well-separated sites are essential. A minimum of three detectors is required to locate the position of the source, and there is general agreement that only with at least four detectors can we speak of a veritable gravitational wave *astronomy*, based on a close collaboration in the exchange and analysis of the experimental data.

4 Noise and sensitivity

The measurement of gravitational wave signals is a constant struggle against the many types of noise entering the detectors. Four such noise sources, the most prominent ones, will be discussed below.

4.1 Laser noise

The requirements on the quality ('purity') of the laser light used for the GW interferometry are a great technological challenge. As it happens, the light sources for the ground-based and the space-borne interferometers will both be Nd:YAG lasers, in the form of non-planar ring oscillators, see Figure 5 [8].

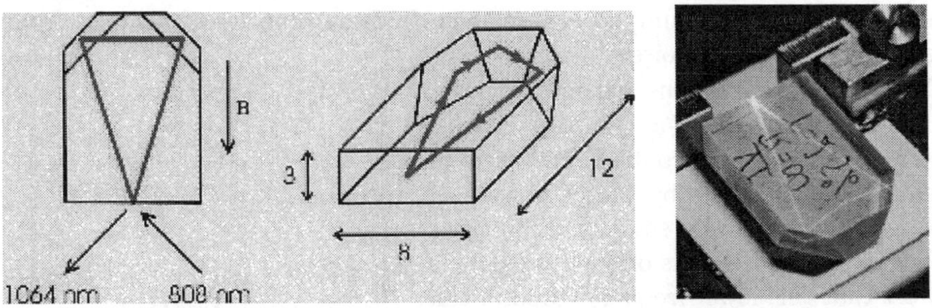

Figure 5. NPRO laser, scheme, dimensions in mm (left), photo (right)

Frequency stability A perfect Michelson interferometer (with exactly matching arms) would be insensitive to frequency fluctuations of the light used. The detectors will, however, by necessity have unequal arms, the ones on the ground due to civil engineering tolerances and a particular modulation scheme chosen, the space detector due to unavoidable imperfections in the orbits of the individual spacecraft.

Therefore, a very accurate control of the laser frequency is required, with (linear) spectral densities of the frequency fluctuations of the order $\widetilde{\delta\nu} = 10^{-4}\,\mathrm{Hz}/\sqrt{\mathrm{Hz}}$. Control schemes have been devised to reach such extreme stability, albeit only in the frequency band required, and not all the way down to DC (which would set an all-time record in frequency stability: $\widetilde{\delta\nu}\,/\,\nu = 3\times10^{-19}\,/\,\sqrt{\mathrm{Hz}}$).

Power stability Again due to asymmetries of the interferometer, the incoming laser beam needs to be closely controlled as to its power, in the frequency band of interest. Here, however, a power stability in the order of 10^{-7} is seen to be sufficient.[8]

Beam purity The illumination of the Michelson interferometer is required to be an almost pure TEM_{00} mode. For small light power, below 1 W as in the space project, this purity can be gotten by passing the light through a single-mode fiber. For the laser powers needed in the ground-based interferometers, however, a 'mode-cleaner' is used, a non-degenerate cavity that is tuned for the TEM_{00} mode, but suppresses the higher lateral modes that represent fluctuations in position, orientation, and width of the beam.[9]

4.2 Seismic noise

The mirrors between which the distances are to be monitored are suspended as pendulums in vacuum, to isolate them from extraneous vibrations: from seismic and acoustic noise.
Combinations of various schemes (pendulum suspension, lead-and-rubber stacks, even active position control) are used to reduce seismic noise by many powers of 10, which is relatively easy for frequencies above, say, 100 Hz. It is only with extreme effort that this lower frequency bound can be lowered to 10 Hz or less. Not only does the natural noise rise drastically towards low frequencies, but also the pendulum isolation becomes less effective. This causes the very steep rise to low frequencies in the righthand sensitivity curve in Figure 2.
VIRGO has developed an extremely powerful seimic isolation system, the "super-attenuator", consisting of a series of 6 successive pendulum stages, in conjunction with an 'inverted pendulum', and an active isolation stage. This suspension will allow VIRGO to extend GW search to lower frequencies than other terrestrial detectors. Figure 6 shows such an attenuator, with a total height of about 10 m. Preloaded cantilever springs in each stage provide excellent vertical isolation.

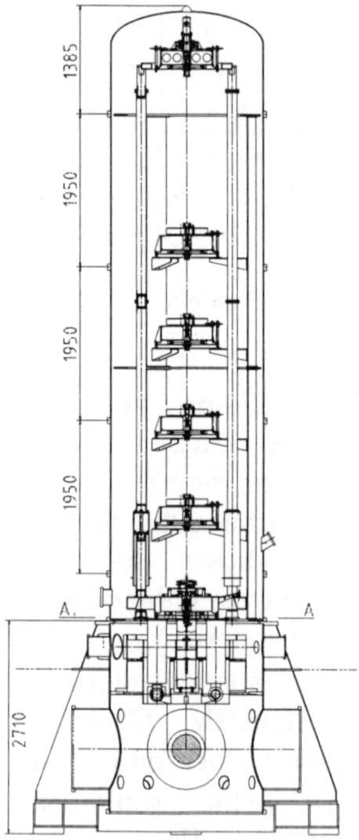

Figure 6. Super-attenuator suspension in VIRGO

4.3 Thermal noise

All optical components – and in particular the mirrors – will cause fluctuations in the optical paths also due to their thermal vibrations, their *Brownian motion.* The noise coming from the pendulum modes of motion is most prominent at low frequencies, rolling off steeply towards higher frequencies. The noise due to the vibrational modes of the substrate rolls off less steeply and is thus a serious disturbance at intermediate frequencies.

By choice of materials (high mechanical Q) and appropriate shaping of the substrates (to keep their resonant frequencies above our kHz range) the effect of these thermal motions can be reduced.

Intensive research is going into the development and choice of appropriate materials for the mirror substrates (pure fused silica, sapphire), and the proper treatment for attaining the highest mechanical Q, e.g. several 10^7. Such high values can be maintained only if the bonding to the suspension 'wires' does not introduce losses. Special bonding techniques are required using fibers of material identical to the substrate (monolithic suspension). Efficient collaboration between the European groups has given very promising results.

But for both of the thermal noise effects, the internal vibrations of the mirrors as well as the pendulation mode, and similarly also for the seismic disturbances, the sensitivity goal in strain $h = 2\delta\ell/\ell$ can only be reached if we choose the armlength ℓ long enough. This is where our need for kilometer dimensions comes from. The steep rise at the left-hand side of the sensitivity curve "Earth" in Figure 2 is mainly due to the seismic and vibrational noise.

4.4 Shot noise

Particularly at higher frequencies, the sensitivity is limited by another fundamental source of noise, the so-called shot noise, a fluctuation in the measured interference coming from the "graininess" of the light. These statistical fluctuations fake apparent fluctuations in the optical path difference ΔL that are inversely proportional to the square root of the light power P used in the interferometer. For the very ambitious aims of the "advanced" detectors, about 10 kW of light power, in the visible or the near infrared, would be required. This is not as unrealistic as it may sound; it can be realized by the concept of "power recycling".

The laser interferometers are planned to monitor the (gravitational-wave induced) changes δL of the light path by observing the dark fringe of the interferometer in one output port. The (unused) light going out at the other port of the beam splitter can be fed back, via a mirror M_P, and in correct phase with the incoming light (Figures 3, 4), so that the circulating light power is significantly enhanced. This scheme was proposed by Ron Drever in 1981, at the same time as Roland Schilling saw it come as a natural consequence in the Garching 30 m prototype, where the appropriate feedback had already been implemented for an efficient frequency stabilisation of the laser.

Shot noise is a 'white' noise, but with the response rolling off as $1/f\mathcal{T}$ at frequencies above the inverse storage time, the apparent strain noise rises with frequency, as shown in the curves 'Space' and 'Earth' in Figure 2.

4.5 Advanced interferometry configurations

An additional "recycling" scheme was later proposed by Brian Meers, and now forms the baseline for the GEO 600 interferometer: 'signal recycling (SR)'. A further mirror, M_S, is added to the interferometer, this one in the output port. The microscopic position of this mirror can be adjusted such that the signal sideband is also resonant in the interferometer, providing an enhancement of the signal, with possibly reduced measuring bandwidth. Schemes like this "signal recycling (SR)", or the related "resonant sideband extraction (RSE)", are expected to be employed in future upgrades also of the other detectors. They can be used to optimize and tune the detector bandwidth independently of the carrier storage time in the arm cavities.

The curve marked "*Earth*" in Figure 2 indicates the sensitivities that will eventually be reached with the current large interferometers, at least in their advanced versions.

5 Next-generation ground-based detectors

Even though the current detectors are not yet in operation, it is essential to develop a next generation of detectors. The study of new technologies to be employed, of new materials, of advanced interferometric configurations has to be pushed forward, so that the necessary new implementations can be undertaken in or around the year 2005.

Three plans for such next-generation detectors have been put forward, which are entered in the lower part of Table 1: Advanced LIGO, LCGT, and EURO. The status of these three future projects will be sketched below.

5.1 Advanced LIGO

Among these, the proposed US project is furthest progressed and well documented.[10] Advanced LIGO makes full use of the common efforts in the LIGO Scientific Collaboration, LSC. For locations, Advanced LIGO will rely on the existing facilities at the sites of Hanford and Livingston. The advantage is clear: no cost for new sites, for civil and vacuum engineering. One drawback is that the incorporation of more "aggressive" approaches (cryogenics, all-refractive optics, Sagnac) is not so easy to realize, and that the option of lower seismic noise of underground sites is forfeited.

The Advanced LIGO groups of LSC have come up with simulations of the expected sensitivity that indicate that an operation limited only by the optics noise (shot noise, radiation pressure noise) appears possible. The suspension would have to be modeled after the GEO 600 triple pendulum concept, mirrors be made from large substrates of sapphire (or YAG), and the schemes of SR or RSE have to be used.

As was recently shown by Buonanno and Chen,[11] the 'detuned' implementation of SR/RSE can even lead to a (moderate) reduction of what is usually termed the 'Standard Quantum Limit'.

5.2 LCGT

The concept of the Japanese project "Large Cryogenic Gravitational-Wave Telescope" (LCGT) is also rather well defined; it will use super-cooled (cryogenic) mirrors. The location of LCGT will be deep inside the mountain that houses the famous neutrino detector Super-Kamiokande. The ground noise is by nearly two orders of magnitude lower than at ground level. The armlength will be 3 km, and an existing tunnel can be used for one arm. Funding is not yet secured.

5.3 EURO

Even more ambitious is the concept of the European detector EURO. The four funding agencies (CNRS, MPG, INFN, PPARC) of France, Germany, Italy, and the UK, agreed to pursue the definition of a common European high-sensitivity detector. However, the completion and commissioning of the current projects, GEO 600 and VIRGO, has the highest priority. Thus, the actual beginning of the project may be as late as 2008. A site deep underground (as for LCGT) is preferred, but not yet decided upon.

Simulations using parameter sets from optimistic but not unreasonable assumptions, verified that an operation limited only by the (quantum-)optical noise, i.e. solely by shot noise and radiation pressure noise, seems possible, using classical techniques, but going to the ultimate frontier of current technologies.

6 Space interferometer LISA

Only a space mission allows us to investgate the gravitational wave spectrum at very low frequencies. For all ground-based measurements, there is a natural, insurmountable boundary towards lower frequencies. This is given by the (unshieldable) effects due to varying gravity gradients of terrestrial origin: moving objects, meteorological phenomena, as well as motions inside the Earth. To overcome this "brick wall", the only choice is to go far enough away, either into a wide orbit around the Earth, or better yet further out into interplanetary space. The European Space Agency (ESA) and NASA have agreed to collaborate on such a space mission called LISA, "Laser Interferometer Space Antenna".[12]

6.1 The LISA configuration

Once we have left our planet behind and find ourselves in outer space, we have some great benefits for free: to get rid of terrestrial seismic and gravity gradient noise, to have excellent vacuum along the arms, and in particular to be able to choose the arm length large enough to match the frequency of the astrophysical sources we want to observe.

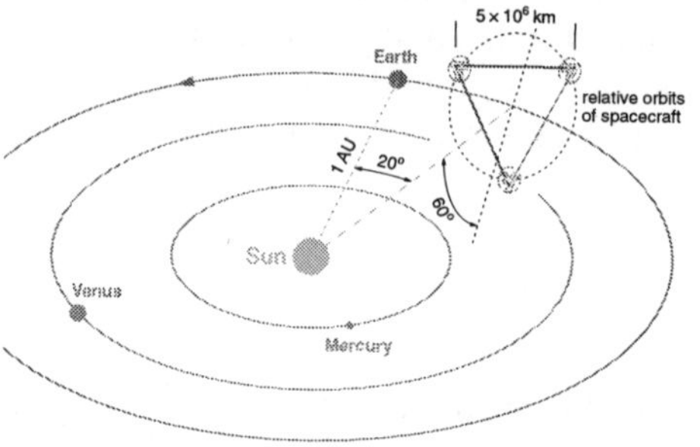

Figure 7. Orbits of the three spacecraft of LISA, trailing the Earth by 60°

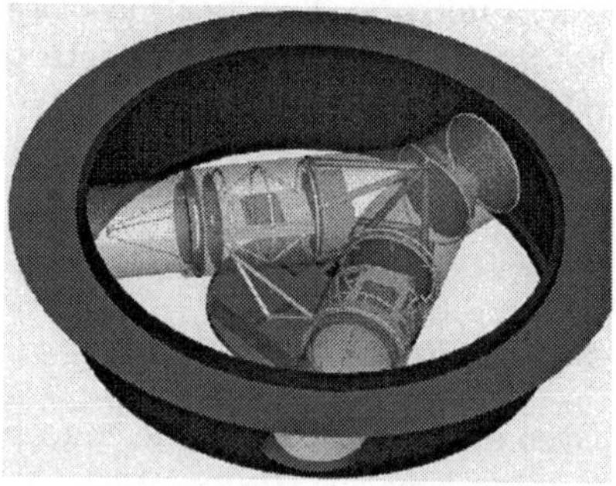

Figure 8. View of one LISA spacecraft, housing two optical assemblies

LISA consists of three identical spacecraft, placed at the corners of an equilateral triangle (Figure 7). The sides are to be 5 million km long ($5{\times}10^9$ m). This triangular constellation is to revolve around the Sun in an Earth-like orbit, about 20° (i.e. roughly 50 million km) behind the Earth. The plane of this equilateral triangle needs to have an inclination of 60° with respect to the ecliptic to make the common rotation of the triangle most uniform. The small orbit correction manoeuvres required can be made with field-effect Cs-ion thrusters. The three spacecraft form a total of three, but not independent, Michelson-type interferometers, here of course with 60° between the arms.

The spacecraft at each corner will have two optical assemblies that are pointed, subtending an angle of 60°, to the two other spacecraft (indicated in Figure 8, with the Y-shaped thermal shields shown semi-transparent). An optical bench, with the test-mass housing in its center, can be seen in the middle of each of the two arms, and a telescopes of 30 cm diameter at the outer ends. Each of the spacecraft has two separate lasers that are phase-locked so as to represent the "beam-splitter" of a Michelson interferometer.

6.2 Gravitational sensors

The distances are measured from test masses housed *drag-free* in the three spacecraft. The three LISA spacecraft each contain two test masses, one for each arm forming the link to another LISA spacecraft. The test masses, 4 cm cubes made of an Au/Pt alloy of low magnetic susceptibility, reflect the light coming from the YAG laser and define the reference mirror of the interferometer arm. These test masses are to be freely floating in space.

For this purpose these test masses are also used as inertial references for the drag-free control of the spacecraft that constitutes a shield to external forces. Development of these sensors is done at various institutions. Figure 9 shows a sensor modelled after already space-proven developments at ONERA [13], other configurations are being discussed.

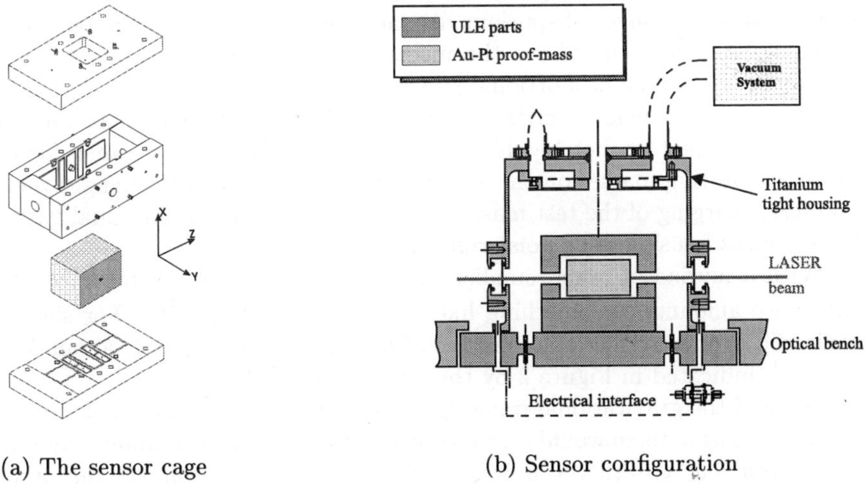

(a) The sensor cage (b) Sensor configuration

Figure 9. Layout of gravitational sensor: (a) exploded view, (b) with housing

These sensors feature a three-axis electrostatic suspension of the test mass with capacitive position and attitude sensing. A resolution of $10^{-9}\,\mathrm{m}/\sqrt{\mathrm{Hz}}$ is needed to limit the disturbances induced by relative motions of the spacecraft with respect to the test mass: for instance the disturbances due to the spacecraft self-gravity or to the test-mass charge.

6.3 FEEP thrusters

The very weak forces required to keep up drag-free operation, less than $100\,\mu\mathrm{N}$, are to be supplied by field-effect electrical propulsion (FEEP) devices: a strong electrical field forms the surface of liquid metal (Cs or In) into a cusp from which ions are accelerated to propagate into space with a velocity (of the order 60 km/s) depending on the applied voltage. Such FEEP thrusters have been developed at various European institutions, and their characteristics will be studied in a technology demonstration mission (Section 6.6).

6.4 Noise in LISA

With the 30 cm optics planned, from 1 W of infrared laser power transmitted, only some 10^{-10} W will be received after 5 million km, and it would be hopeless to have that light reflected back to the central spacecraft. Instead, also the distant spacecraft are equipped with lasers of their own, phase-locked to the incoming laser beam.

Due to the low level of light power received, shot noise plays a major role in the total noise budget of spurious displacements. Again, with the response rolling off at frequencies above the inverse round-trip time, this shot noise leads to the frequency-proportional rise in the sensitivity 'Space' in Figure 2.

At frequencies below 1 mHz, the noise is mainly due to accelerations of the test mass that cannot be shielded even by the drag-free scheme: forces due to gravitating masses on the spacecraft when temperature changes their distances, charging of the test masses due to cosmic radiation, residual gas in the test mass housing. The noise rolls off as $1/f^2$.

With a myriad of other, smaller, noise contributions the total apparent path noise amounts to something like $\widetilde{\delta L} \approx 20 \times 10^{-12}\,\mathrm{m}/\sqrt{\mathrm{Hz}}$. For signals monitored over a considerable fraction of a year, the sensitivity is about $h \approx 3 \times 10^{-24}$, indicated in Figure 2 by the curve marked "Space".

Some of the gravitational wave signals are guaranteed to be much larger. Failure to observe them would cast severe doubts on our present understanding of the laws that govern the universe. Successful observation, on the other hand, would give new insight into the origin and development of galaxies, existence and nature of dark matter, and other issues of fundamental physics.

6.5 *Status of LISA*

LISA is approved by ESA as a cornerstone mission under Horizons 2000. A System and Technology Study [12] has substantiated that improved technology, lightweighting, and collaboration with NASA can lead to a considerable reduction of cost. Thus, a new, ***faster, cheaper, and better*** approach, together with NASA, is being pursued, under the auspices of an international LISA Science Team. Launch is foreseen for 2011, not very long after first operation of the next-generation ground based detectors.

LISA has a nominal lifetime of 2 years, but the equipment and thruster supply are chosen to allow even 10 years of operation.

6.6 *Technology demonstrator*

Some of LISA's essential technologies (gravitational sensor, micronewton thrusters, interferometry) are to be tested in a mission LTP (LISA Technology Package) on board an ESA SMART-2 satellite. This package is to be launched into an orbit relatively far away from Earth in August 2006.

6.7 *LISA follow-on*

Even as early as now concepts are being discussed for a successor to LISA, on the possible enhancements in sensitivity and/or frequency band. One scheme would try to bridge the frequency gap between ground and space detectors, by reducing the arm lengths, leaving the general configuration unchanged. Another concept is to have a square constellation instead of the triangle, providing independent interferometers. These can be used to detect and measure a stochastic background of gravitational waves, similar to, but reaching much further back than the 3 K electromagnetic background radiation.

7 Conclusion

The difficulties (and thus the great challenge) of gravitational wave detection stem from the fact that gravitational waves have so little interaction with matter (and space), and thus also with the measuring apparatus. Great scientific and technological efforts, large detectors, and a working international collaboration are required to detect and to measure this elusive type of radiation.

And yet – just on account of their weak interaction – gravitational waves (just as neutrinos) can give us knowledge about cosmic events to which the electromagnetic window will be closed forever. This goes for the processes in the (millisecond) moments of a supernova collapse, as well as of the many mergers of binaries that might be hidden by galactic dust. And it is also true for the distant, but violent, mergers of galaxies and their central (super)massive black holes. A LISA follow-on mission, but also combinations of terrestrial detectors, might probe the GW background from the very beginning of our universe (10^{-14} s or even only 10^{-22} s after the big bang).[14]

In this way, gravitational wave detection can be regarded as a new window to the universe, but to open this window we must continue on our way in building and perfecting our antennas. It will only be after these large interferometers are completed (and perhaps even only after the next generation of detectors) that we can reap the fruits of this enormous effort: a sensitivity that will allow us to look far beyond our own galaxy, perhaps to the very limits of the universe.

References

1. B.F. Schutz, in *Detection of Gravitational Waves*, ed. D. Blair, Cambridge University Press (1991), p. 406
2. A. Buonanno, T. Damour, Phys. Rev. D **59** (1999) 084006
3. B.F. Schutz, in *Lighthouses of the Universe*, ESO Astrophysics Symposia, Springer (2002) (in press)
4. A. Abramovici et al., Science **256** (1992) 325–333
5. A. Vicere, The VIRGO experiment, in AIP Conference Proc., **555**, AIP, 2001, 138–145
6. B. Willke, The GEO 600 gravitational wave detector, accepted Class. Quantum Grav. (2002)
7. M.K. Fujimoto, Overview of the TAMA Project, in: Gravitational Wave Detection II, Universal Academy Press (Tokyo, 2000) 41–43
8. I. Zawischa et al., The GEO 600 laser system, accepted Class. Quantum Grav. (2002)
9. A. Rüdiger et al., Opt. Acta **28** (1981) 641–658
10. LIGO Laboratory document M000352-00-M (Dec 2000) 117-162
11. A. Buonanno and Y. Chen, Class. Quantum Grav. **18** (2001) L95
12. LISA: System and Technology Study Report, ESA document ESA-SCI(2000)11, July 2000
13. V. Josselin, M. Rodrigues, P. Touboul, Acta Astronautica **49/2** (2001) 95–103
14. B. Allen, gr-qc/9604033, (p. 9/10)

AN INTERFEROMETER WITH A MESOSCOPIC BEAM SPLITTER: AN EXPERIMENT ON COMPLEMENTARITY AND ENTANGLEMENT

J.M. RAIMOND, P. BERTET, S. OSNAGHI, A. RAUSCHENBEUTEL,
G. NOGUES, A. AUFFEVES, M. BRUNE AND S. HAROCHE

Laboratoire Kastler Brossel, Département de Physique, Ecole Normale Supérieure
24 rue Lhomond, F-75005, Paris, France
E-mail: jmr@lkb.ens.fr

Bohr, in his famous discussion with Einstein on complementarity, has stressed that the nature - quantum or classical - of the interferometer parts plays an essential role to account for the fringes visibility. We have performed a Ramsey interferometry experiment with very excited Rydberg atoms, in which one "beam splitting" element is a classical field and the other - a small field stored in a superconducting cavity - continuously evolves from a microscopic quantum device into a macroscopic classical system. When microscopic, it records an unambiguous information on the atomic "path" in the interferometer and no fringes show up. When classical, it is not changed by the interaction with the atom and Ramsey fringes are observed. This experiment illustrates the complementarity concept and its intimate link with the notion of entanglement. It sheds light on the quantum-classical boundary in fundamental measurement processes.

1 Complementarity in an optical interferometry experiment

Complementarity is one of the most fundamental aspects of the quantum world. It can be illustrated in a vivid way in simple situations. In a famous discussion with Einstein, Bohr imagined an optical Young's slits interferometer, with one movable slit [1]. When deflecting the interfering particle, this slit receives a momentum kick. When the slit is a macroscopic object, this kick is negligible and the interferometer produces fringes. When the slit is a very light object, the kick modifies appreciably its motion. The slit then records an unambiguous information about the path followed by the particle in the interferometer ("which-path" information). No fringes show up then, since the two interfering paths are distinguishable. The "wave" (interferences) and "particle" (well defined trajectory) aspects of the quantum object are both necessary to understand the experiment but are never revealed simultaneously.

Let us discuss this simple scheme in more quantitative terms, with a Mach Zehnder interferometer (see figure 1). The two interfering paths are separated and recombined by two beam splitting elements B_1 and B_2 (50% transmission each) and two mirrors M and M'. The relative phase ϕ of the interfering amplitudes is varied by a retarding element. The photons are

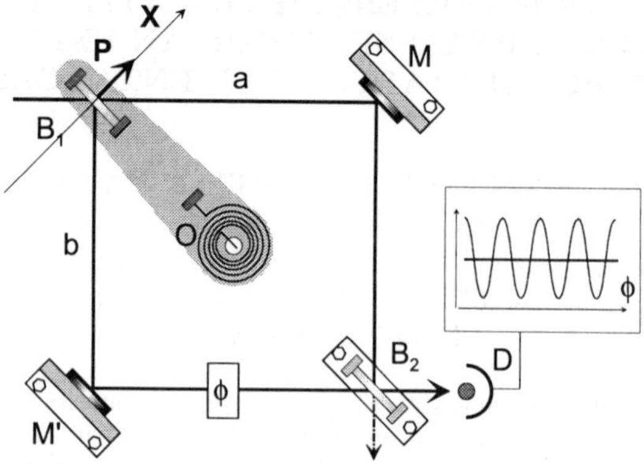

Figure 1. Gedankenexperiment illustrating complementarity with an optical Mach Zehnder interferometer. It involves two beam splitting elements B_1 and B_2 and two interfering paths labeled a and b. The beam splitter B_1 moves around the center O of the interferometer. When it is light enough, its motional state records "which-path" information about the interfering photon and fringes are washed out.

finally detected in the upper output port. The beam splitter B_2 is rigidly attached to the interferometer, while B_1 is mounted on a rotating assembly, free to move around the center of the interferometer. A spring provides a restoring force. When the amplitude of motion is small enough, B_1 undergoes an harmonic oscillator motion along the X axis. We assume that it is initially in its ground state $|0\rangle$.

When the photon is reflected in path b, B_1 receives a momentum kick $p \simeq h/\lambda$, λ being the wavelength of the incoming radiation. The resulting motional state of B_1 is a coherent state $|\alpha\rangle$, with an amplitude α proportional to p. The kick provides an unambiguous which path information when p is much larger than the initial momentum spread ΔP in state $|0\rangle$ ($\Delta P = \sqrt{\hbar\sqrt{mk}/2}$, m and k being the effective mass and spring stiffness). This condition writes also $\Delta X > \lambda$ since, in the ground state, $\Delta X \Delta P \simeq \hbar$. When the "mass" m is small, ΔP is smaller than p, and a which-path information is obtained. This small momentum indeterminacy corresponds to an initial position spread $\Delta X > \lambda$. The fringes are thus washed out, since their phase is undetermined. If m is "big", ΔP is much larger than p. No which-path information is stored, $\Delta X \ll \lambda$ and fringes show-up. Intermediate situations can also be considered, where partial which-path information is recorded by a mesoscopic beam splitter. Fringes then appear with a reduced contrast.

There is a very strong link between complementarity and entanglement, not fully revealed by the discussion in terms of the Heisenberg uncertainty relations. The photon, before hitting B_2, is entangled with B_1, in an EPR-like [2] state:

$$|\Psi\rangle = \frac{1}{\sqrt{2}}\left(|a\rangle|0\rangle + |b\rangle|\alpha\rangle\right) , \quad (1)$$

where the first ket in each term refers to the photon ($|a,b\rangle$ corresponding to paths a or b), and the second to B_1. The final detection probability is:

$$P(\phi) = \frac{1}{2}\left(1 + \Re e^{i\phi}\langle 0|\alpha\rangle\right) . \quad (2)$$

The fringe contrast is the modulus of the scalar product of the B_1 states corresponding to the two paths, which measures directly the degree of entanglement between B_1 and the interfering photon. An unambiguous which-path information corresponds to a maximum entanglement, a zero scalar product and a vanishing contrast. This discussion of fringes visibility in terms of entanglement is quite general [4]. The entanglement can take place with the interferometer itself, as in the gedankenexperiment described here, with an external measuring device recording the photon's path, or with an environment. As soon as the environment states corresponding to the two paths are orthogonal, the fringes are washed out in this decoherence process [3].

2 A Ramsey interferometer to probe complementarity

Many experiments have been proposed [4,5,6,7] or realized [8,9,10,11,12] to illustrate complementarity. None so far, to our knowledge, has addressed the progressive micro- to macroscopic transition in the design of the interferometer itself. Following former theoretical proposals [13,14], we realized such an experiment [15], with an atomic Ramsey interferometer [16] instead of an optical Mach Zehnder one. We describe here only the main aspects of the experiment sketched in figure 2. More details can be found elsewhere [17,18,19]. Rubidium atoms prepared in the circular Rydberg state of principal quantum number 51 (level e) are sent one by one through the interferometer with a velocity of 500 m/s. Their position is known within 1 mm at all times during their transit across the apparatus. The atoms undergo a transition between e and the lower energy circular Rydberg level g (principal quantum number 50), at frequency $\nu_{eg} = 51.1$ GHz. This transition is induced by two coherent pulses R_1 and R_2 mixing with equal weights e and g ($\pi/2$ pulses), at two times separated by a delay $T = 24$ μs. The final atomic energy state is detected in D. Between R_1 and R_2, a two microsecond electric field pulse Stark shifts the atomic levels and produces a controlled phase shift ϕ of the atomic state superposition.

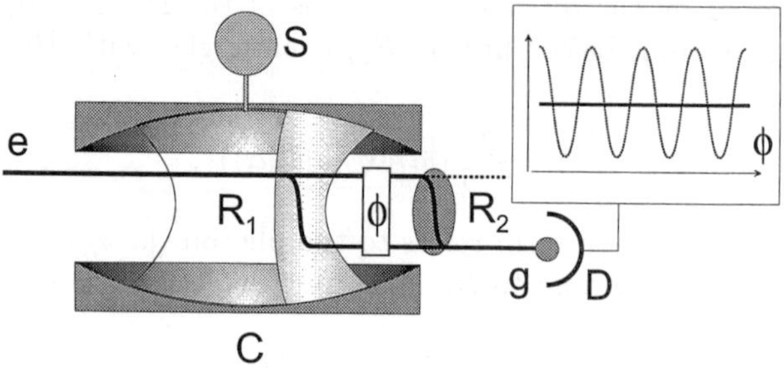

Figure 2. Ramsey interferometer. A Rydberg atom's state is split into two levels e and g by a microwave pulse R_1 induced by a small coherent field stored in the superconducting cavity C. The two paths are later recombined by a second classical microwave pulse R_2. The atom is then detected by field-ionization in D.

The transition probability P_g is reconstructed as a function of ϕ by averaging a large number of single atom events. The final atomic state results from the interference between two quantum paths, depicted in figure 2. In this pictorial representation, mixing atomic states and real space, the analogy between the Ramsey interferometer and the Mach Zehnder one is apparent.

The pulse R_1 is produced by a small coherent field $|\alpha\rangle = \sum_n c_n|n\rangle$, with $c_n = \exp(-|\alpha|^2/2)\alpha^n/\sqrt{n!}$, injected by the source S in the cavity C before the atom enters it. The cavity C is an open Fabry Perot resonator with superconducting niobium mirrors (photon damping time $T_r = 1$ ms). An electric field applied across the mirrors is used to control the atomic transition frequency through Stark effect. The atom is tuned on resonance with the cavity mode during the last part of its transit trough the 6 mm mode waist. The interaction time is set to fulfill a $\pi/2$ pulse condition. Note that this condition can be met even when $\alpha = 0$, the vacuum Rabi frequency at cavity center, $\Omega/2\pi = 49$ kHz, being larger than the reciprocal of the cavity transit time [17] (strong coupling regime of cavity QED). After R_1, the atom and the field mode are in an entangled state which writes:

$$|\Psi\rangle = \frac{1}{\sqrt{2}}\left(|e\rangle|\alpha_e\rangle + |g\rangle|\alpha_g\rangle\right) , \tag{3}$$

in a form similar to Eq.(1). Here, the cavity states corresponding to the two paths are given by:

$$|\alpha_e\rangle = \sqrt{2}\left(\sum_n c_n \cos(\Omega\sqrt{n+1}t_\alpha)|n\rangle\right) \tag{4}$$

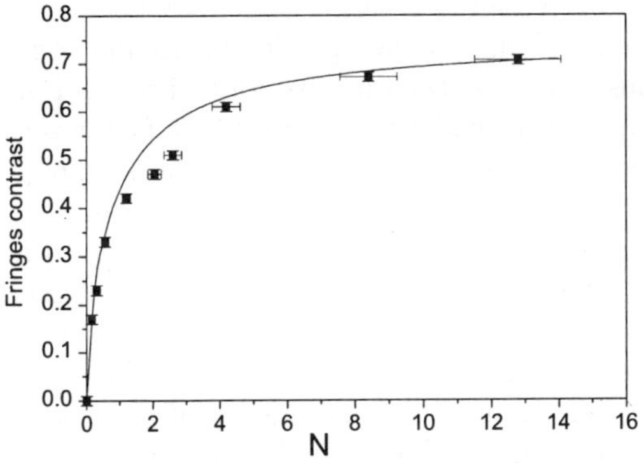

Figure 3. Contrast of the Ramsey fringes as a function of the photon number N in R_1. The solid line is theoretical.

$$|\alpha_g\rangle = \sqrt{2}\left(\sum_n c_n \sin(\Omega\sqrt{n+1}t_\alpha)|n+1\rangle\right) , \tag{5}$$

where the effective atom-cavity interaction time t_α is a decreasing function of α defined by the $\pi/2$ pulse condition. When the photon number $N = |\alpha|^2$ is zero, $|\alpha_e\rangle = |0\rangle$ and $|\alpha_g\rangle = |1\rangle$. The atom and the cavity are thus in a maximally entangled state. When N is very large, $|\alpha_e\rangle \simeq |\alpha_g\rangle \simeq |\alpha\rangle$. The cavity is thus disentangled from the atom, since $|\Psi\rangle$ is a product state.

The pulse R_2 is a classical field, fed by S in a low-Q standing wave mode, independent from C. The damping time of this mode is extremely short and the photons renewed many times during the atom-field interaction. The absorption or the emission of a single photon by the atom in this huge photon flux does not leave any indication about the atomic transition and the R_2 field does not get entangled with the atomic state. The pulse R_2 behaves thus as a classical Ramsey pulse [20]. The final detection probability P_g exhibits oscillations versus ϕ, with a contrast given by $|\langle\alpha_e|\alpha_g\rangle|$ – see Eq. (2). In figure 3, we plot the Ramsey fringes contrast as a function of the photon number N in R_1. This number is determined in an auxiliary experiment, based on the light shifts experienced by a non-resonant atom in the cavity field [21].

When $N = 0$, the cavity mode records an unambiguous which-path information (0 photon when the intermediate state is e, 1 photon when it is g). The cavity mode plays then the role of the microscopic beam splitter B_1 in the Mach Zehnder arrangement, the photon number being equivalent to the beam splitter mass m. When N is large, the cavity field is not entangled with the atom, and Ramsey fringes show up. The maximum contrast is limited to 75%, due to various experimental imperfections. The solid line in figure 3 corresponds to $|\langle\alpha_e|\alpha_g\rangle|$ multiplied by a constant 0.75 factor accounting for these imperfections. The agreement with the experiment is quite good. These data thus reveal clearly the progressive apparition of the fringes when the "size" of the beam splitter element is increased from microscopic to macroscopic.

Note that this experiment can also be interpreted in terms of the "photon number – phase" uncertainty relation. The first Ramsey pulse R_1 imprints a phase information on the atomic state superposition. This information is "read out" by R_2. The Ramsey fringes signal thus reveals the phase correlations between the two pulses. When the photon number N is large, the phase of R_1 is a well-defined quantity and the fringes show up with unit contrast. When the photon number decreases, the phase fluctuations increase accordingly. The contrast of the fringes thus decreases. Ultimately, the vacuum state in C contains no phase information and no fringes show up. This approach could deceptively lead to the conclusion that the fringes disappearance is the simple effect of a noisy classical phase perturbation. This is not the case, however. The exchange of which-path information is a fully reversible process, as illustrated in the next section.

3 A quantum eraser experiment

The disappearance of the fringes for a small photon number in R_1 results from the entanglement of the interfering atom with the field in C. By measuring this field in a basis mixing the states correlated to the interferometer's paths, it is possible to restore fringes, conditioned to the result of this measurement. This is the principle of the "quantum eraser" [4,11,14]. A direct measurement of the state of C is not possible in our set-up. Such an experiment requires first a copy of the cavity mode state onto an atomic state.

The experiment thus involves two atoms. The first one, the "interferometer" atom A_i undergoes the R_1 $\pi/2$ pulse in the cavity and, later, the Stark phase shift ϕ, followed by the classical R_2 pulse. We consider in this section only the simple case when C is initially empty, and where no fringes show up in the preceding experiment. After their interaction, A_i and C are thus in the maximally entangled state $(|e_i,0\rangle + |g_i,1\rangle)/\sqrt{2}$. The second ("eraser") atom, A_e, crosses C after A_i. It is prepared in state g and undergoes a π Rabi oscillation pulse in the field of a single photon stored in C. When a photon is present in C, it is absorbed with unit probability and the eraser atom exits in level e. After this interaction, the cavity is left in the vacuum state, which factors out, and the two atoms are in the EPR entangled state [2,24]:

$$|\Psi_{EPR}\rangle = \frac{1}{\sqrt{2}} (|e_i, g_e\rangle - |g_i, e_e\rangle) \ , \tag{6}$$

where the indices in the kets identify the atoms.

At this point, the which-path information has been merely transferred from C to A_e, but is not yet "erased". If the state of A_e were directly measured, A_i would be projected onto states e or g accordingly, due to the non-local correlations. The pulse R_2 applied on A_i would thus prepare it in a superposition of e and g with equal weights, irrespectively of ϕ. No Ramsey fringes would show up. The entanglement between A_i and A_e is incompatible with the interferences, as well as the entanglement between A_i and C in the first experiment. A state transformation, followed by a measurement on A_e is required to recover the fringes.

Atom A_e undergoes a $\pi/2$ classical microwave pulse R'_2 after it exits C. This pulse produces the transformations $|e\rangle \rightarrow (|e\rangle + |g\rangle)/\sqrt{2}$ and $|g\rangle \rightarrow (-|e\rangle + |g\rangle)/\sqrt{2}$. The $A_e - A_i$ state is then:

$$|\Psi'\rangle = \frac{1}{2} [|g_e\rangle(|e_i\rangle - |g_i\rangle) - |e_e\rangle(|e_i\rangle + |g_i\rangle)] \ . \tag{7}$$

A detection of A_e in e_e (g_e) then projects A_i onto the coherent superposition $(|e_i\rangle - |g_i\rangle)/\sqrt{2}$ $[(|e_i\rangle + |g_i\rangle)/\sqrt{2}]$. The path ambiguity is thus restored. The final populations of levels e and g for A_i at the exit of the interferometer are then oscillating functions of the phase ϕ. Ramsey fringes show up on A_i, conditioned by the result of the detection of A_e. The fringes corresponding to A_e in e or g are in phase opposition.

The scheme described above is precisely the one of a previous experiment aimed at preparing an EPR atomic pair, through the successive interaction of the two atoms with an initially empty cavity mode [24]. The test of the non-local atomic correlations involves two experiments. In the first, the atomic state populations are directly detected and reveal energy correlations (see Eq. (6)). In the second, $\pi/2$ resonant pulses R_2 and R'_2 are applied onto the two atoms to observe the atomic entanglement in a "transverse" basis, realizing the scheme proposed above. In this early experiment, the interference pattern was revealed by scanning the relative phase ϕ' of the two Ramsey pulses R_2 and R'_2, instead of scanning the phase ϕ of the atomic superposition. The two procedures are clearly equivalent. The conditional probabilities Π_{ee} and Π_{ge} for detecting A_i in e provided A_e has been detected in e or g respectively are plotted in figure 4 as a function of ϕ'. Fringes show up in the conditional probabilities, in phase opposition for the two states of A_e, demonstrating the

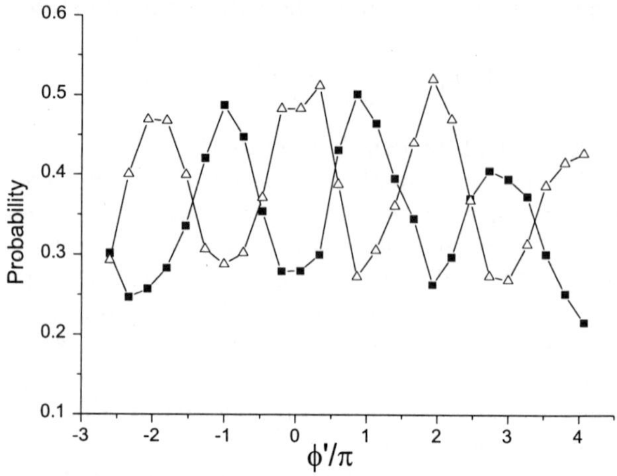

Figure 4. Quantum eraser experiment. The conditional probabilities Π_{ee} (solid squares) and Π_{ge} (open triangles) are plotted as a function of the relative phase ϕ'/π of pulses R_2 and R'_2. The lines joining the data points have been added for visual convenience. The Ramsey fringes on A_i conditioned to the detection of A_e are clearly apparent.

quantum eraser procedure. Summing the two Π_{ee} and Π_{ge} signals is equivalent to not sending A_e: the resulting signal exhibits no fringes.

4 Conclusion

Rydberg atoms and superconducting cavities make it possible to illustrate in a vivid way fundamental quantum properties such as complementarity, entanglement and non-locality. They are also well suited to tailor and process the complex entangled states required for demonstrations of quantum information processing functions. The experiment described above prepares an atomic EPR pair via the successive interactions of the two atoms with the same cavity mode [24]. More complex sequences made it possible to prepare a three-particle entangled state [19]. Finally, a "cavity-assisted collision" between two Rydberg atoms interacting at the same time with an off-resonance cavity mode can be used to directly entangle these atoms in a process insensitive to cavity imperfections [25]. The realization of these experiments opens interesting perspectives for complex entanglement manipulations in cavity QED.

Acknowledgments

Laboratoire Kastler Brossel is Unité Mixte de Recherche of Ecole Normale Supérieure, Université Pierre et Marie Curie and Centre National de la Recherche Scientifique. This work was supported by the Commission of the European Community and by the Japan Science and Technology Corporation (International Cooperative Research Project, Quantum Entanglement Project).

References

1. Bohr, N. in *Quantum Theory and Measurement*, J.A. Wheeler and W.H. Zurek eds., p.9, Princeton University Press, (1983).
2. A. Einstein, B. Podolsky and N. Rosen, Phys. Rev. **47**, 777 (1935).
3. W.H. Zurek, Physics Today, **44**, 10 p. 36 (1991).
4. M. O. Scully and M. S. Zubairy, *Quantum optics*, Cambridge University Press (1997).
5. S. Haroche et al., Appl. Phys. B **54**, 355 (1992).
6. P. Storey et al., Nature **367**, 626 (1994).
7. B. Englert et al, Nature **375**, 367 (1995).
8. U. Eichmann et al., Phys. Rev. Lett. **70**, 2359 (1993).
9. T. Pfau et al., Phys. Rev. Lett. **73**, 1223 (1994).
10. M. S. Chapman et al., Phys. Rev. Lett. **75**, 3783 (1995).
11. T. J. Herzog et al., Phys. Rev. Lett. **75**, 3034 (1995).
12. S. Dürr et al., Phys. Rev. Lett. **81**, 5705 (1998).
13. B.G. Englert, H. Walther and M. O. Scully, Appl. Phys. B **54**, 366 (1992).
14. S.B. Zheng, Opt. Comm. **173**, 265 (2000).
15. P. Bertet et al, Nature **411**, 166 (2001).
16. N.F. Ramsey, *Molecular Beams*, Oxford Univ. Press, New York (1985).
17. M. Brune et al., Phys. Rev. Lett. **76**, 1800 (1996).
18. G. Nogues et al., Nature **400**, 239 (1999).
19. A. Rauschenbeutel et al., Science **288**, 2024 (2000).
20. J.I. Kim et al., Phys. Rev. Lett. **82**, 4737 (1999).
21. M. Brune et al., Phys. Rev. Lett. **72**, 3339 (1994).
22. M. O. Scully and K. Drühl, Phys. Rev. A. **25**, 2208, (1982).
23. P. G. Kwiat et al., Phys. Rev. A. **45**, 7729 (1992).
24. E. Hagley et al., Phys. Rev. Lett. **79**, 1 (1997).
25. S. Osnaghi et al., Phys. Rev. Lett. **87**, 037902 (2001).

CAVITY QED WITH COLD ATOMS

H. J. KIMBLE AND J. MCKEEVER

Norman Bridge Laboratory of Physics 12-33, California Institute of Technology, Pasadena, CA 91125, USA

We report experimental progress toward real-time observation and control of atomic motion with optical fields in the strong coupling regime of cavity QED. One experiment utilizes a conventional, far-detuned field inside a high-finesse resonator to provide a trapping potential. Our investigation into the limitations to atomic storage times in this trap has led to an interesting discovery, namely that trap lifetime is sensitive to mirror motion driven by thermal excitation of vibrational modes in the mirror substrates. Calculations of this mirror motion and its effect on intensity fluctuations of intracavity fields are compared to measured noise spectra. In another experiment, single atoms are trapped by intracavity fields with average intensities corresponding to single photons. Detection of this light transmitted through the cavity allows real-time tracking of the atomic motion and reconstruction of atomic trajectories.

1 Introduction

Cavity quantum electrodynamics (QED) offers powerful possibilities for the deterministic control of atom-photon interactions quantum by quantum[1]. Indeed, modern experiments in cavity QED have achieved the exceptional circumstance of strong coupling, for which single quanta can profoundly impact the dynamics of the atom-cavity system. The diverse accomplishments of this field set the stage for advances into yet broader frontiers in quantum information science for which cavity QED offers unique advantages, including the creation of quantum networks to implement fundamental quantum communication protocols and for distributed quantum computation[2].

The primary technical challenge on the road toward such scientific goals is the need to trap and localize atoms within a cavity in a setting suitable for strong coupling. Since the initial work of Mabuchi et al. in 1996[3], several groups have been pursuing the integration of the techniques of laser cooling and trapping with those of cavity quantum electrodynamics (QED)[4]. Two separate experiments in our group have achieved significant milestones in this quest, namely the real-time trapping and tracking of single atoms in cavity QED[5,6,7].

As illustrated in Figure 1, in both these experiments the arrival of a single atom into the cavity mode can be monitored with high signal-to-noise ratio in *real time* by a near resonant field with mean intracavity photon number $\bar{n} < 1$. These atomic transit signals are subsequently used to trigger optical confining potentials which trap the atom in a single well of a standing wave. Both experiments exploit the dipole forces of red-detuned optical fields to provide confining potentials by attracting an atom to regions of high intensity. However, the two traps operate in very different regimes from experimental and theoretical points of view. In one experiment, we employ a far-off-resonance trap (FORT)[6] to excite a second longitudinal mode of the resonator located 16 nm below the primary longitudinal mode utilized for strong coupling in cavity QED. At the large detuning used for this trap, atomic spontaneous emission is strongly suppressed so that the force is largely conservative and is well modeled by spatially dependent shifts of the atomic ground and excited state energies. In a second experiment without any "classical" FORT field, we exploit excitation tuned near the primary cavity QED mode, with a near-resonant laser field that provides initial atomic detection for intracavity photon number $\bar{n} = 0.05$ being ramped up to a higher intensity with $\bar{n} \simeq 1$ to provide a trapping potential. In this case, the trapping and diffusive forces arise from the spatial dependence of the coupling coefficient for the atom-cavity system, and are qualitatively different from analogous forces in free space.[8] In the next section, we first discuss the FORT experiment and describe our investigations into the limits on the trap lifetime. In the ensuing section, we go into more detail on the single-photon trapping experiment.

2 A FORT inside a Cavity: Present Limitations and Future Directions

We begin by mentioning some experimental details and parameters of the experiment. Cesium atoms are collected in a magneto-optical trap (MOT) a few mm above the cavity, and then released to fall freely toward the cavity mode after a few ms of polarization-gradient cooling. The cavity length is 44.6 μm with a finesse of 4.2×10^5, so that the relevant rates for cavity QED are $(g_0, \gamma_\perp, \kappa)/2\pi = (32, 2.6, 4)$ MHz, meaning that the ratio of coherent coupling to damping is well within the strong coupling regime. Here, g_0 is half the single-photon Rabi frequency, $\gamma_\perp$ is the transverse atomic decay rate (half the Einstein-A coefficient), and κ is the decay rate for the cavity field in the absence of the atom. This regime of strong coupling enables real-time detection of a single atom in the cavity mode by monitoring the pW-scale cavity transmission, in this case using heterodyne detection. By injecting light at 869 nm, another longitudinal mode of the cavity is driven on resonance, creating a dipole-force trapping potential, which has a standing-wave structure along the cavity axis and a Gaussian profile in the transverse direction. The cavity's finesse at this wavelength is 3.5×10^5, so that in a typical experiment,

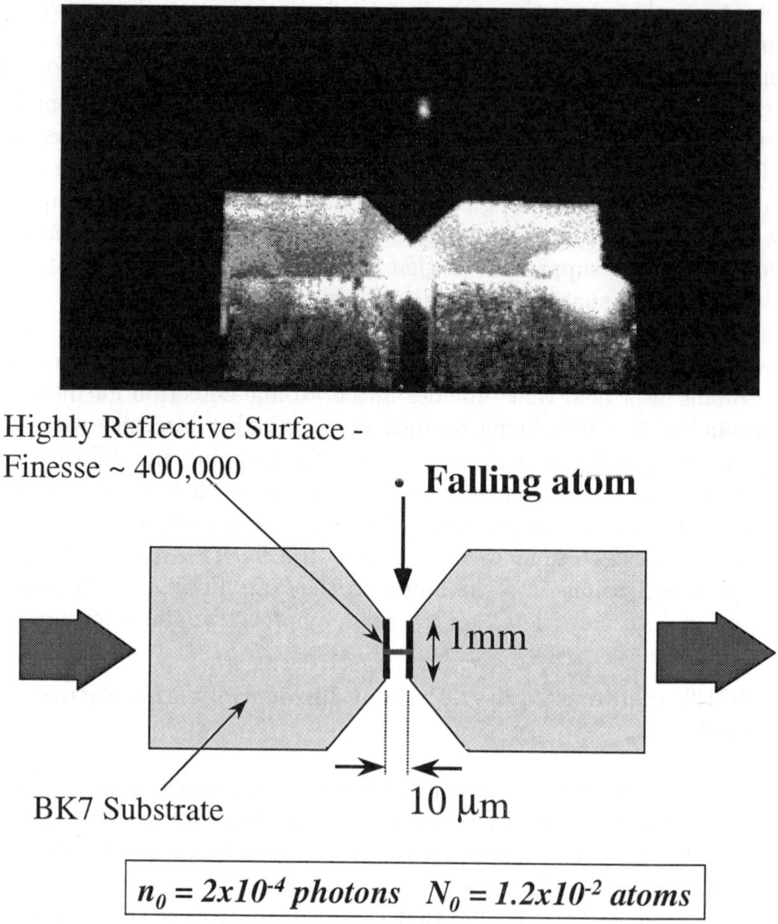

Figure 1. Experimental setup for delivering cold atoms to high finesse optical cavities. The parameters are relevant to the experiment of Hood et al.[7]

we need only inject approximately 7 μW of power into the cavity, to result in a peak ground state shift of about 75 MHz.

The timing of a typical experiment starts with the FORT being triggered on by the detection of a single atom in the cavity mode. After a fixed FORT trapping interval Δt, a near-resonant probe is used again to determine whether the atom stayed in the cavity (the probe light is turned off during the entire trapping period). As reported in Ref.[6], measurement of the mean trapping success ratio for many values of the FORT duration Δt yields a lifetime $\tau = (28 \pm 6)$ ms. Over the past two years, we have spent considerable effort to understand the mechanisms that limit the trap lifetime and which heat atomic motion within the FORT. Many potential culprits have been investigated as suspects in generating heating or loss mechanisms, of which here we mention only a few. Our vacuum chamber has a pressure of 10^{-10} Torr, implying a lifetime τ of about 100 s from collisions with background gas. Photon scattering events from the FORT light occur at a rate on the order of 50 s^{-1}, but the recoil energy is only 2 kHz for Cs. Of course, the impact of this far-off resonance scattering can be greatly increased due to imperfections in polarization[9], so we have explored both circularly polarized and linearly polarized intracavity FORTs, with similar results for the lifetime. Our conditions of strong coupling in the cavity also imply that stray fields near the Cesium resonance must be reduced to levels well below the critical photon number, $n_0 = 0.003$ in order to avoid extraneous heating. We have identified and eliminated several possible sources of stray intracavity light, such that we believe this is no longer a limiting factor.

Another well known source of heating in dipole force traps is intensity noise at twice the trapping frequency ν_{tr} causing parametric heating at a rate $\tau_e^{-1} = \pi^2 \nu_{tr}^2 S_e(2\nu_{tr})$[10]. $S_e(f)$ is the power spectral density of fractional intensity noise evaluated at frequency f. In our case, the high finesse of the cavity at the FORT wavelength of 869 nm, $F = 350,000$, leads to conversion of FM to AM. That is, what might otherwise be viewed as relatively harmless jitter of a few kiloHertz in the detuning of the FORT laser from cavity resonance (of FWHM $\approx$ 11 MHz) manifests itself as intensity noise in the intracavity (and hence transmitted) field. This noise has been the primary focus of our investigations[11], and our attempts to understand the source of the measured noise has led us to a somewhat surprising conclusion. At frequencies above 500 kHz, those most relevant to the atom's axial motion within the intracavity FORT, the noise can, in large measure, be attributed to thermally excited vibrations of the mirror substrates.[12,13,14] More specifically, the elastic modes for mechanical vibrations of the mirror substrates are each excited with $k_B T$ thermal energy (experiments are done at room temperature). This results in displacement of the mirror surfaces, which in turn we observe as intensity noise on the FORT light at the cavity output.

Figure 2 shows a comparison of the measured intensity noise on the transmitted field with the predicted light noise obtained from the calculated mirror motion. The noise measurement was carried out with the laser frequency detuned a half linewidth from cavity resonance, with a transmitted power of 2.8

μW. For the theoretical trace, the frequencies and amplitudes of the peaks are obtained by numerical computation of the normal modes of a right-circular cylinder, with energy $k_B T$ imparted to each mode. Following Gillespie and Raab,[12] each mode amplitude is then weighted by its effective mass to account for the actual mode displacement integrated over the cavity mode's field distribution on the mirror's surface. The resulting intensity noise on the light transmitted by the cavity is directly related to the overall power spectrum for the motion of the mirror faces. The laser power and detuning for the theoretical trace in Figure 2 are the same as the experimental values. The only adjustable parameter in the comparison is the quality factor for the mechanical modes. Here, we assume structure damping,[15] and set the quality factor $Q = 90$ to match the observed value for the mode at 800 kHz. Note that the lack of correspondence between many of the calculated and observed mode frequencies presumably is associated with the fact that the theoretical computation is for a *right circular cylinder* while the actual mirror substrates have a somewhat more complex shape (as can be seen in the photograph in Figure 1).[6,7] In collaboration with Dennis Coyne of the LIGO project, we are carrying out a finite element analysis corresponding to the actual substrate shapes.

We are pursuing several technical avenues to mediate FORT heating arising from the thermally induced motion of the cavity mirrors. The first is improvement to the servo system that locks the frequency of the FORT laser to the cavity-QED resonator, which is somewhat difficult because only a few μW of FORT light transmit through the high-reflectivity cavity output mirror. Another option is to use a different wavelength for the FORT trapping. Going much further to the red of atomic resonance (to approximately 940 nm to produce a *zero-net shift FORT*[16]) would lower the cavity finesse by 10^2, and thereby greatly reduce the FM to AM conversion problem (by roughly the square of the change in finesse).

However, quite apart from the issue of cavity QED with single atoms, this thermal noise may lead to interesting scientific directions in its own right.[13,14] Possibilities include selective cooling of a single mode of vibration using feedback, or even reaching the standard quantum limit for observation of the motion of a test mass.

Once the heating rates are understood and reduced as much as possible, there are several additional steps to be taken in this experiment to achieve true atomic center-of-mass control in cavity QED. Two technical hurdles to be overcome are the needs to cool the atom in the trap and to control which antinode of the standing wave the atom becomes trapped in (for the purpose of determining the coupling strength at the trapping location). In order to address

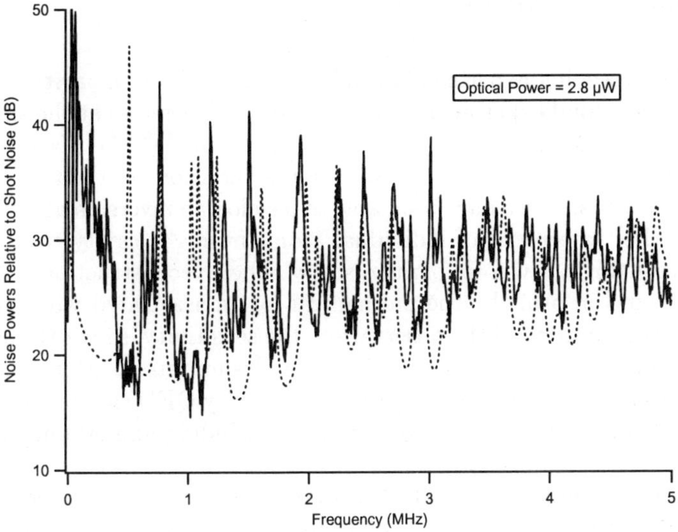

Figure 2. Comparison of the measured intensity noise power spectrum (solid trace) to the spectrum predicted by calculation of thermally excited vibrational motion of the mirror substrates (dashed trace). For the lowest frequency peak of the theoretical trace, the corresponding spectral density of vibrational motion is about $5 \times 10^{-17} m/\sqrt{Hz}$. Note that the theoretical trace is for a right circular cylinder, whereas the actual mirror substrates are of a more complex shape.

these issues, we have performed numerical simulations of three-dimensional quasi-classical atomic motion using solutions of the master equation[17]. By calculating the spatial dependence of the friction and diffusion coefficients, as well as the conservative forces due to both the cavity QED and FORT optical fields, we have found that axial cooling to the Doppler limit should be readily achievable with a probe beam of appropriate detuning and sufficiently weak intensity. In addition, the calculations show that it should be possible to determine the coupling strength at the center of the relevant trapping well by way of an appropriate set of measurements of the intensity of a transmitted probe beam while the atom is trapped.

3 The Atom-Cavity Microscope

Apart from the preceding experiments combining a FORT with cavity QED, we are pursuing a second experiment that relies upon light forces at the single-photon level to trap a single atom within the cavity mode[5,7,8]. Because an atom moving within the resonator generates large variations in the transmission of a weak probe laser, we have been able to develop an inversion algorithm to reconstruct the trajectories of individual atoms from the cavity transmission, thereby realizing a new form of microscopy. These reconstructions reveal single atoms bound in orbit by the mechanical forces associated with single photons. Over the duration of the observation, the sensitivity is near the standard quantum limit for sensing the motion of a Cesium atom. More details about the reconstruction of atomic trajectories in cavity QED can be found at *http://www.its.caltech.edu/~qoptics/atomorbits/*, including animations of the orbits.

Using extensive numerical simulations, we have also analyzed the trapping dynamics in the limit of single photons and atoms, focusing on two points of interest[8]. Firstly, we investigate the extent to which light-induced forces in these experiments are distinct from their free-space counterparts and whether or not there are qualitatively different effects of optical forces at the single-photon level within the setting of cavity QED. Secondly, we explore the quantitative features of the resulting atomic motion and how these dynamics are mapped onto experimentally observable variations of the intracavity field. Not surprisingly, qualitatively distinct atomic dynamics arise as the coupling and dissipative rates are varied.

Acknowledgements

We thank the many collaborators who have been involved in the efforts to trap single atoms in cavity QED in a regime of strong coupling that we have described here, including Joseph Buck, Kevin Birnbaum, Andrew Doherty, Theresa Lynn, Christina Hood, Christoph Naegerl, Dan Stamper-Kurn, Ron Legere, David Vernooy, and Jun Ye. We gratefully acknowledge critical interactions with Dennis Coyne and Kent Blackburn of the LIGO Lab at Caltech concerning thermal noise. This work has been funded by the National Science Foundation, by the Caltech MURI on Quantum Networks administered by the US Army Research Office, and by the Office of Naval Research.

References

1. For a recent review, see contributions in the Special Issue of Physica Scripta **T76** (1998).
2. J. I. Cirac, S. J. van Enk, P. Zoller, H. J. Kimble, and H. Mabuchi, Physica Scripta **T76**, 223 (1998).
3. H. Mabuchi, Q. A. Turchette, M. S. Chapman, and H. J. Kimble, Opt. Lett. **21**, 1393 (1996).
4. See also the contribution by G. Rempe in this meeting.
5. J. Ye *et al.*, IEEE Trans. Instru. & Meas. **48**, 608 (1999).
6. J. Ye, D. W. Vernooy, and H. J. Kimble, Phys. Rev. Lett. **83**, 4987 (1999).
7. C. J. Hood, T. W. Lynn, A. C. Doherty, A.S. Parkins, and H. J. Kimble, Science **287**, 1447 (2000).
8. A. C. Doherty, T. W. Lynn, C. J. Hood, and H. J. Kimble, Phys. Rev. A **63**, 013401 (2001); available as quant-ph/0006015.
9. K. L. Corwin, S. J. M. Kuppens, D. Cho, and C. E. Wieman, Phys. Rev. Lett. **83**, 1311 (1999).
10. T. A. Savard, K. M. O'Hara and J. E. Thomas, Phys. Rev. A **56**, R1095 (1997).
11. C. W. Gardiner, J. Ye, C. Nägerl and H. J. Kimble, Phys. Rev. A **61**, 045801 (2000).
12. A. Gillespie and F. Raab, Phys. Lett. A **178**, 357 (1993).
13. Y. Hadjar Y, P. F. Cohadon, C. G. Aminoff, M. Pinard, and A. Heidmann, Europhys. Lett. **47**, 545 (1999); F. Cohadon, A. Heidmann, and M. Pinard, Phys. Rev. Lett. **83**, 3174 (1999).
14. I. Tittonen, G. Breitenbach, T. Kalkbrenner, T. Muller, R. Conradt, S. Schiller, E. Steinsland, N. Blanc, and N. F. de Rooij, Phys. Rev. A **59**, 1038 (1999).
15. P. Saulson, Phys. Rev. D **42**, 2437 (1990).
16. H. J. Kimble, C. J. Hood, T. W. Lynn, H. Mabuchi, D. W. Vernooy, and J. Ye in *Laser Spectroscopy: XIV International Conference*, eds. Rainer Blatt, et al. (World Scientific, Singapore, 1999), 80.
17. S. J. van Enk, J. McKeever, H. J. Kimble, and J. Ye, Phys. Rev. A **64**, 013407 (2001); available as quant-ph/0005133.

SINGLE-ATOM MOTION IN OPTICAL CAVITY QED

G. REMPE, T. FISCHER, P. MAUNZ, P.W.H. PINKSE AND T. PUPPE
Max-Planck-Institut für Quantenoptik, Hans-Kopfermann-Str. 1, D-85748 Garching, Germany
E-mail: gerhard.rempe@mpq.mpg.de

Recent experiments performed with individual atoms in a microscopic high-finesse optical cavity are reviewed. By monitoring the cavity transmission with a weak probe laser, moving atoms can be observed with high spatial and temporal resolution. This allows one to obtain detailed knowledge about the cavity-mediated light forces, which can be exploited to manipulate the motion of a single atom in real time. For many atoms in the cavity, long-range light forces mediated by the common cavity field lead to collective motional effects.

1 Introduction

Laser cooling and trapping techniques make possible to perform experiments in which individual particles are coupled in a controlled way. For example, stored ions can interact via a single mode of a vibrational field. Here, single phonons can be created or annihilated on demand. Another example is cavity quantum electrodynamics, where an atom interacts with a single mode of a light field. Such a system can be used as a triggered source of single photons [1,2], and a first experiment in this direction with single atoms falling through a high-finesse cavity has recently been performed [3]. But in contrast to the system with a trapped ion, where the phonon has a long lifetime, a photon generated inside an optical cavity can quickly be transmitted through the cavity mirrors with near-unity efficiency.

This indicates a fundamental difference between the two systems, namely that dissipation cannot be neglected in the atom-photon experiments. Nevertheless, experiments can now be performed in the so-called strong-coupling regime, where a single quantum of energy is exchanged between the atom and the cavity many times before escaping into the environment. Two conditions must be fulfilled to achieve such a novel situation: first, the characteristic (vacuum Rabi) oscillation frequency determined by the atom-cavity coupling constant, g, has to be large. This requires a large electric field per photon and, hence, a cavity with a small mode volume. Second, the energy loss rate must be small. Two decay channels are important: one is spontaneous emission of the atom into the free-space continuum of modes not covered by the cavity. This gives rise to a decay of the atomic dipole at rate γ, equal to half the spontaneous linewidth for a radiatively broadened atom. This decay channel is usually not affected in a Fabry-Perot type cavity supporting a mode with a small solid angle, typically of the order of $\Delta\Omega/4\pi\approx10^{-5}$. The other loss channel is the escape of a photon from the cavity due to mirror transmission or absorption. This leads to a decay of the cavity field at rate κ which can be large when the mirror separation is small. Hence, ultra-high reflectivity mirrors are prerequisite for experimental research in the regime of strong coupling.

Dissipation has the advantage that it allows one to observe the system in real time. In this context, the strong-coupling regime is most interesting, as here the presence of an atom changes the cavity transmission dramatically. In particular, a single atom can be detected with a light field containing less than one photon on average, as long as the critical atom number, $N_C=2\gamma\kappa/g^2$, and the saturation photon number, $n_S=\gamma^2/2g^2$, are small [4,5]. Moreover, the light force from a single photon in the cavity can significantly influence the motion of a cold atom, if the kinetic energy is smaller than the vacuum-Rabi energy, g. Indeed, it was shown experimentally that it is possible to channel single atoms through the nodes and antinodes of a standing wave cavity field containing much less than one photon on average [6,7]. A single atom could even be trapped in a light field containing about one photon on average [8,9]. An interesting application is a high-resolution microscope for single atoms [9], in particular when higher-order transverse modes of the cavity are employed to obtain two-dimensional position information in a plane transverse to the cavity axis [10]. Moreover, the dynamical interplay between the motion of the atom and the intensity of the light field in the cavity can be used to cool the atom by means of cavity dissipation [11,12]. We now briefly summarize some experiments where the motion of single atoms in a high-finesse cavity containing at most a few photons on average was investigated in detail.

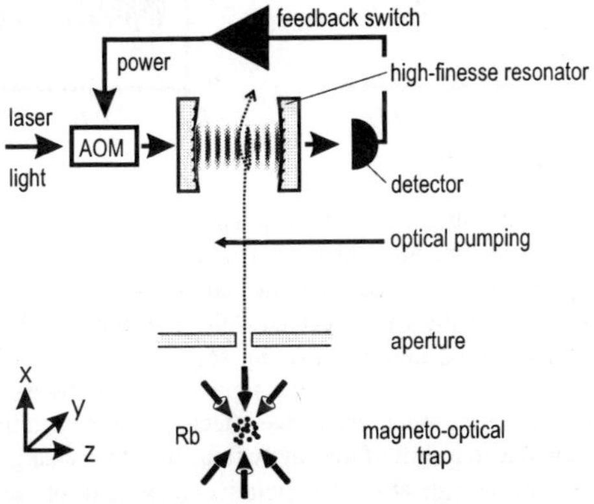

Figure 1. Experimental setup [5]. ^{85}Rb atoms are collected from a background vapor, cooled by polarization gradient light forces and subsequently launched into a high-finesse optical cavity. On their way up, the atoms are pumped into the F=3, m_F=3 Zeeman level of the $5S_{1/2}$ ground state. For a mirror separation of 116 μm, a mode waist of 29 μm and right circularly polarized light resonant with the $5S_{1/2}$ F=3 – $5P_{3/2}$ F=4 transition at 780 nm, the atom-cavity coupling constant at an antinode of the standing-wave cavity mode is $g=2\pi \times 16$ MHz. The decay rates of the atomic dipole, γ, and the cavity field, κ, are $(\gamma, \kappa) = 2\pi \times (3, 1.5)$ MHz, respectively. The intensity of the laser pumping the cavity field can be switched to a higher value upon the detection of an atom in the cavity.

2 Detection of single atoms

In our experiments, laser-cooled rubidium atoms are first collected in a magneto-optical trap (see figure 1) and then cooled in an optical molasses. Next, they are launched upwards in an atomic fountain and enter the high-finesse cavity ($F \approx 430{,}000$) which is placed close to the point where the atoms turn around under the influence of gravity. This has several advantages: first, it allows us to perform experiments with atoms whose kinetic energy is smaller than the vacuum-Rabi energy. Hence, it can be expected that forces of a light field containing about one photon on average induce significant motional effects. Second, a slowly moving atom can be observed for a long time even in a cavity with a small beam waist. Third, the ability to control the power of the laser pumping the cavity field allows us to manipulate in real time the motion of an isolated atom. Depending on the flux from the fountain we can study single atoms in transit, trap an atom or study many atoms in simultaneous transits.

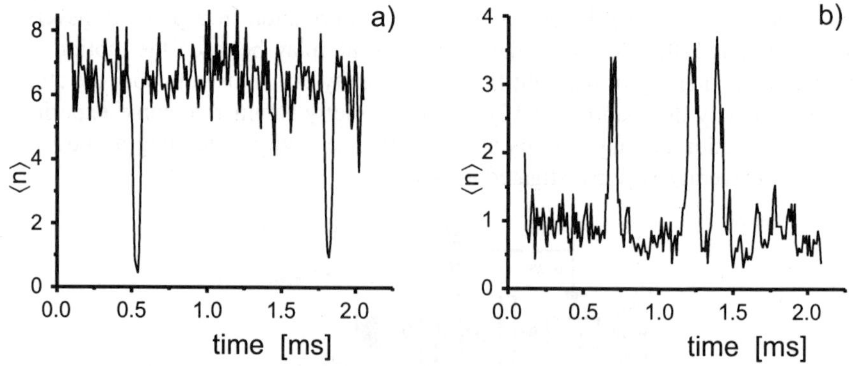

Figure 2. Real-time detection of individual atoms [5]. The intracavity photon number is plotted while several atoms pass the cavity mode one after the other with a speed of about 80 cm/s. a) the laser is on-resonance with the cavity, but red detuned from the atom by $\Delta_A = -2\pi \times 26.5$ MHz. For these parameters, the two atoms passing the cavity decrease the transmission. b) for the same atom-laser detuning, but the cavity is blue shifted from the laser by $-\Delta_C = 2\pi \times 4$ MHz. This results in an increase of the transmission upon passage of an atom. In the figure, three atoms pass the cavity one after the other. As the laser is detuned from the atom, dispersive effects dominate and the signals shown in a) and b) can be interpreted as being due to a shift of the cavity frequency by a change of the index of refraction of the cavity medium, caused by a single atom. The (relative) magnitude of this effect is $|\delta n/n| = |\delta\omega/\omega| \approx g/\omega \approx 4 \times 10^{-8}$, with ω the optical frequency.

3 Cavity-mediated light forces

If the light is red or blue detuned relative to the atomic transition, the atoms are attracted towards or repelled, respectively, from the antinodes of the standing-wave in the cavity. This effect can be observed by counting the number of atoms detected close to an antinode, where the atom-cavity coupling constant is largest. These atoms lead to deep dips in the cavity transmission (see figure 2a)) when the laser is in resonance with the empty cavity. When scanning this frequency relative to the atomic frequency, it was found that the number of deep dips is much larger for red detuning than for blue detuning. The experimental data shown in figure 3 agree well with the result of a simulation which takes into account the full motional dynamics induced by the position- and velocity-dependent intracavity light field.

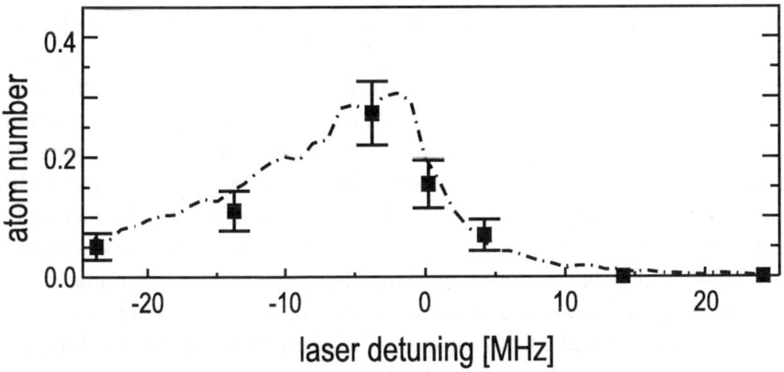

Figure 3. Atom counting [7]. For the laser on resonance with the empty cavity, and for different detunings from the atomic frequency, the relative number of atoms at an antinode is determined by counting the number of deep dips in the cavity transmission. Squares represent data. The dashed-dotted line is the result of a Monte-Carlo simulation based on an analytic solution of the quantum master equation, taking into account the dipole force, momentum diffusion and the velocity-dependent force.

Other information about light forces is obtained from the intensity autocorrelation function of the transmitted light, which reveals a fast oscillation of the intracavity field. This is caused by atoms channeling through the standing light wave, thereby modulating the light intensity at twice the oscillation frequency. We emphasize that the dependence of the intracavity intensity on the atomic position is highly nonlinear. In particular, the intensity change induced by the oscillatory motion of the atom is more pronounced for blue detuning, where the atoms channel through narrow nodes, than for red detuning, where the atoms channel through wide antinodes. When comparing the experimental data with the results of a numerical simulation, good agreement could only be obtained when the dipole force, momentum diffusion and a novel velocity-dependent force were taken into account [7]. The latter force is particularly interesting. It vanishes for small atom-cavity coupling constant, and has the novel feature that it allows to cool an atom without relying on spontaneous emission of the atom [11,12]. In fact, the dissipation required for cooling can be provided by the decay of the cavity field. In principle, this allows one to cool a two-level atom to a temperature well below the Doppler limit, or cool particles where a cycling transition is not available.

4 Single-photon optical tweezers

Mechanical binding of a single atom and a single photon was first envisioned about a decade ago for long-lived Rydberg atoms in a microwave cavity [13,14]. It has recently been observed in the optical domain where light forces are much larger [8,9]. In order to trap an atom in the conservative dipole potential in the cavity, the laser light coupled into the cavity is suddenly switched to a higher power when an atom is detected near an antinode. This increases the depth of the optical dipole potential and holds the atom in a cavity field containing about one photon on average. The subsequent motion of the trapped atom induces pronounced changes of the cavity transmission, which is monitored in real time. An example is displayed in figure 4. The observed signals are in good qualitative agreement with predictions based on a quantum jump Monte Carlo simulation of the trajectory of a point-like atom. Moreover, the fourth-order intensity correlation function displays a periodic structure, which we attribute to sudden flights of the atom between different antinodes of the standing wave. These flights are initiated by fluctuations of the dipole force heating the atom in the direction of the cavity axis. They are terminated by the velocity-dependent force cooling the atom for the parameters of the experiment. The cooled atom can be recaptured in one of the neighboring not-too-distant antinodes (instead of hitting one of the mirrors). Hence, the long trapping times observed in the experiment are further evidence for the existence of a cavity-mediated friction force.

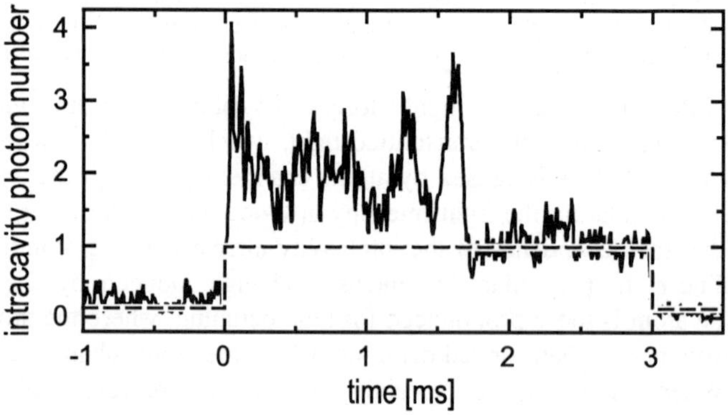

Figure 4. Single-atom trapping [8]. The number of photons transmitted through the cavity is counted for time intervals of 10 μs. At time t=0, an atom traversing the cavity triggers an 8-fold increase of the pump power (dashed line). The increase in the dipole potential causes the atom to remain in the cavity for about 1.7 ms. The pump power is switched back to its original value after 3 ms. The laser is red detuned from the atom and the cavity by $\Delta_A = -2\pi \times 45$ MHz and $\Delta_C = -2\pi \times 5$ MHz, respectively. The average trapping time of 280 μs is limited by transverse heating caused by spontaneous emission events. A single photon in the cavity corresponds to a transmitted power of 0.9 pW, leading to about 20 detected photons in 10 μs. Apart from field fluctuations, the changes of the transmitted power reflect the motion of the trapped atom. A nearest-neighbor smoothing algorithm is applied to the data.

5 Long-range light forces between strongly coupled atoms

If two or more atoms are placed inside the mode of the cavity, every atom can change the light field dramatically. This also changes the light forces on the other atoms. In other words, a light force between atoms is mediated by the cavity. This force is quite different from that in free-space, because the interaction strength is not a function of the inter-atomic distance, but rather of the position of the atoms in the cavity, which is much larger than a wavelength. We have found evidence for such a long-range dipole-dipole interaction in the spectra of a high-finesse cavity filled with ultra-cold rubidium atoms [15]. The forces between these strongly coupled atoms influence the spatial distribution of the atoms, which manifests itself in an asymmetric normal-mode spectrum, as displayed in figure 5. The asymmetry occurs already for photon numbers well below 0.05. Good agreement between the observed data and calculated spectra could only be achieved when including in the Monte-Carlo simulation the full motional dynamics of the many-atom system. For this purpose, we extended the existing work on the dipole force and the diffusion coefficient for a single atom [16] to that in the presence of many atoms [17].

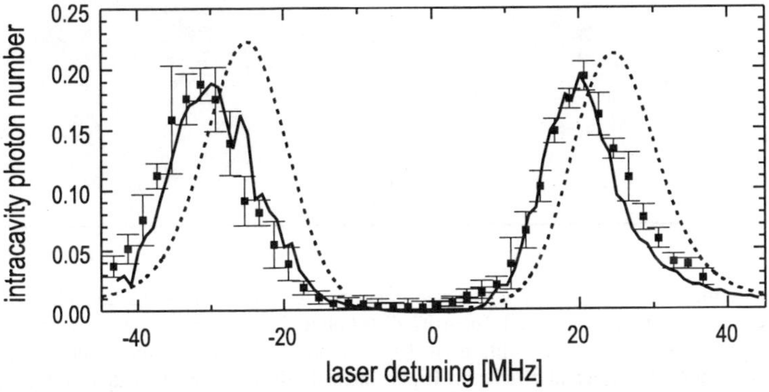

Figure 5. Normal mode spectrum [15]. The number of photons in the cavity is measured as a function of the detuning, Δ_C, of the laser from the cavity, which is resonant with the free-space atoms. Symbols represent measured data. The dashed line comes from a semi-classical simulation ignoring light forces. The solid line is the result of a Monte-Carlo simulation including the motion of several strongly coupled atoms simultaneously.

6 Atomic kaleidoscope

An interesting application of the strongly coupled atom-cavity system is to observe a single moving atom with high spatial (μm) and temporal (μs) resolution. For example, an atom moving along the standing-wave induces a fast modulation of the intracavity light field [8,18]. In principle, this allows one to measure the (relative) position of the atom in the direction of the cavity axis. In the transverse plane, however, position information is much more difficult to extract because of the rotational symmetry of the fundamental TEM_{00} cavity mode [9]. This disadvantage can be overcome by using higher-order frequency-degenerate transverse modes [10]. Here, the atom couples different modes depending on its position. This produces a pattern of bright and dark regions at the cavity output, as plotted in the right part of figure 6. The pattern changes when the atom moves around, thereby resembling a toy kaleidoscope. A measurement of this intensity distribution with, e.g., a two-dimensional detector array should allow one to reconstruct the atomic position and, hence, the atomic trajectory. An important feature of this scheme is that it does not require any knowledge of the light forces to trace a trajectory. Instead, the position of the atom can directly be deduced from the measured intensity distribution.

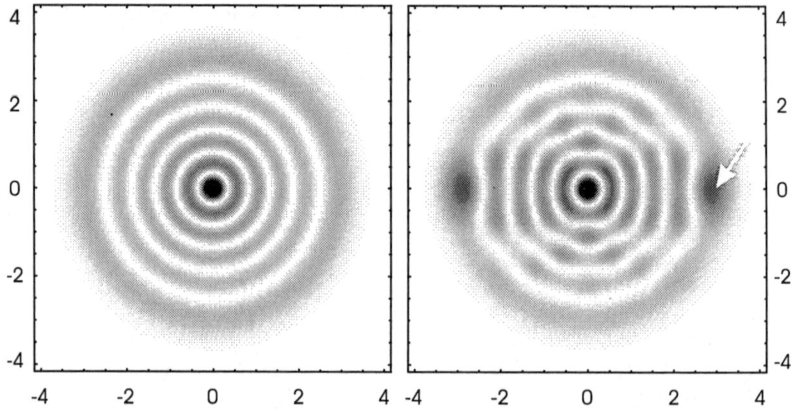

Figure 6. Atomic kaleidoscope [10]. Calculated intensity distribution in a transverse plane of area 8×8 w_0^2, where w_0 is the mode waist. Left: intensity distribution for the manifold of frequency-degenerate Laguerre-Gaussian TEM_{mn} modes with m+n=10. A plane input wave excites all possible modes simultaneously, but with different efficiency. Right: intensity distribution calculated for the same pump conditions, but with a single atom at a position indicated by the arrow. Parameters are $(g, \gamma, \kappa) = 2\pi \times (16, 3, 1.5)$ MHz. The laser is detuned from the atom and the cavity by $\Delta_A = -2\pi \times 50$ MHz and $\Delta_C = -2\pi \times 5$ MHz, respectively. Note that the pattern induced by the atom displays a 180 degree symmetry.

References

1. Law C. K. and Kimble H. J., Deterministic generation of a bit-stream of single-photon pulses, *J. Mod. Opt.* **44** (1997) pp. 2067-2074.
2. Kuhn A., Hennrich M., Bondo T. and Rempe G., Controlled generation of single photons from a strongly coupled atom-cavity system, *Appl. Phys.* B **69** (1999) pp. 373-377.

3. Hennrich M., Legero T., Kuhn A. and Rempe G., Vacuum-Stimulated Raman Scattering Based on Adiabatic Passage in a High-Finesse Optical Cavity, *Phys. Rev. Lett.* **85** (2000) pp. 4872-4875.
4. Mabuchi H., Turchette Q. A., Chapman M. S. and Kimble H. J., Real-time detection of individual atoms falling through a high-finesse optical cavity, *Opt. Lett.* **21** (1996) pp. 1393-1395.
5. Münstermann P., Fischer T., Pinkse P. W. H. and Rempe G., Single slow atoms from an atomic fountain observed in a high-finesse optical cavity, *Opt. Comm.* **159** (1999) pp. 63-67.
6. Hood C. J., Chapman M. S., Lynn T.W. and Kimble H. J., Real Time Cavity QED with Single Atoms, *Phys. Rev. Lett.* **80** (1998) pp. 4157-4160.
7. Münstermann P., Fischer T., Maunz P., Pinkse P. W. H. and Rempe G., Dynamics of Single-Atom Motion Observed in a High-Finesse Cavity, *Phys. Rev. Lett.* **82** (1999) pp. 3791-3794.
8. Pinkse P. W. H., Fischer T., Maunz P. and Rempe G., Trapping an atom with single photons, *Nature* **404** (2000) pp. 365-368.
9. Hood C. J., Lynn T. W., Doherty A. C., Parkins A. S. and Kimble H. J., The Atom-Cavity Microscope: Single Atoms Bound in Orbit by Single Photons, *Science* **287** (2000) pp. 1447-1453.
10. Horak P., Ritsch H., Fischer T., Maunz P., Puppe T., Pinkse P. W. H. and Rempe G., An optical kaleidoscope using a single atom, quant-ph/0105048.
11. Horak P., Hechenblaikner G., Gheri K. M., Stecher H. and Ritsch H., Cavity-Induced Atom Cooling in the Strong Coupling Regime, *Phys. Rev. Lett.* **79** (1997) pp. 4974-4977.
12. Vuletić V. and Chu S., Laser Cooling of Atoms, Ions, or Molecules by Coherent Scattering, *Phys. Rev. Lett.* **84** (2000) pp. 3787-3790.
13. Haroche S., Brune M. and Raimond J. M., Trapping atoms by the vacuum field in a cavity, *Europhys. Lett.* **14** (1991) pp. 19-24.
14. Englert B.-G., Schwinger J., Barut A. O. and Scully M. O., Reflecting slow atoms from a micromaser field, *Europhys. Lett.* **14** (1991) pp. 25-31.
15. Münstermann P., Fischer T., Maunz P., Pinkse P. W. H. and Rempe G., Observation of Cavity-Mediated Long-Range Light Forces between Strongly Coupled Atoms, *Phys. Rev. Lett.* **84** (2000) pp. 4068-4071.
16. Hechenblaikner G., Gangl M., Horak P. and Ritsch H., Cooling an atom in a weakly driven high-Q cavity, *Phys. Rev.* A **58** (1998) pp. 3030-3042.
17. Fischer T., Maunz P., Puppe T., Pinkse P. W. H. and Rempe G., Collective light forces on atoms in a high-finesse cavity, *New Journal of Physics* **3** (2001) pp. 11.1-11.20.
18. Pinkse P. W. H., Fischer T., Maunz P., Puppe T. and Rempe G., How to catch an atom with single photons, *J. Mod. Opt.* **47** (2000) pp. 2769-2787.

OPTICAL COOLING IN HIGH-Q MULTIMODE CAVITIES

HELMUT RITSCH, PETER DOMOKOS, PETER HORAK, MARKUS GANGL
Institut für Theoretische Physik
Universität Innsbruck
Technikerstr. 25, A-6020 Innsbruck, Austria
E-mail:Helmut.Ritsch@uibk.ac.at

Using a semiclassical description we study the motion of cold atoms in the field of a driven high-Q optical cavity, which supports several radiation field modes close to the atomic transition frequency. We present stochastic differential equations for the coupled particle and field dynamics. While the atoms influence the field as a time varying refractive index and loss the motion is governed by the light forces induced by the cavity field. Kinetic energy can be extracted from the particles through cavity dissipation. For sufficiently large detuning spontaneous emission plays a minor role and the cooling scheme should work for any particles with a suitable optical dipole moment. We derive analytic expressions for the friction force and the momentum diffusion to estimate the cooling rate and temperature and exhibit the scaling properties of the cooling as function of atom and mode number. For the case of degenerate modes we demonstrate the possibility to accurately reconstruct atomic trajectories from the measured output field distributions. Finally, as a second example we show that there are good prospects for implementation of the scheme using evanescent fields appearing in microoptical devices.

1 Introduction

Laser manipulation of atomic gases has seen dramatic improvements towards lower temperatures and higher phase space densities [1,2,3,4] up to the degeneracy regime in recent years. Nevertheless there are limitations in the achievable final phase space density and the atomic species to which the most advanced methods can be applied are limited. Most schemes rely on spontaneous emission and as an essential ingredient this creates the need of a closed optical level scheme and poses the problem of light reabsorption, which limits the applicability to molecules (rotational and vibrational couplings) and atoms with no suitable transition.

This motivates the search for new cooling methods. Raizen and coworkers[5] proposed stochastic cooling of an ensemble of trapped atoms. Alternatively we suggested the use of a strongly coupled high-Q optical cavity to extract motional energy from a cloud of trapped polarizable particles. Here the dissipation channel is realized by cavity damping and the scheme allows final average kinetic energies well below the Doppler limit [6,7]. Experimental evidence for the realizability of the scheme has been obtained recently for single atoms [8,9] and a few photons in mode in cavity QED setups. An application of cavity induced cooling to atoms and molecules far off resonance in a confocal resonator has been studied recently [10].

Starting from our previously derived semiclassical model we present here a general formulation for cavity induced cooling involving many atoms in a multimode field configuration. Due to their momentary positions the atoms experience a certain lightshift, the gradient of which gives the dipole force acting on the atoms. At the same time the sum of the lightshifts of all atoms is equivalent to the frequency shift of the cavity resonance induced by the atoms. Hence this frequency shift is a collective property of the atomic ensemble and can be used for backaction on the atoms via the pump field. In its simplest form one uses a constant pump of fixed frequency. The atom cavity system is then dynamically shifted in and out of resonance with the pump depending on the momentary atomic positions. By choosing the right parameters, motional energy can be extracted from the atoms via this feedback [7].

If several field modes are involved in the dynamics, the atoms influence the intensity as well as the relative phases of the modes and photons can be redistributed between the various modes. In the simplest multimode case, i.e., a single particle in a ring cavity, a strong enhancement of the cooling has been predicted [11], as small shifts of the particle positions couple more efficiently to relative phase shifts of the two counter propagating fields than to intensity changes. For several particles simultaneously coupled to the same radiation modes besides cooling one also finds particle-particle interactions, which can either lead to a buildup of correlations or fast thermalization without direct particle interaction.

2 Semiclassical model

Consider a dilute gas of N identical two-level atoms of mass m, transition frequency ω_a, and spontaneous emission rate Γ interacting with M modes of a high-finesse optical cavity with resonance frequencies ω_k, cavity decay rates κ_k, and mode functions $f_k(x)$, $k = 1..M$, respectively. The mode functions are mutually orthogonal and fulfill the normalization condition $\int |f_k|^2\, dx = V_{00}$, where $V_{00} = dw_0^2\pi/4$ is the TEM_{00} mode volume of a Fabry-Perot cavity of length d and waist w_0. The maximum single photon Rabi frequency of the TEM_{00} mode is denoted by g. The cavity modes are driven with pump strengths η_k .

As a central approximation we treat the atomic center-of-mass motion classically. This is a good approximation as long as the atom is well localized in both position and momentum space and can be fulfilled when $\hbar k^2/M\gamma < 1$. Such a semiclassical description of the system implies a set of coupled stochastic differential equations (SDEs) for the atomic positions x_n, momenta p_n, and mode amplitudes α_k [12,13],

$$dx_n = \frac{p_n}{m}\, dt, \tag{1}$$

$$\begin{aligned} dp_n = & -U_0\Big[\mathcal{E}(x_n)\nabla_n\mathcal{E}^*(x_n) + \mathcal{E}^*(x_n)\nabla_n\mathcal{E}(x_n)\Big]\, dt \\ & +i\gamma\Big[\mathcal{E}(x_n)\nabla_n\mathcal{E}^*(x_n) - \mathcal{E}^*(x_n)\nabla_n\mathcal{E}(x_n)\Big]\, dt + dP_n, \end{aligned} \tag{2}$$

$$\begin{aligned} d\alpha_k = & -\eta_k^* dt + i\Big[\Delta_k\alpha_k - U_0\sum_n \mathcal{E}(x_n) f_k^*(x_n)\Big] dt \\ & -\Big[\kappa_k\alpha_k + \gamma\sum_n \mathcal{E}(x_n) f_k^*(x_n)\Big] dt + dA_k, \end{aligned} \tag{3}$$

where $\mathcal{E}(x) = \sum_k f_k(x)\alpha_k$ is the field amplitude at position x, $U_0 = \Delta_a g^2/(\Delta_a^2+\Gamma^2)$ the light shift per photon ($\Delta_a = \omega_p - \omega_a$), $\gamma = \Gamma g^2/(\Delta_a^2+\Gamma^2)$ the photon scattering rate, and $\Delta_k = \omega_p - \omega_k$ the detuning of the kth mode from the pump laser.

The derivation of the above SDE is based on an adiabatic elimination of the internal atomic degrees of freedom, for which we must ensure a small atomic saturation parameter, that is, $s = |\alpha|^2 g^2/(\Delta_a^2 + \Gamma^2) \ll 1$. To this end we assume a sufficiently large atomic detuning $\Delta_a \gg \Gamma$. Qualitatively the interpretation of the various terms in eqs. (1)-(3) is rather simple. The two terms in eq. (2) correspond to the dipole force and the radiation pressure force, respectively. Equation (3) describes the pumping of the cavity modes, the frequency detuning from the pump laser shifted according to the coupling to the atoms, and the mode damping due to cavity decay and photon scattering by the atoms. dP_n and dA_k are white noise increments which can be characterized by a diffusion matrix $D_{ij} = \langle dF_i\, dF_j\rangle$, where $F_i = \{P_n, A_k\}$. The diffusion matrix exhibits nontrivial cross correlations between momentum diffusion and cavity amplitude and phase fluctuations [12]. The precise form of all noise terms is given in Ref. [13].

The set of SDEs (1)-(3) is well suited for a numerical implementation. Note that the number of equations grows only linearly with the number of atoms and modes and the inclusion of the quantum noise terms allows to obtain reliable estimates for steady-state temperatures and trapping times.

3 Friction, diffusion and temperature for a single atom in a single mode

In the following section we consider a high Q standing wave cavity with only a single mode function $f(x) = \cos(kx)$ excited. In order to observe a significant effect we require $\Delta \approx -\kappa$ and $N|U_0| \approx \kappa$. For atomic velocities v fulfilling $kv < \kappa$ we obtain the position-averaged friction force and diffusion coefficient:

$$\overline{F_1} = -k^2 \frac{\eta^2 U_0^2}{4\kappa^4}, \; \overline{D} = k^2 \kappa \frac{\eta^2 U_0^2}{8\kappa^4} \tag{4}$$

and hence the steady-state temperature can be estimated as

$$k_B T = -\frac{\overline{D}}{\overline{F_1}} = \frac{\kappa}{2}. \tag{5}$$

We clearly can get sub-Doppler cooling for $\kappa < \Gamma$. From the friction coefficient we also obtain the cooling time τ_c defined from the exponential decay time of the kinetic energy $E(t)$ of the form

$$E(t) = \left[E(0) - \frac{k_B T}{2}\right] \exp(-t/\tau_c) + \frac{k_B T}{2}, \quad \text{with} \quad \tau_c = \frac{m}{2|\overline{F_1}|} = \frac{\kappa^4}{\eta^2 U_0^2} \omega_R^{-1}. \tag{6}$$

Here $\omega_R = \hbar k^2/(2m)$ is the recoil frequency. In general we see that very good mirrors (small κ), a small cavity volume (large U_0) and light atoms are required for optimum performance.

4 Scaling properties for many atoms in a single mode

Let us now consider N atoms. As they all couple to the same mode they will interact and perturb each other via this mode [14,15]. In fact, for a few atoms and a few photons inside a high-finesse cavity such effects have recently been demonstrated experimentally [16]. Here we numerically integrate the SDEs (1)abd calculate the time evolution of the kinetic energy.

In Fig. 1 we plot the mean kinetic energy as a function of time for a single and for ten atoms and rescale the parameters such that NU_0 and the optical potential depth (proportional $U_0\eta^2/\kappa^2$) are constant. The curves can be well approximated by an exponential fit as in eq. (6) with cooling times $\tau_c = 142\kappa^{-1}$ $(N = 1)$ and $\tau_c = 1110\kappa^{-1}$ $(N = 10)$. For the given parameters, the analytical estimate given by eq. (6) yields $\tau_c = 118.6\kappa^{-1}$ and $1186\kappa^{-1}$, respectively, which is surprisingly good. Therefore, the single atom cooling time appears to make sense for the many atom case. However, the operating

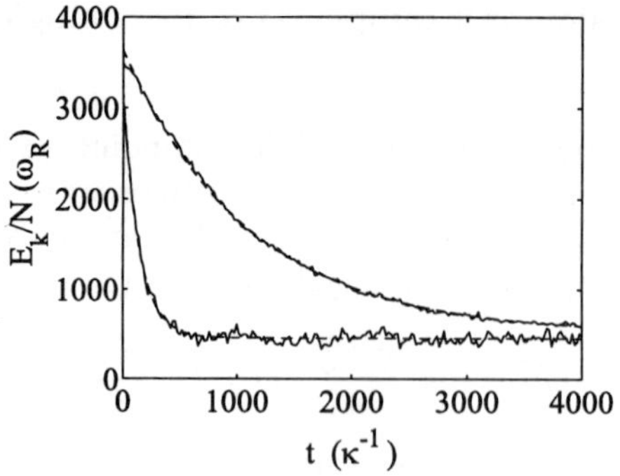

Figure 1. Kinetic energy per atom obtained from numerical simulations averaged over 100 realizations (solid curves). The dashed curves are exponential fits. Lower curves: $N = 1$, upper curves: $N = 10$. The parameters are $NU_0 = -0.6\kappa$, $N^2\gamma = 0.03\kappa$, $\Delta = -0.6\kappa$, $\eta = 3\sqrt{N}\kappa$, $\kappa = 415\omega_R$.

conditions differ for varying atom numbers, according to the condition that NU_0 remains constant.

The steady-state kinetic energies obtained from the exponential fits of Fig. 1 are $466\omega_R$ ($N = 1$) and $510\omega_R$ ($N = 10$). While these are approximately of the same size, they are significantly larger than the value of $\kappa/4 \approx 100\omega_R$ obtained from eq. (5). However, eq. (5) was derived assuming $U_0 \ll \kappa$ and a flat spatial distribution in the steady state. Both of these conditions are violated for the present parameters.

It should also be emphasized that an exponential fit as in Fig. 1 only works for sufficiently hot atoms where the friction force can be replaced by its average value. For very cold and well localized atoms the position dependence of the friction must be taken into account. In fact, the friction force vanishes completely at the bottom of the potential wells. Hence, if in the course of the cooling process the localization increases, the time scale of the cooling increases and the kinetic energy no longer decreases according to an exponential law [17].

For the same scaling of the parameters with N as discussed above, we plot the steady-state kinetic energy and the cooling time for particle numbers from one to 100 in Fig. 2. We find an approximately constant final temperature and a cooling time proportional to N. Taking the scaling of U_0 with the particle number N into account, these results agree with the predictions of the single-atom estimates of the previous section.

In an alternative operating regime where $NU_0 < \kappa$ the numerical simula-

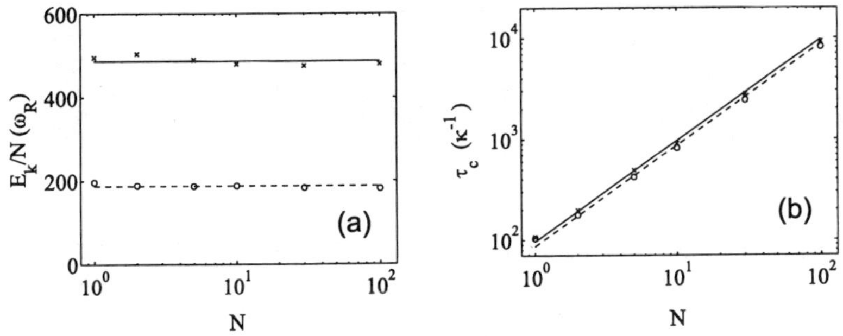

Figure 2. Steady-state kinetic energies per atom (a) and cooling times τ_c (b) for atom numbers from 1 to 100. Crosses show the numerical results for $\Delta = -0.6\kappa$, circles for $\Delta = -\kappa$. All other parameters are the same as in Fig. 1. The solid and dashed lines are linear fits.

tions show that the cooling time for an ensemble of N atoms is of the same order of magnitude as the cooling time for a single atom and individual atoms in the cloud are cooled independently from each other as it has also been predicted by Vuletic and Chu [10]. The cavity simply enhances backscattering of photons by moving atoms according to their Doppler shift analogous to free-space Doppler cooling. This is, however, only valid up to a certain limiting photon and atom number. For a very far off resonance field in a macroscopic cavity this would be the typical operating regime. Here the cooling is slow but it involves only very few spontaneous emission events.

5 Atom kaleidoscope: one atom coupled to a mode family

As can be seen easily from eqs. (1)-(3), besides creating an energy shift and a loss for each field mode, an atom will also induce a coupling to other modes. Hence for a multimode situation the field pattern will strongly depend on the atomic positions with constraints posed by the common symmetries of all modes. The effect reminds of a toy-kaleidoscope in which small objects in a symmetric arrangement of mirrors create beautiful patterns. We will see that this allows to obtain accurate information on the atom's position. As concrete example we consider higher-order frequency-degenerate transversal modes of an optical resonator. Typical examples are the Hermite-Gaussian (HG) or the Laguerre-Gaussian (LG) modes. The generated patterns allow to determine the atomic position directly from a field measurement. For simplicity, we assume that all modes have a common eigenfrequency ω and decay rate κ.

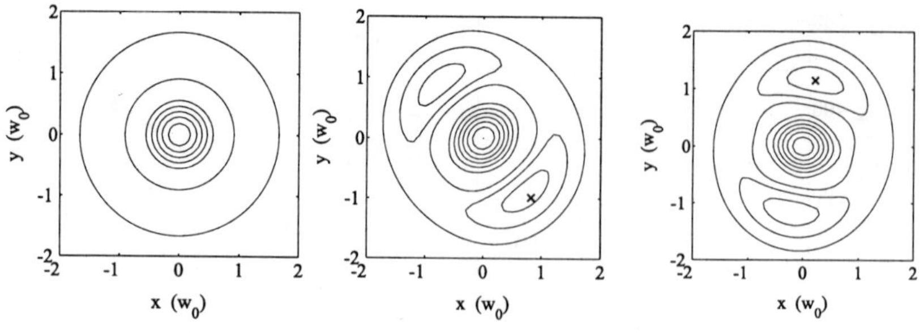

Figure 3. Steady state field intensity for different positions of the atom.

As an example, we consider an atom at rest at position $\mathbf{r}_a$ coupled to the three degenerate cavity modes with $(p, m) = (1, 0), (0, -2), (0, 2)$. It is then possible to solve eqs. (1)-(3) for the mean stationary field amplitudes $\alpha_{pm}^{\text{stat}}(\mathbf{r}_a)$:

$$\alpha_{pm}^{\text{stat}} = \frac{\eta_{pm}^*}{i\Delta - \kappa} + \frac{iU_0 + \gamma}{i\Delta - \kappa} u_{pm}^*(\mathbf{r}_a)\mathcal{E}_0(\mathbf{r}_a), \tag{7}$$

where $\mathcal{E}_0$ is the electric field at the position of the atom,

$$\mathcal{E}_0(\mathbf{r}_a) = \frac{\sum_{p,m} u_{pm}(\mathbf{r}_a)\eta_{pm}^*}{(i\Delta - \kappa) - (iU_0 + \gamma)\sum_{p,m} |u_{pm}(\mathbf{r}_a)|^2}. \tag{8}$$

Figure 3 shows the steady state field intensities for the empty cavity and for two different atomic positions. For the chosen parameters (rubidium atoms, $(g_0, \Gamma, \kappa) = 2\pi \times (16, 3, 1.5)\,\text{MHz}$, $(\eta_{10}, \eta_{0-2}, \eta_{02}) = 2\pi \times (6.4, 0, 0)\,\text{MHz}$, $\Delta = -2\pi \times 2.25\,\text{MHz}$ and $\omega_p - \omega_a = -2\pi \times 114\,\text{MHz}$, leading to $U_0 = \Delta$, $\gamma = 2\pi \times 60\,\text{kHz}$) the atom distributes photons between the cavity modes in such a way that a local maximum of the total field intensity is created near the position of the atom, see Fig. 3. This dependence of the cavity field on the atomic position suggests that measuring the cavity output field distribution yields ample information on the atomic motion. In fact, it can be shown that the functions $\alpha_{pm}^{\text{stat}}(\mathbf{r}_a)$ can be inverted to yield the atomic position. Of course one is limited by the common symmetries of all modes. For instance, for the system considered in Fig. 3, a 180° rotation around the cavity axis forms a symmetry operation.

As an example we show in Fig. 4 that this method in fact allows to reconstruct the trajectory of an atom passing through the cavity by spatially resolved observation of the cavity output. First, we numerically generate a sample trajectory (solid line in Fig. 4) and calculate the corresponding

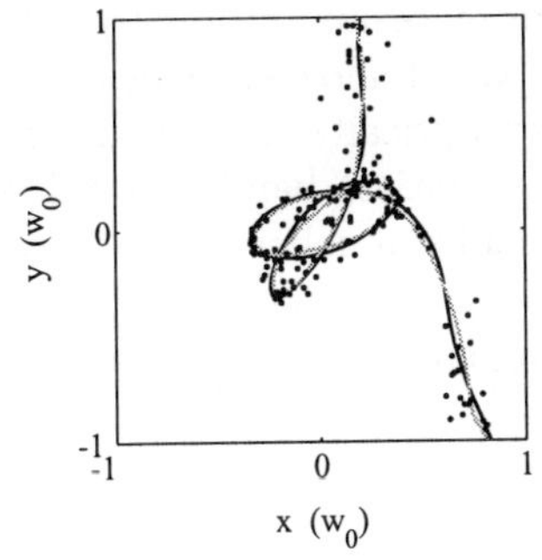

Figure 4. Simulated and reconstructed positions and trajectory of the atom.

cavity output taking into account spatial distribution, counting statistics, and quantum noise. Next, we use this "measurement data" to reconstruct the atomic position at each time step (single dots). Out of these points we can finally form a continuous curve as a function of time by appropriate smoothing (dotted line). Comparing the reconstructed with the original trajectory we note that for the depicted area close to the cavity axis the reconstruction works well.

6 Conclusions

We investigated the possibility of cooling a cloud of neutral particles by a cavity-enhanced cooling scheme which does not rely on atomic spontaneous emission. Numerical simulations show that the friction force and the cooling time is of the same order of magnitude as for a single atom for the same system parameters. However, the operating regime which is most ideal for one atom cannot be used for many atoms, so that the best achievable cooling time increases approximately linearly with the atom number. Although the result might be somewhat different for the case of strongly localized atoms, the general trend should be similar. Comparison with the analogous expressions for free-space Doppler cooling indicate that cavity-induced cooling is mainly favorable for high-finesse optical cavities fulfilling $g > \kappa$. Numerical simulations also suggest that the cooling efficiency is increased in cavities which support many nearly degenerate optical modes. A confocal setup seems a good choice in terms of atom number scaling and time scale [10]. Using the parameters of the high-Q resonators used in recent cavity-QED experiments [8,9] one obtains spontaneously scattered photon numbers per cooling time as low as 5-10. Replacing the heavy rubidium or cesium atoms by lithium even yields photon numbers below one, which would allow efficient cooling without spontaneous emission. Alternatively microoptically confined fields as in microspheres, thin high index surface layers or optical band gap guides on surfaces could be used to trap and cool a smaller number of particles at well defined spatial positions[18]. In addition the fields provide ample information of the particle position and motion allowing for spatially resolved single particle detection in the context of cavity QED.

References

1. M. Morinaga *et al.*, Phys. Rev. Lett. **83**, 4037 (1999).
2. A.J. Kerman, V. Vuletic, C. Chin, and S. Chu, Phys. Rev. Lett. **84**, 439 (2000).
3. M.T. DePue *et al.*, Phys. Rev. Lett. **82**, 2262 (1999).
4. M.D. Barrett, J.A. Sauer, and M.S. Chapman, Phys. Rev. Lett. **87**, 010404 (2001).
5. M.G. Raizen, J. Koga, B. Sundaram, Y. Kishimoto, H. Takuma, and T. Tajima, Phys. Rev. A**58**, 4757 (1998).
6. P. Horak, G. Hechenblaikner, K.M. Gheri, and H. Ritsch, Phys. Rev. Lett. **79**, 4974 (1997).
7. G. Hechenblaikner, M. Gangl, P. Horak, and H. Ritsch, Phys. Rev. A**58**, 3030 (1998).
8. P.W.H. Pinkse, T. Fischer, P. Maunz, and G. Rempe, Nature **404**, 365 (2000).
9. C.J. Hood, T.W. Lynn, A.C. Doherty, A.S. Parkins, and H.J. Kimble, Science **287**, 1447 (2000).
10. V. Vuletic and S. Chu, Phys. Rev. Lett. **84**, 3787 (1999); V. Vuletic, H.W. Chan, and A.T. Black, Phys. Rev. A**64**, 033405 (2001).
11. M. Gangl and H. Ritsch, Phys. Rev. A**61**, 043405 (2000).
12. P. Domokos, P. Horak, and H. Ritsch, J. Phys. B: At. Mol. Opt. Phys. **34**, 187 (2001).
13. P. Horak and H. Ritsch, Phys. Rev. A**64**, 0334xx (2001).
14. A. Hemmerich, Phys. Rev. A**60**, 943 (1999).
15. A.S. Parkins and H.J. Kimble, J. Opt. B: Quantum Semiclass. Opt. **1**, 496 (1999).
16. P. Münstermann, T. Fischer, P. Maunz, P. W. H. Pinkse, and G. Rempe, Phys. Rev. Lett. **84**, 4068 (2000).
17. M. Gangl and H. Ritsch, Phys. Rev. A**61**, 011402 (2000).
18. P. Domokos and H. Ritsch, Europhys. Lett. **54**, 306 (2001).

SINGLE IONS INTERFERING WITH THEIR MIRROR IMAGES

JÜRGEN ESCHNER, CHRISTOPH RAAB, PAVEL BOUCHEV,
FERDINAND SCHMIDT-KALER, RAINER BLATT

*Universität Innsbruck, Institut für Experimentalphysik,
Technikerstr. 25, 6020 Innsbruck, AUSTRIA
E-mail: Juergen.Eschner@uibk.ac.at*

The spontaneous emission of an atom is inhibited and enhanced when the atom is placed into a structured dielectric environment [1]. When another atom of the same kind is nearby, their spontaneous emission may exhibit cooperative sub- and superradiance [2]. We study these most fundamental quantum optical processes by recording single photons emitted by a single trapped atom which interacts with its mirror image over a distance of 50 cm: By retroreflecting the fluorescence of a single trapped Ba^+ ion with a high-quality lens and a mirror 25 cm away, we observe interference fringes with 72% visibility as the mirror distance varies. Simultaneous observation of the light transmitted through the mirror shows the population of the upper level to vary in anticorrelation with the interference fringes, which indicates inhibition and enhancement of a single atom's single spontaneous emission events. When two ions are trapped, they interfere with each other's mirror images, which indicates super- and subradiance mediated by the distant mirror. In this case the fringe visibility is 5%. The experiment allows to study variations in the vacuum fluctuations around a trapped ion on a sub-optical scale and to determine its position with respect to the mirror with nanometer-resolution.

1 Introduction

Since the first single ion was experimentally prepared and observed [3], single trapped atoms have found numerous applications in various areas. These range from precision measurements of physical constants and frequency standards [4], over experiments on fundamental quantum mechanics, to their application for the storage and processing of quantum information.

The lasting interest in single trapped ions is based on two main experimental features which become possible through the combination of an ion trap, in particular of the Paul type, with laser cooling. Together these techniques result in a localization of the single particle to typically a few ten nanometers or even below, in temperatures in the sub-millikelvin regime, in a high degree of isolation of the ion from its environment, and in quasi unlimited interaction time.

During the last few years experiments with single atoms have moved on towards the coherent manipulation of their internal and motional quantum state. These prospects have opened another rich field of applications of single trapped atoms because such manipulations, when applied to several ions in the same trap, form the basis of one of the promising implementations of quantum information processing. Indeed, the preparation of pure quantum states [5,6], their unitary rotation with high fidelity [6,7], conditional dynamics [7], as well as deterministic entanglement of a trapped ion string [8] have already been demonstrated.

The results reported in this paper belong to the field of fundamental studies with single trapped particles which earlier have revealed such prominent effects as quantum jumps [9] and antibunching [10]. In the experiments described here, we investigate interference of the light emitted by single Barium ions in a Paul trap. In an initial study, we send spontaneously emitted photons through a Mach-Zehnder interferometer and observe interference fringes with high contrast. In the second experiment which is the main topic of this report [11], one or two ions are trapped and, by a special optical arrangement, their light is retroreflected and focused back on the ion(s). We show that this light not only exhibits interference but also acts back on the ion or, in the case of two ions, creates a coupling between them. The analysis of the experimental observations with simple model considerations shows that the fundamental effects behind the observations are, in the one-ion case, inhibition and enhancement of spontaneous emission due to a modification of the electromagnetic mode structure and, in the two-ion case, sub- and superradiant emission due to reabsorption of photons emitted by the other ion. While modified spontaneous emission has been observed in several experimental [12] systems and sub- and superradiant emission has been studied with trapped ions [13], our experiments together with their model description highlight in particular the intimate relation between these fundamental phenomena.

2 Experimental setup

A single Ba^+ ion is trapped in a Paul trap of 1.4 mm diameter; its oscillation frequencies ω_z (ω_r) in the trap potential are between 1.2 and 2 (0.6 and 1) MHz. The ion is laser-cooled by continuous excitation on its $S_{1/2} \leftrightarrow P_{1/2}$ and $P_{1/2} \leftrightarrow D_{3/2}$ resonance lines at 493.4 nm and 649.7 nm, respectively. See Fig. 1 for a schematic of the experiment and the relevant levels of Ba^+; more details are described in earlier publications [14]. Both lasers have linewidths well below 100 kHz. The laser beams are combined on a dichroic beamsplitter before they are focused into the trap, and both light fields are linearly polarized. The laser intensities at the position of the ion are set roughly to saturation. The 650 nm laser is tuned close to resonance, the 493 nm laser is red-detuned by about the transition linewidth ($\Gamma = 15.1$ MHz) for Doppler cooling. A 2.8 Gauss magnetic field which is orthogonal to both the laser wave vector and the laser polarization defines a quantization axis and lifts the degeneracy of the Zeeman sublevels. The precise parameters are determined by fitting an 8-level Bloch equation calculation to a scan of the fluorescence intensity vs. the detuning of the 650 nm laser [15], see Fig. 2 for an example. A high-quality lens (L1), oriented at 90° to the excitation beams and situated 12.5 mm away from the ion, collects the fluorescence light of the ion in 4% solid angle and transforms it into a parallel beam of 21.4 mm diameter.

In the initial experiment, this light is analysed with a Mach-Zehnder interferometer as sketched in Fig. 1b. In the main experiment displayed in Fig. 1a, a mirror 25 cm away retroreflects the 493 nm part of the light collimated with L1, while transmitting the 650 nm part. The mirror is angle-tuned for 180° back-reflection with a precision mirror mount and, for fine adjustment, with two piezo translators (PZTs). The retroreflected light is focused by L1 to the position of the ion and, together with the light emitted directly into that direction, it is collected with a second lens (L2) at -90° to the excitation beams and recorded with a photomultiplier (PM1). Coarse alignment, i.e. superposition of the ion and its mirror image, is controlled visually through L2 while fine adjustment is done by optimizing the signal. The distance between mirror and ion is varied by an amount d (in the range of ±1 μm) by shifting the mirror along the optical axis with another PZT. The 650 nm light transmitted through the mirror is recorded by a second photomultiplier PM2.

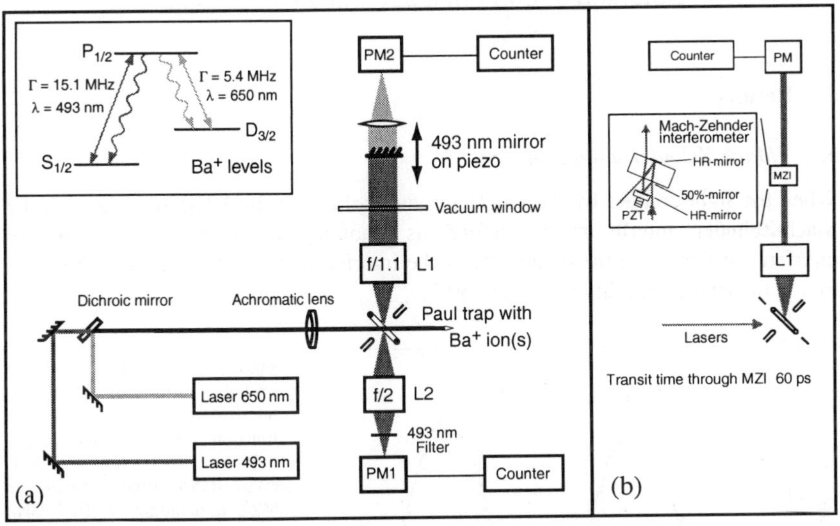

Figure 1. (a) Setup of the backreflection experiment and level diagram for Ba^+. See text for details. (b) Setup of Mach-Zehnder experiment.

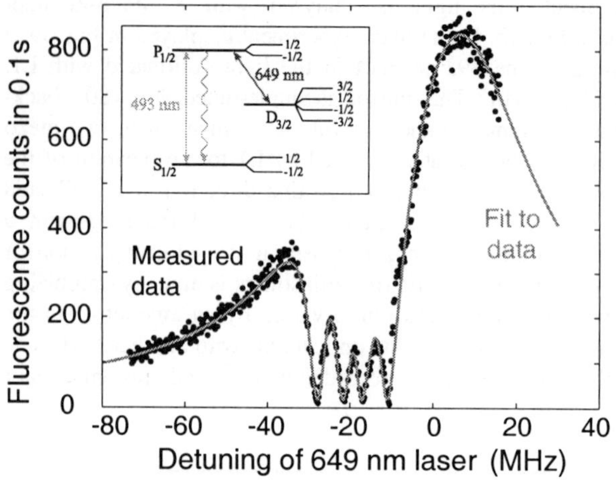

Figure 2. Inset: Levels of Ba^+ with Zeeman structure indicated. Onset: Measured excitation spectrum (points), i.e. fluorescence counts vs. detuning of 650 nm laser, and fit calculated from 8-level optical Bloch equations (line). With the geometry used (quantization axis, laser beam, and polarization mutually orthogonal), four dark resonances are observed when the lasers excite a Raman resonance between an $S_{1/2}$ substate and a $D_{3/2}$ substate.

3 Results

3.1 Mach-Zehnder interferometer

When the light emitted by the single ion and collimated by L1 is sent through the Mach-Zehnder interferometer before its photons are counted, we observe interference fringes in the count rate as the interferometer is scanned. As shown in Fig. 3, the fringe visibility is close to 60%.

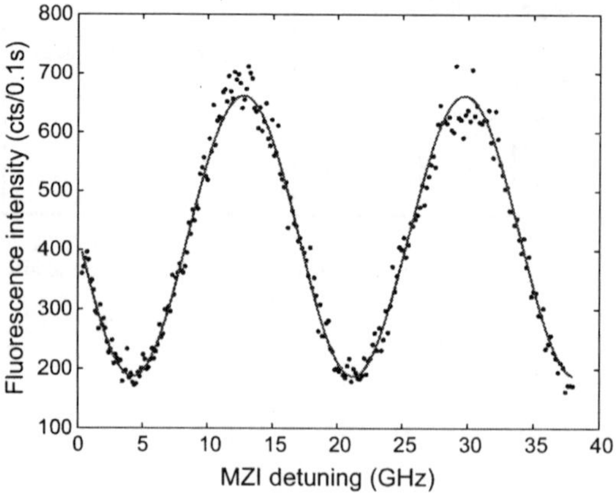

Figure 3. Photon count rate behind the Mach-Zehnder interferometer (MZI), as a function of the path length difference (or detuning). With 60 ps transit time through the MZI, a detuning of 16.7 GHz corresponds to a path length difference of λ = 493 nm.

It is instructive to look at the number of photons inside the interferometer: Its path length corresponds to a transit time of 60 ps, such that at the maximum count rate of $4.6 \cdot 10^3\ s^{-1}$ and with ~ 20% counting efficiency, the average number is ~ $1.4 \cdot 10^{-6}$. The probability for two photons to be in the interferometer simultaneously is correspondingly smaller and it is even further reduced, by a factor $2 \cdot 10^{-3}$, due to the antibunching property of the single ion's resonance fluorescence. Therefore the interference fringes clearly demonstrate that every single photon interferes with itself, and the experiment combines nicely the wave and the particle character of the ion's resonance fluorescence.

3.2 Backreflection experiment with one ion

In this experiment, as displayed in Fig. 1a, a direct and a retroreflected part of the resonance fluorescence of a single Ba^+ ion are recorded together on PM1 while the distance between mirror and ion is varied. A scan of fluorescence vs. mirror shift is shown in Fig. 4. Interference fringes appear which repeat when the mirror is shifted by half the 493 nm wavelength. The interference contrast (or visibility V) in this example is 72%. We have identified various sources of visibility reduction: Residual thermal motion of the ion limits it to 93%, spectral broadening due to inelastic scattering reduces it by another 2%. The remaining reduction is caused by acoustic noise and abberations in the optical system.

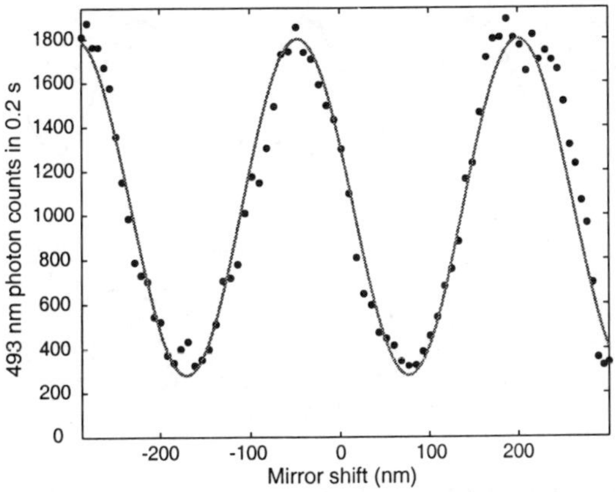

Figure 4. Interference of direct and back-reflected parts of the fluorescence of a single ion: Photon count rate at PM1 vs. mirror displacement (points). The fit (line) accounts for the nonlinear expansion of the PZT with applied voltage.

The observation shows clearly that light from the ion and from its mirror image, i.e. light scattered by the same atom into opposite directions, is coherent and can therefore interfere. While such interference would also be observed if the two light fields were superimposed on a beam splitter, the particular feature of this experiment is that the two fields are superimposed *at the position of the ion*. Thereby, our retroreflecting lens-mirror setup creates a *back-action* on the atom which is a fundamentally different effect. In an intuitive picture this back-action is explained by a modification of the electromagnetic vacuum at the position of the ion: The mirror creates nodes and antinodes in those modes which are collimated by the lens and then retroreflected, among them the modes which are analyzed by the detector. Since the spontaneous emission into any of these modes is proportional to the mode intensity at the position of the ion, we observe reduced or increased fluorescence depending on whether the ion is at a node or antinode, i.e. depending on its distance from the mirror.

If some fraction of the total fluorescence is suppressed or enhanced, we also expect the total rate of fluorescence to vary at roughly the same percentage level. An observation of such a variation would verify that a back-action takes place. Therefore we recorded, simultaneously with the interference fringes, the fluorescence at 650 nm which is transmitted through the mirror (see Fig. 1a) and which is directly proportional to the population of the excited $P_{1/2}$ level of the ion. The result is shown in Fig. 5. The 650 nm fluorescence exhibits a clear ~ 1% sinusoidal variation anticorrelated with the interference signal, indicating that an interference minimum (maximum) at 493 nm leads to higher (lower) population of the excited state. This shows that the mirror 25 cm away in fact acts on the internal atomic dynamics of the ion.

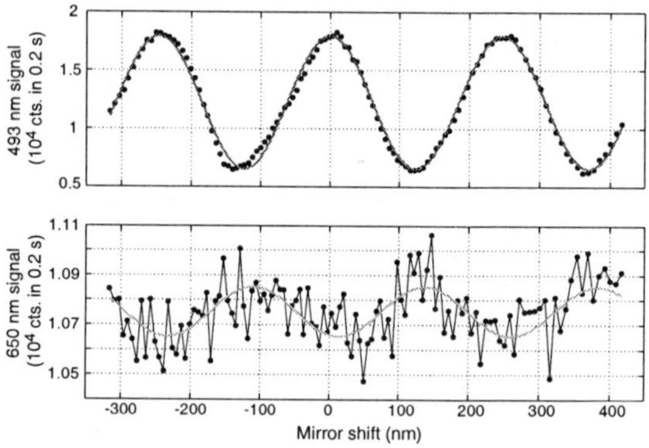

Figure 5. Interference fringes at 493 nm (top) and simultaneously recorded fluorescence at 650 nm transmitted through the mirror (bottom). Points are experimental data, bold lines are fits showing sinusoidal oscillations at the same frequency. The visibility of the modulation is 47% (top) and 0.9% (bottom). Within the experimental error the relative phase of the fits is in agreement with anticorrelation.

3.3 *Backreflection experiment with two ions*

With the same setup as before but with *two* laser-cooled ions in the trap, we adjust the mirror such that the mirror image of each ion is superimposed with the real image of *the other* ion. When we scan the mirror we find a result as displayed in Fig. 6. Again, interference fringes appear with the same period as before and with about 5% contrast. However, their interpretation must be clearly different since it is not light from the same atom that interferes, neither is there a back-action of an atom on itself. Instead, the two indistinguishable processes which create the interference are emission by one ion towards the detector and emission by the other towards the mirror, and the two atoms interact *with each other.*

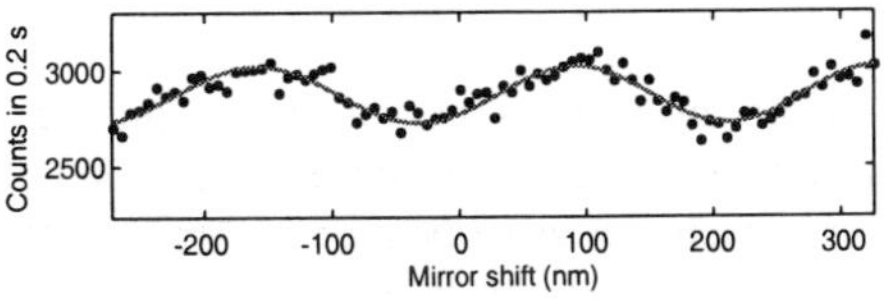

Figure 6. Interference fringes as in Fig. 4 but now with *two* ions, each interfering with the mirror image of the other. The visibility is ~ 5%; the main reason for its reduction, compared to the one-ion experiment, is the strong driven (micro-) motion of the ions in the Paul trap when their mutual repulsion displaces them from the trap center.

4 Model description

To include the back-action (or interaction) created by the mirror into the description of the system we have to take into account that when a photon reaches the detector, the two directions into which it can have been emitted are indistinguishable. This is represented in the Optical Bloch Equations (OBEs) for the atomic dynamics by adding coherently the two decay processes, one of them delayed by the travel time τ to the mirror and back [11].

A corresponding model calculation for the one-ion case, using the parameters of Fig. 5, indeed predicts a variation of the total fluorescence with the ion-mirror distance by 0.9% when the effective fraction of the total emitted light which can be brought to interference is set to 1.7%. Such a variation of the total fluorescence rate due to mirrors or other dielectric boundary conditions is usually called inhibited and enhanced spontaneous emission; it can also be regarded to result from reabsorption or stimulated emission induced by the back-reflected photons.

In the 2-ion case, the two indistinguishable processes which interfere are emission by one ion towards the detector and emission by the other ion towards the mirror. Using the same tools as for a single ion, we now modify the OBEs for the 2-atom density matrix correspondingly, finding that a new term appears in the dynamics which describes simultaneous emission by one ion and absorption by the other and which is modulated with the distance between the ions via the mirror. This shows that in fact reabsorption (and its inhibition) of the emitted photons goes along with the observed interference in the two-ion case. A slightly different viewpoint is that, depending on the delay τ, either the symmetric or the antisymmetric two-atom wave function is preferentially populated, which leads to enhanced or suppressed collective spontaneous emission, respectively. This is sub- and superradiance as originally described by Dicke [2].

In an earlier experiment [13] the corresponding lifetime modification was studied with two ions whose spacing was reduced to about 1.5 μm by a strongly confining trap. In our case, their interaction is mediated by the lens-mirror system over a distance of 50 cm.

5 Discussion

Apart from its fundamental aspects, the experiment has some interesting practical implications: We can regard the setup as a microscope to determine the position of the ion relative to the mirror, the precision of such a measurement being limited only by the noise in the photon counting signal. With the parameters of Fig. 4, we find that the Poissonian counting error translates into an error of the position measurement of only 1.7 nm. This means that within a typical measurement time of 0.1 to 1 s the center position of the ion can be determined more precisely than the extension of its ground state wave packet in the trap (~ 7 nm), which opens up exciting possibilities of measuring and even manipulating the position and motion of the ion on a scale below its position uncertainty. In the same sense our interference signal reveals spatial variations in the electromagnetic mode structure around the ion on a sub-optical scale; here the resolution is set by the thermal wave packet (~ 35 nm). A related study, using a Ca^+ ion in an optical resonator, has been presented in another contribution to this conference [16].

We have also observed that the interference signal can be used as a very sensitive detector for the driven (micro-) motion of the ion in the Paul trap. In turn, observation of that motion reveals more details of the electromagnetic mode structure around the ion. This will be the scope of future studies.

In summary, we have used a retroreflecting mirror at a distance of 25 cm to suppress or enhance the spontaneous emission events of a single trapped atom into the retroreflected modes by up to 72%. The total spontaneous emission rate is modified by ~ 1%. When two atoms are trapped, their spontaneous decay can be correlated via the mirror to produce sub- and superradiant emission. The experiments highlight the intimate relation between these fundamental quantum optical one- and two-atom effects. They are also very encouraging in the view of currently ongoing efforts to couple one or two single atom(s) to the mode of a high-finesse cavity, which is an important step in experimental quantum information processing.

6 Acknowledgements

We thank Dietrich Leibfried, Peter Zoller, Giovanna Morigi, Ignacio Cirac, and Uwe Dorner for stimulating discussions and helpful comments. We gratefully acknowledge support by the European Commission (TMR network QSTRUCT, ERB-FMRX-CT96-0077), by the Austrian Science Fund (FWF, P11467-PHY and SFB15), and by the Institut für Quanteninformation GmbH.

References

1. E. M. Purcell, Phys. Rev. **69**, 681 (1946); P. W. Milonni, *The Quantum Vacuum* (Academic, San Diego, 1994), in particular Ch. 6.
2. R. H. Dicke, Phys. Rev. **93**, 99-110 (1954).
3. W. Neuhauser, M. Hohenstatt, P. Toschek, H. Dehmelt, Phys. Rev. A **22**, 1137–1140 (1980).
4. See contribution to this book by J. Bergquist and coworkers from NIST Boulder.
5. F. Diedrich, J. C. Bergquist, W. M. Itano, and D. J. Wineland, Phys. Rev. Lett. **62**, 403-406 (1989); C. Monroe, D. M. Meekhof, B. E. King, S. R. Jefferts, W. M. Itano, D. J. Wineland, P. Gould, Phys. Rev. Lett. **75**, 4011–4014 (1995); B. E. King, C. S. Wood, C. J. Myatt, Q. A. Turchette, D. Leibfried, W. M. Itano, C. Monroe, and D. J. Wineland, Phys. Rev. Lett. **81**, 1525-1528 (1998).
6. Ch. Roos, Th. Zeiger, H. Rohde, H. C. Nägerl, J. Eschner, D. Leibfried, F. Schmidt-Kaler, R. Blatt, Phys. Rev. Lett. **83**, 4713-4716 (1999); H. Rohde, S. T. Gulde, C. F. Roos, P. A. Barton, D. Leibfried, J. Eschner, F. Schmidt-Kaler, R. Blatt, J. Opt. B **3**, S34-S41 (2001).
7. C. Monroe, D. M. Meekhof, B. E. King, W. M. Itano, D. J. Wineland, Phys. Rev. Lett. **75**, 4714–4717 (1995).
8. C. A. Sackett, D. Kielpinski, B. E. King, C. Langer, V. Meyer, C. J. Myatt, M. Rowe, Q. A. Turchette, W. M. Itano, D. J. Wineland, C. Monroe, Nature **404**, 256-259 (2000).
9. W. Nagourney, J. Sandberg, and H. Dehmelt, Phys. Rev. Lett. **56**, 2797-2799 (1986); Th. Sauter, W. Neuhauser, R. Blatt, and P. E. Toschek, Phys. Rev. Lett. **57**, 1696-1698 (1986); J. C. Bergquist, R. G. Hulet, W. M. Itano, and D. J. Wineland, Phys. Rev. Lett. **57**, 1699-1702 (1986).
10. F. Diedrich and H. Walther, Phys. Rev. Lett. **58**, 203-206 (1987).
11. J. Eschner, C. Raab, F. Schmidt-Kaler, R. Blatt, accepted for publication in Nature.
12. K. H. Drexhage, in *Progress in Optics* **12**, 163-232, ed. E. Wolf (North-Holland, Amsterdam, 1974); F. DeMartini, G. Innocenti, G. R. Jacobovitz, P. Mataloni, Phys. Rev. Lett. **59**, 2955-2958 (1987); W. Jhe, A. Anderson, E. A. Hinds, D. Meschede, L. Moi, S. Haroche, Phys. Rev. Lett. **58,** 666-669 (1987); D. J. Heinzen, J. J. Childs, J. F. Thomas, M. S. Feld, Phys. Rev. Lett. **58**, 1320-1323 (1987); C. J. Hood, T. W. Lynn, A. C. Doherty, A. S. Parkins, H. J. Kimble, Science **287**, 1447-1453 (2000); P. W. H. Pinkse, T. Fischer, P. Maunz, G. Rempe, Nature **404**, 365-368 (2000); P. Goy, J. M. Raimond, M. Gross, S. Haroche, Phys. Rev. Lett. **50**, 1903-1906 (1983); R. G. Hulet, E. S. Hilfer, D. Kleppner, Phys. Rev. Lett. **55**, 2137-2140 (1985); G. Rempe, H. Walther, N. Klein, Phys. Rev. Lett. **58**, 353-356 (1987); G. Gabrielse, H. G. Dehmelt, Phys. Rev. Lett. **55**, 67-70 (1985).
13. R. G. DeVoe, R.G. Brewer, Phys. Rev. Lett. **76**, 2049-2052 (1996).
14. C. Raab, J. Bolle, H. Oberst, J. Eschner, F. Schmidt-Kaler, R. Blatt, Appl. Phys. B **67**, 683-688 (1998) and Appl. Phys. B **69**, 253 (1999); C. Raab, J. Eschner, J. Bolle, H. Oberst, F. Schmidt-Kaler, R. Blatt, Phys. Rev. Lett. **85**, 538-541 (2000).
15. M. Schubert, I. Siemers, R. Blatt, W. Neuhauser, P. E. Toschek, Phys. Rev. A **52**, 2994-3006 (1995).
16. G. R. Guthöhrlein, M. Keller, W. Lange, H. Walther, K. Hayasaka, see their proceedings article in this book.

ADVANTAGES AND LIMITS TO LASER COOLING IN OPTICAL LATTICES

DAVID S. WEISS

Penn State Physics Dept., 104 Davey Lab., University Park, PA 16802, USA

E-mail: dsweiss@phys.psu.edu

Atoms laser cooled in optical lattices have reached higher densities and lower temperature than other laser cooling methods have for comparable numbers of atoms. We will review why this is so, and discuss the limitations of laser cooling in optical lattices. We will also discuss how to compare the phase space density of cold atoms in deep lattice potentials to atoms in free space or simple traps.

1 Introduction

The laser cooling of optically thick samples of atoms works better when they are confined in deep, far-off-resonant 3D optical lattices. Once atoms are isolated at separate lattice sites, they do not collide with each other. For this reason, and because they neither diffuse nor ballistically expand, the density is completely maintained during the cooling. For laser cooling methods that do not involve cooling to dark states, like polarization gradient cooling, a large fraction of atoms can be shelved in dark states, so that cooling is minimally affected by the rescattering of spontaneously emitted photons [1]. For methods that laser cool to a dark state, like 3D Raman sideband cooling [2-3], the tight binding of atoms at lattice sites can suppress the heating from rescattered photons [4]. After cooling, the localization of atoms at sites can be adiabatically converted to a lower temperature when the lattice is shut off, without compromising the bulk density of the sample [5].

Optical lattices also present a way to increase the spatial density of atomic samples, by dynamically changing the lattice dimension [6]. Combining increased density with decreased temperature, laser cooling in optical lattices has reached a phase space density of 1/30 for $5\text{x}10^7$ [3], or 1/100 for $3\text{x}10^8$ atoms [2]. These results exceed by several orders of magnitude what has been accomplished for this many atoms without optical lattices.

A natural question is: what limits the phase space density obtained from these methods? We will discuss the three principal issues related to this larger question: what peak lattice occupation can be obtained; how close to the vibrational ground state can atoms in an optical lattice be cooled; and what is the phase space density of a particular atomic distribution in an optical lattice.

Before proceeding, it is useful to point out that a high 3D lattice phase space density can also be obtained by starting with a high phase space density sample outside of a lattice, like a Bose Einstein condensate, and transferring those atoms into a 3D optical lattice. Such an experiment has already been accomplished with a

1D lattice [7], and the extension to 3D presents exciting prospects. It will almost certainly will be accomplished somewhere soon. The focus of this paper, however, is more closely related to the reverse problem. Can one create a Bose-Einstein condensate by cooling atoms in an optical lattice?

2 What is the highest obtainable optical lattice filling fraction?

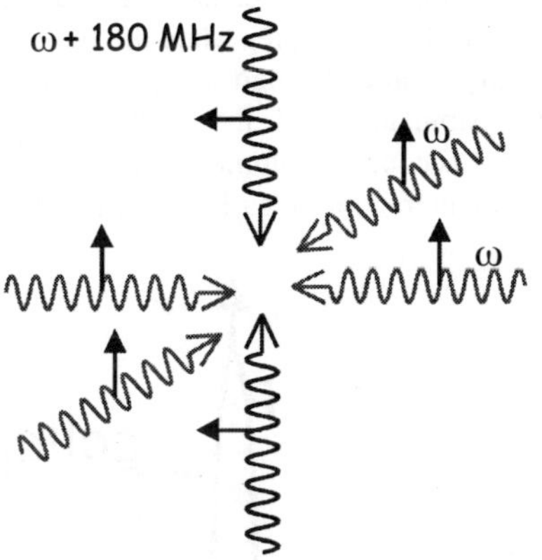

Fig. 1: Schematic of our 3D optical lattice, which consists of a vertically oriented 1D lattice overlapped with a horizontally oriented 2D lattice of somewhat different optical frequency.

The answer to this question requires a brief review of how the highest occupation to date has been obtained [6]. Cs atoms are first polarization gradient cooled in a far-off-resonant 3D optical lattice, which is made by overlapping a vertically oriented 1D lattice with a horizontally oriented 2D lattice of somewhat different frequency (see Fig. 1). Since the two sets of beams do not interfere, the total potential is just the sum of the two [1]. Our lattice is made with 2 W of laser light in orthogonal, nearly collimated 0.5 mm waist beams, red detuned from the $6S_{1/2}$ to $6P_{3/2}$ transition by 30,000 linewidths. The lattice potential is hundreds of microkelvin deep, and essentially the same for all ground states. Polarization gradient cooling in the far-off-resonant lattice is as simple as briefly turning back on the MOT laser beams. It achieves 70% vibrational ground state occupation in each direction.

After cooling, the 2D lattice is shut off adiabatically, which leaves atoms still tightly confined in the vertical direction by the 1D lattice, but relatively free to move in horizontal planes. The atomic distribution is like a stack of pancakes. The atomic temperature in the plane is less than 500 nK, which is much less than the ~250 μK depth of the trap in the horizontal direction due to the Gaussian profile of the 1D lattice beams. In the horizontal plane the atomic distribution is momentum squeezed, as illustrated in Fig. 2. After the 2D lattice is shut off the atoms start to collapse into the center of the 1D lattice beams. The pancakes decrease their radii while maintaining their thickness. Because only the central part of the Gaussian is occupied, the potential is nearly harmonic, so if there were no atom-atom interactions the density in the center of the 1D lattice beams would peak one quarter of the horizontal oscillation period later. The density does, in fact, peak at that time, increasing by an order of magnitude over its initial value, to a density that is 10% higher than the density of 3D lattice sites when the full 3D lattice is on, or $7.5\text{x}10^{12}$ atoms/cm^3. This density is already high compared to what laser cooled systems typically achieve, but simulations of the collapse that take into account trap anharmonicity but do not account for atomic collisions, predict a peak density seven times higher. Atom-atom interactions clearly limit the peak density.

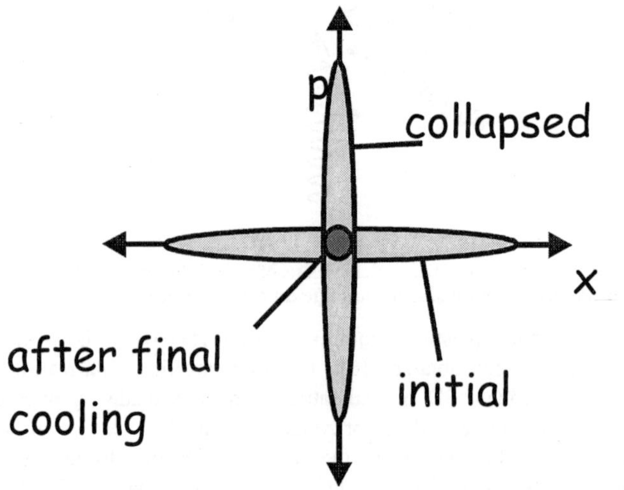

Figure 2: Cartoon to illustrate the distribution of atoms before and after the pancake collapse. Upon shutoff of the 2D lattice, the atoms are momentum squeezed in the horizontal plane. At the peak of collapse they are position squeezed. The final cooling returns them to their low initial temperature, without changing the final spatial distribution. In reality, primarily because of collisions, the phase space density actually drops during the collapse.

Less than 10% of the atoms are lost to inelastic collisions. The details of how elastic collisions limit the density have not been fully studied. The densities are sufficiently high that the sample is hydrodynamic, i.e., each atom is involved in several collisions during the collapse. Although the two-dimensional character of the potential may be important to the dynamics, the system is not strictly two dimensional, because there is non-negligible occupation of excited transverse vibrational levels and because they have enough energy by the time they reach the center to cause further transverse excitation. We can report some qualitative observations about the collapse that are consistent with the gas having finite bulk compressibility. The more initial kinetic energy the atoms have, the higher the peak density in the collapse. For instance, small dense clouds do not collapse to as high a peak density in the same potential as large dilute clouds, and the peak density is lower when the same initial distribution collapses in a shallower potential. It is also interesting to note that when the atoms are made to collapse in 1D filaments, which occurs when the 2D lattice is kept on and the 1D lattice is shut off, the peak density they reach is less than half of the pancake collapse. Since the peak density depends on collision cross sections, it no doubt depends on the type of atom. To date, we have only studied Cs collapse, and the largest peak density during collapse that we have obtained is $1\text{x}10^{13}$ atoms/cm^3. It seems likely that atoms with smaller scattering lengths will collapse to much higher densities.

To convert whatever spatial density is obtained after a collapse into higher lattice site occupation, the full 3D lattice potential can be adiabatically turned back on. The turn-on can be slow compared to typical lattice oscillation frequencies of 200 kHz, while still fast compared to the oscillation in the beam waist scale trap, which is ~100 Hz. When this was done we were able to infer that 90% of the atoms became immediately bound at sites, while the rest of the atoms only became bound during subsequent laser cooling. We found that when the laser cooling light was turned back on there was a rapid loss of atoms, in pairs, at multiply occupied sites, so that in the end, 44% of the sites were occupied by a single atom. If atoms at the same site are always lost in pairs during laser cooling, clearly the maximum obtainable occupation after cooling is 50%. In this procedure, therefore, higher initial densities would not translate into a higher site occupation after cooling.

There are conceivable ways to overcome this limit. The most straightforward would be to reduce the loss of atoms during cooling, so that multiple atoms could be cooled at the same site. For Cs, loss of pairs that were not even already near the vibrational ground state occurred in about 10 photon scatters during polarization gradient cooling. When both atoms have been cooled to near the ground state the inelastic collision rate would be even higher. Since the frequency of the optical pumping light in Raman cooling is not particularly constrained, one could imagine that it could be done at a frequency where the photo-assisted collision rate is very small. It is not clear that such a magic frequency exists, although one might guess that if it did it would be close to resonance, so that molecular states would tend to form where the 6S+6P potential is flattest. Perhaps the photo-associative rate is already small enough for other atoms.

Another possibility would be if the initial distribution of lattice bound atoms after the collapse were non-Poissonian. Mean field energies cannot compete with the ~100 μK potentials that are required to rebind atoms at 3D lattice sites, but perhaps if the collapse was more gentle, it would be possible to bring these two energy scales closer together. A more gentle collapse does not lead to a sufficiently high density in Cs, but perhaps it would in other species. Non-Poissonian distributions might also be obtained by dramatically increasing the mean field energy after the collapse but just before the full 3D lattice is turned back on. This might be accomplished by optical excitation into strongly repulsive states.

Of course, if the size of the ground vibrational state were sufficiently large, multiple atoms could be cooled at a single site. In fact, this has been accomplished with CO_2 laser-based lattices [8] and by using small angles between shorter wavelength lattice beams [9]. It is more difficult, although certainly not impossible, to reach the Lamb-Dicke limit in these experiments, which limits the laser cooling effectiveness. To the extent that the Lamb-Dicke limit is reached, the density may still be too high for laser cooling multiple atoms at a site. Were atoms to be cooled in large scale lattices, then adiabatically shutting off the lattice (which is discussed in section 4) might take a prohibitively long time, unless steps are taken to adiabatically reduce the lattice spacing after laser cooling.

3 How close to the vibrational ground state can lattice bound atoms be cooled?

The highest vibrational ground state occupations have been obtained using Raman sideband cooling. In brief, Raman sideband cooling works by first optically pumping to a single internal energy state, which can be the lowest energy state of the system [2-3,10]. Next, stimulated transitions are made that reduce the vibrational energy while changing the atoms' internal state. Then the atoms are optically pumped back to the initial internal state, and the cycle is repeated. In the Lamb-Dicke limit, which is naturally obtained in optical lattices, the vibrational state tends not to change during optical pumping, so repeated application of these two steps steadily decreases vibrational energy. Because atoms in the initial internal state and the ground vibrational level are dark to both the stimulated transition between ground states and the optical pumping, atoms accumulate in that state.

Since Raman sideband cooling is a dark state cooling method, there is in principle no limit to the achievable temperature, and the ground state occupation can be complete. But there are practical limitations. Off resonant excitation on other hyperfine transitions can depopulate the dark state to degrees that vary with the particular transitions being used. The dominant problem is imperfect polarization of the optical pumping light, which can provide a path out of the desired dark state. This harmful rate can be made less than a hundredth of the desired optical pumping rate, which at first glance would lead one to expect on the order of 99% ground state occupation. However, the problems are significantly compounded by rescattering at high densities, because it can require many optical pumping attempts to get a net increase of one atom in the dark state. This was the limiting factor in our previously published 3D Raman sideband cooling result, where we obtained 37% 3D vibrational ground state occupation with 25% lattice occupation and 5×10^7 atoms, and 55% occupation with half that number and density [3]. Our subsequent studies of suppression of rescattering by weakly optical pumping in optical lattices suggest that with the appropriate optical pumping parameters, the rescattering problems will be overcome [4]. Ultimately, it should be possible to cool even densely packed atoms almost entirely to the vibrational ground state.

4 What is the phase space density of cold atoms in an optical lattice?

For this discussion, we will consider a distribution of lattice bound atoms where each site has either zero or one atom, and each atom is in its vibrational ground state. This is precisely the type of state that can be obtained by the methods discussed above. We will consider a uniform lattice with a well-defined boundary, which is essentially equivalent for our purposes to an infinite lattice. The atoms are bosons with positive scattering length, so that they have a repulsive long range interaction. Work is in preparation and will be published elsewhere to deal with more realistic lattice profiles and with imperfect cooling at each site.

There are

$$\Omega = \frac{N_L!}{(N - N_L)!N!}$$

ways to distribute N atoms among N_L lattice sites. For a given implementation of this cooling method, atoms are left tightly bound at a particular set of sites. Since the exact location of these atoms can in principle be determined and does not change over time, it has zero entropy. We can, however, assign an effective entropy to the distribution based on the number of ways that such a distribution can be realized,

$$S = k_B \ln \Omega \approx k_B \mathrm{N}(\mathrm{x} \ln \mathrm{x}\text{-}(\mathrm{x}-1)\ln(\mathrm{x}-1),$$

where $x=N_L/N$. Now imagine that the lattice potential is slowly lowered until it is shut off. By slow, we mean that the changes are quasi-reversible, so that if we were to turn back on the lattice, the atoms would return to one of the Ω possible lowest energy states, with all sites singly occupied and all atoms in the local vibrational ground state. While the potential is so deep that the mean field energy of two atoms at a site greatly exceeds the characteristic tunneling energy, reversibility simply requires that changes be made slowly compared to the oscillation period at individual sites. When these energies become comparable, quasi-reversibility requires that changes be made much more slowly than tunneling times across the lattice. Roughly speaking, once atoms are free to tunnel among sites, they must be allowed to do so before the potential is changed too much. If x=1, this transition is exactly the Mott insulator to superfluid transition that is discussed in Ref. [11]. A lattice with one atom in the vibrational ground state at every site transforms into a pure BEC.

In general, the entropy of a degenerate Bose gas is given by

$$S = 1.28 k_B \mathrm{N} \left(\frac{\mathrm{T}}{\mathrm{T}_0} \right)^{3/2}$$

where T_0 is the critical temperature [12]. At the BEC transition point, $S=1.28k_BN$. Setting that entropy equal to the effective entropy of the initial lattice state, and solving for x, shows that a lattice occupancy of 0.538 corresponds to the BEC transition point. Interestingly, this is just outside the range of the lattice-based cooling methods described above.

With further refinement, it looks possible to use these methods to exceed the BEC threshold. It is not the total entropy that is ultimately important for determining whether condensation will occur, but the peak phase space density. The trick will be to use a flat top bulk distribution in the lattice, which tends to occur naturally when the spatial density before the final laser cooling significantly exceeds the density of lattice sites. When such a distribution is adiabatically transformed into a Gaussian trap, the phase space density at trap center is higher than the average phase space density in the trap, and the initial occupancy required to reach BEC can be much lower. These calculations will appear elsewhere.

5 Conclusion

In summary, we have reviewed how far-off-resonant optical lattices can aid laser cooling. The present inability to laser cool two Cs atoms at micron-scale optical lattice site leads to a maximum laser cooled site occupancy of 50%. We have suggested ways to overcome this limit with different atomic species or with lattices with larger spacing. Although ground vibrational state occupations have been limited to around 50% for more than $3\text{x}10^7$ atoms in 3D lattices, more recent experiments suggest that it is possible to significantly exceed 90% [4]. We have showed that atoms cooled to the ground state in a uniform lattice with 53.8% occupancy will be at the threshold of BEC when the lattice is shut off slowly enough. BEC might be reached by these methods by either increasing the site occupancy or by starting with a non-Gaussian distribution in the optical lattice.

1. S.L. Winoto, M.T. DePue, N.E. Bramall, and D.S. Weiss, *Phys. Rev. A*, **59**, R19 (1999).
2. A.J. Kerman, V. Vuletic, C. Chin, and S. Chu, *Phys. Rev. Lett.* **84**, 439 (2000).
3. D.J. Han, S. Wolf, S. Oliver, C. McCormick, M.T. DePue, and D.S. Weiss, *Phys. Rev. Lett.* **85**, 724 (2000).

4. S. Wolf, S.J, Oliver and D.S. Weiss, *Phys. Rev. Lett.* **85**, 4249 (2000).
5. A. Kastberg, W.D. Phillips, S.L. Rolston, R.J.C. Spreeuw, and P.S. Jessen, *Phys. Rev. Lett.* **74**, 2253 (1995).
6. M.T. DePue, C. McCormick, S.L. Winoto, S. Oliver, and D.S. Weiss, *Phys. Rev. Lett.* **82**, 2262 (1999).
7. C. Orzel, A. Tuchman, M. Fenselau, M. Yasuda, and M. Kasevich, *Science* **291,** 2386 (2001); B.Anderson and M. Kasevich, *Science* **282**, 1686 (1998).
8. S. Friebel, R. Scheunemann, J. Walz, T. Hansch, and M. Weitz, *Appl. Phys. B* **67**, 699 (1998).
9. D. Boiron, A. Michaud, J.M. Fournier, L. Simard, M. Sprenger, G. Grynberg, and C. Salomon, *Phys. Rev. A* **57**, R4106 (1998).
10. C. Monroe, D. Meekhof, B. King, S. Jefferts, W. Itano, D. Wineland, and P. Gould, *Phys. Rev. Lett.* **75**, 4011 (1995); S. Hamann, D. Haycock, G. Klose, P. Pax, I. Deutsch, and P. Jessen, *Phys. Rev. Lett.* **82**, 4149 (1998); H. Perrin, A. Kuhn, I. Bouchoule, and C. Salomon, *Europhys. Lett.* **42**, 395 (1998); V. Vuletic, C. Chin, A. Kerman, and S. Chu, *Phys. Rev. Lett.* **81**, 5768 (1998).
11. D. Jaksch, C. Bruder, J. Cirac, C. Gardiner, and P. Zoller, *Phys. Rev. Lett.* **81**, 3108 (1998).
12. Landau and Lifshitz, *Statistical Physics, 3rd Edition Part 1,* (Pergamon: New York, 1980) p.182.

COHERENT TUNNELING AND QUANTUM CONTROL IN AN OPTICAL DOUBLE-WELL POTENTIAL

P. S. JESSEN, D. L. HAYCOCK, G. KLOSE AND G. SMITH
Optical Sciences Center, University of Arizona, Tucson, AZ 85721, USA
E-mail: poul.jessen@optics.arizona.edu

P. M. ALSING, I. H. DEUTSCH, J. GRONDALSKI AND S. GHOSE
Dept. of Physics and Astronomy, University of New Mexico, Albuquerque, NM 87131, USA
E-mail: ideutsch@tangelo.phys.unm.edu

Quantum coherence and tunneling of a particle in a double-well potential is central to our understanding of many physical phenomena. We have developed a laboratory realization of this "toy model", based on laser cooled Cs atoms in a one-dimensional optical lattice. The system states are spinor wavepackets with highly entangled internal and external degrees of freedom, allowing us to use the atomic spin as a "meter" to observe the center-of-mass motion. With appropriate initial condition the system Rabi oscillates between right- and left-localized states of the ground doublet. In a separate experiment we have measured the entire density matrix of the atomic angular momentum, thereby extending the use of the spin as a meter and opening up new possibilities to study phenomena such as quantum chaos and decoherence in the double-well system.

1 Introduction

An extrapolation of current trends in modern nanoscience suggests that transistors and other circuit elements will reach the single-electron limit within the next 15 to 20 years, and atomic dimensions in a few additional decades. At some point along this path technology will cross the quantum/classical boundary. What lies ahead, then, is not merely miniature versions of familiar classical devices, but rather a whole new world of quantum systems and coherent quantum dynamics. It is now known that machines whose components and evolution are manifestly quantum can perform tasks which are impossible according to classical physics, such as certain computations and secure communication, teleportation and measurements beyond the standard quantum limit [1]. To harness this power for real-world applications is a formidable challenge, which will require, along the way, the development of methods to accurately prepare, manipulate and measure the state of complex, mesoscopic quantum systems.

Dynamics associated with a particle in a double-well potential has long served as a paradigm for quantum coherent evolution. In this article we describe a particular implementation of this "toy model", based on atomic spinor wavepackets trapped in a magneto-optical lattice [2]. Our system is *mesoscopic*, in the sense that the characteristic action can be varied continuously across the quantum/classical boundary (the range $0.1-10\hbar$). It is also *complex*, in the sense that the spinor wavepackets represent highly entangled states of the atomic center-of-mass and external degrees of freedom, and that the coupling of the two leads to dynamics whose classical counterpart exhibits deterministic chaos [3,4]. Far-off-resonance optical lattices offer an excellent platform for the study of this and similar model systems: the atom-laser interactions can be accurately modeled, the optical potential can be designed with great flexibility, the coherence times can be very long, and dissipation can be added via laser cooling and heating. Last but not least, we can apply a range of powerful experimental tools, including pure state preparation, controlled unitary evolution and quantum state reconstruction.

2 Optical Lattices for Alkali Atoms

In the case of an alkali atom in a laser field tuned close to the D2 resonance line, at a detuning much larger than the excited state hyperfine splitting but less than the fine structure splitting, the light shift potential can be cast in the form

$$\hat{U}(z) = U_J(z) + g_F \mu_B \hat{\mathbf{F}} \cdot \mathbf{B}_{eff}(z),$$

$$U_J(\mathbf{x}) = \frac{2U_1}{3} |\vec{\varepsilon}_L(\mathbf{x})|^2, \quad \mathbf{B}_{fict}(z) = -i\frac{U_1}{3\mu_B}[\vec{\varepsilon}_L^*(\mathbf{x}) \times \vec{\varepsilon}_L(\mathbf{x})], \quad \mathbf{B}_{eff}(z) = \mathbf{B}_{fict}(z) + \mathbf{B} \tag{1}$$

where $\hat{\mathbf{F}}$ is the angular momentum operator for the hyperfine ground state F, U_1 is the light shift for a transition with unit oscillator strength, and $\vec{\varepsilon}_L(\mathbf{x})$ is the local polarization of the light field [5]. Eq. 1 allows for the presence of an externally applied, uniform magnetic field $\mathbf{B}$, in which case $\hat{U}(z)$ is properly described as a *magneto-optical* lattice. In the following we are concerned with a simple one-dimensional configuration of counter-propagating laser beams with linear polarizations forming an angle θ (the so-called 1D lin-θ-lin lattice). The resulting light field can be decomposed into standing waves of positive and negative helicity (Fig. 1), whose nodes are separated by $\Delta z = \theta / k$. One typically defines *diabatic* and *adiabatic* potentials as the diagonal elements (in the basis $\{|m_F\rangle\}$) and eigenvalues of $\hat{U}(z)$ respectively. The former govern the motion when the atomic internal state is independent of z, the latter when it adiabatically follows the direction of $\mathbf{B}_{eff}(z)$ (the Born-Oppenheimer approximation). In a lattice with a transverse field, $\mathbf{B}_\perp \neq 0$, the lowest adiabatic potential forms a periodic array of double-wells. The polarizations on either side of these wells have opposite helicity, so that atomic motion across a well is accompanied by rotation of the spin. This correlation allows us to use the spin as a "meter" to observe the time dependent motion of the atomic wavepacket. To accurately model the system we solve for the eigenvalues (bandstructure) and eigenstates (Bloch spinors) of the full lattice Hamiltonian. For typical experimental parameters the two lowest energy bands of the lattice are flat (indicating negligible tunneling between separate double wells), and split by an energy $\hbar\Omega$ much smaller than the separation to the next excited bands. In this situation it is possible to restrict the dynamics to a subspace spanned

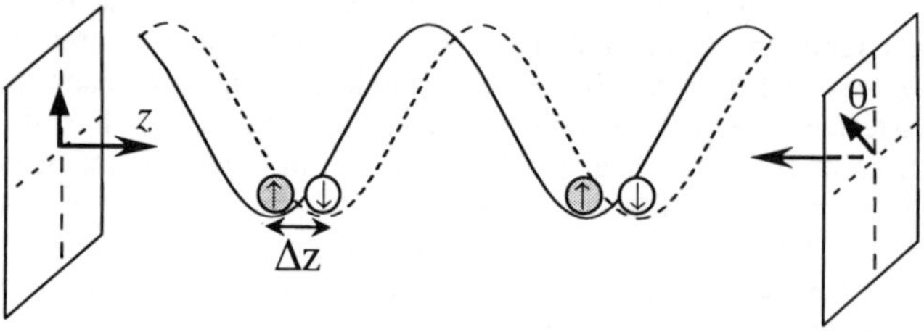

Figure 1. 1D lin-θ-lin lattice. The offset Δz between the two standing wave components is determined by the polarization angle θ. An external magnetic field allows the spin to precess as an atom moves from one side of a double-well to the other.

by a pair of Wannier spinors $\{|\psi_S\rangle, |\psi_A\rangle\}$, which correspond to the symmetric/antisymmetric ground states of individual double-wells. Left and right localized states can then be obtained as the superposition $|\psi_{L/R}\rangle = (|\psi_S\rangle \pm |\psi_A\rangle)/\sqrt{2}$, and one sees immediately that the system will Rabi oscillate between these at frequency Ω. Fig. 2 shows the internal and center-of-mass probabilities for the spinor wavepackets at different times during a Rabi period. Note that the key property of a double-well system is preserved: the spatial probability densities corresponding to $|\psi_{L/R}\rangle$ are localized on the left and right sides of the well, and the minimal overlap between them ensures that the states can be effectively distinguished by the mesoscopic center-of-mass coordinate. Fig. 2 also illustrates the high degree of entanglement between spin and space degrees of freedom, especially in the "Schrödinger Cat"-like superposition found at a quarter Rabi period.

3 Coherent Tunneling in a Magneto-Optical Double-Well Lattice

To observe tunneling we prepare Cs atoms in a well defined initial quantum state, say $|\psi_L\rangle$, and follow their subsequent quantum coherent evolution. We start out by laser cooling in a standard magneto-optical trap/3D molasses followed by further cooling in a near-resonance 1D lin-θ-lin lattice, from which atoms are then transferred to a superimposed far-off-resonance 1D lin-θ-lin lattice. Once in the far-off-resonance lattice the atoms are optically pumped to $|m_F = 4\rangle$, and the motional ground state in the corresponding potential is selected by lowering the depth and accelerating the lattice to allow atoms outside the ground band to escape. Optical pumping and state selection is done in the presence of a large external field B_z to lift degeneracies between the optical potentials and prevent precession of the magnetic moment. When the population outside the ground band has been eliminated we increase the lattice depth, change the lattice acceleration to free-fall, ramp up the transverse field B_x and finally ramp B_z to zero. This sequence adiabatically connects the quantum state prepared by cooling, optical pumping and state selection (the ground state in the $m_F = 4$ potential) to the left-localized state of the optical double well.

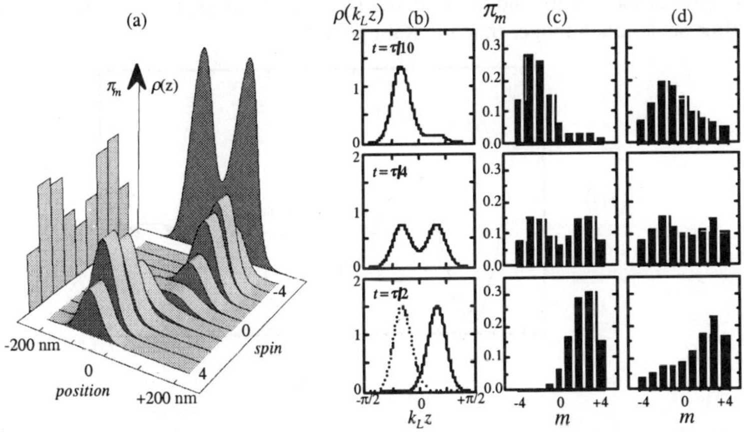

Figure 2. Spinor wavepackets during a Rabi oscillation between left and right localized states. At a quarter period the state is a Schrödinger Cat-like superposition. (a) Center-of-mass probability density distribution for each magnetic sublevel in the "cat" state, showing the correlation between spin and spatial degrees of freedom. The marginal distributions in spin and space are projected onto the walls. (b) Center-of-mass probability densities obtained by tracing over the spin degree of freedom. The dotted curve indicates the wavepacket at $t = 0$. (c) Magnetic populations obtained by tracing over the center-of-mass degree of freedom. (d) Experimentally measured populations.

We detect Rabi oscillations between $|\psi_L\rangle$ and $|\psi_R\rangle$ by measuring the time-dependent magnetization of the atomic ensemble. A Stern-Gerlach measurement is easily accomplished by releasing the atoms from the lattice, quickly applying a bias field B_z to keep the quantization axis well defined, and letting the atoms fall to a probe beam in the presence of an inhomogeneous magnetic field. The magnetic populations can then be extracted from the separate arrival time distributions for different magnetic sublevels. Fig. 3 shows a typical oscillation of the atomic magnetization as a function of time. Our data fits well to a damped sinusoid, and allows us to extract a good measure for the Rabi frequency. We find generally excellent agreement between the measured Rabi frequencies and a bandstructure calculation with no free parameters, over a wide range of experimental conditions. Direct examination of the magnetic populations during Rabi oscillation is consistent with a "Schrödinger Cat"–like superposition at the appropriate time (Fig. 2d). In this first experiment the Rabi oscillations dephase on a timescale of a few hundred microseconds, most likely due to variations in the Rabi frequency across the atomic sample. The probable cause is an estimated ~5% variation of the lattice beam intensities, which is consistent with the observed dephasing times. We estimate the timescale for decoherence due to photon scattering to be of order ~1ms, which is too slow to account for the observed damping. A next generation of the experiment is now underway, in which we hope to increase the dephasing time by an order of magnitude by better control over lattice beam and magnetic field inhomogeneities.

With better homogeneity and larger detuning we hope to explore coherent dynamics on a time scale much longer than the Rabi period.

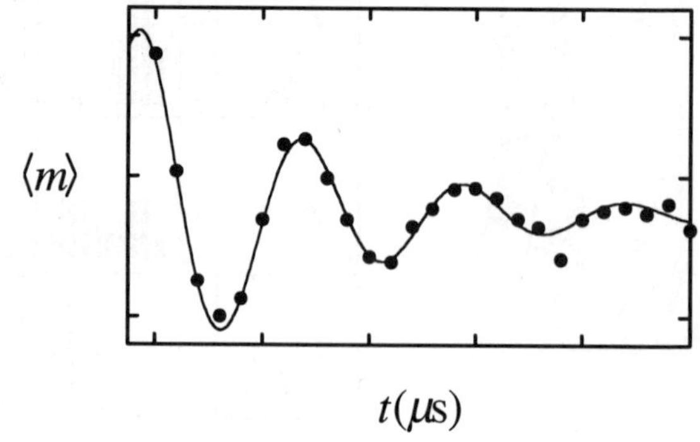

Figure 3. Typical oscillation of the atomic magnetization as a function of time. The solid line is a fit to a decaying sinusoid.

4 Measuring the Quantum State for a Large Angular Momentum

In our experiments so far we have followed the coherent evolution of spinor wavepackets by the simple expedient of measuring the magnetic populations. This type of measurement cannot, in and by itself, distinguish a coherent superposition from an incoherent mixture, and quantum coherence must instead be inferred from the presence of Rabi oscillations at later times. However, the correlation between internal and external degrees of freedom suggests that we extend the use of the spin meter to directly measure coherence in the double-well system. Motivated in part by these considerations we have developed an experimental method to reconstruct the complete density matrix for a large angular momentum, in our case the $6S_{1/2}(F=4)$ hyperfine ground manifold of Cs [6]. We also expect the technique to become a valuable tool to evaluate quantum logic gates for neutral atoms [7].

Reconstruction of a (generally mixed) quantum state is a nontrivial problem with no general solution, though system specific algorithms have been found in a limited number of cases [8]. Our reconstruction scheme relies on repeated Stern-Gerlach measurements with respect to many different quantization axes. These measurements can be performed in a straightforward manner, by choosing different directions for the magnetic bias field which determines the quantization axis for the Stern-Gerlach analyzer. The algorithm requires that one measures the $2F+1$ magnetic populations for $4F+1$ different orientations of the quantization axis, characterized by polar angles θ and azimuthal angles φ. If one chooses a common θ and $4F+1$ evenly separated φ in the interval $[0, 2\pi]$ it is possible to derive analytical expressions for the $(2F+1)^2$ elements of the density matrix in terms of the $(2F+1)(4F+1)$ measured populations [9]. In practice it is better to relate the measured populations to the unknown elements of the density matrix by means of a non-square matrix whose elements are determined by the angles θ and φ, and then invert the problem numerically using the Moore-Penrose pseudoinverse [6]. This gives us the freedom to choose angles that are more suitable for the experimental geometry at hand, while at the same time decreasing the sensitivity to experimental and numerical errors. A drawback of the inversion method is that it occasionally results in a non-positive and therefore non-physical density matrix. If we find this to be the case for a particular data set we instead parameterize the density matrix in terms of $(2F+1)^2-1$ real-valued variables in a way that ensures Hermiticity, unit trace and positivity, and do a best fit to the measured data. This method finds the best *physical* density matrix to fit the measurement, and often gives a better match between the input and reconstructed state.

In order to evaluate the performance of our reconstruction procedure we apply it to a number of input states for which the density matrix is known with good accuracy. As an example we can prepare a test state closely approximating $|m_F=-4\rangle$ by optical pumping with σ_- polarized light tuned to the $6S_{1/2}(F=4)\rightarrow 6P_{3/2}(F'=4)$ transition. The quality of this input can be easily verified through a single Stern-Gerlach measurement along $\hat{z}$, which confirms the near-unit population of $|m_F=-4\rangle$ and therefore implies that all off-diagonal elements must be very close to zero. Both the input and measured density matrices are shown in Fig. 4, and are clearly in excellent agreement. A more complex and therefore interesting input state can be prepared by optical pumping with linear polarization along $\hat{y}$, which prepares a state close to $|m_F=0\rangle$ in a coordinate system with quantization axis along $\hat{y}$. The reconstruction procedure is still performed in a coordinate system with quantization axis along $\hat{z}$, in which case the

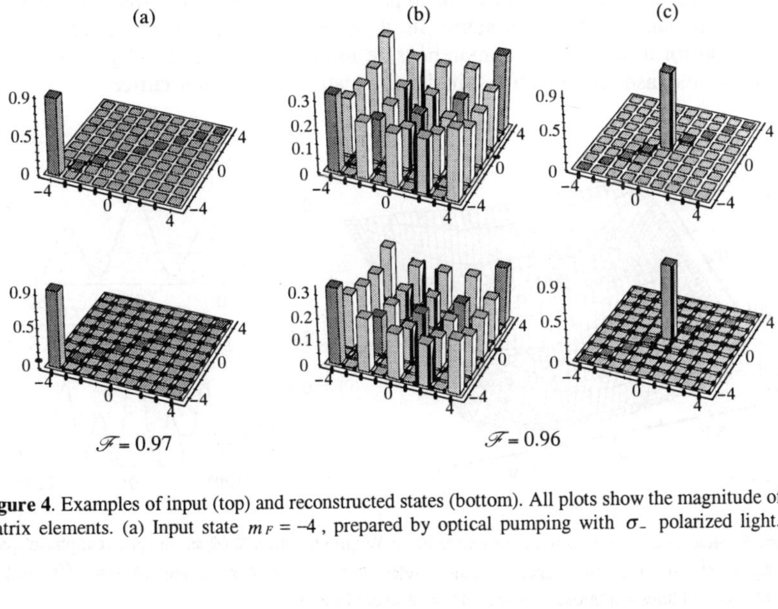

Figure 4. Examples of input (top) and reconstructed states (bottom). All plots show the magnitude of the matrix elements. (a) Input state $m_F=-4$, prepared by optical pumping with σ_- polarized light. (b)

Input state prepared by optical pumping with linear polarization along $\hat{y}$. This produces a state close to $m_F = 0$ in a coordinate system with quantization axis along $\hat{y}$, but with large coherences in the basis used for the reconstruction. (c) Same input and reconstructed state as (b), but shown in the rotated basis where the state is close to $m_F = 0$.

input and reconstructed density matrices have a relatively complex form with large coherences far from the diagonal, as shown in Fig. 4b. Visual comparison is simpler in the representation with quantization axis along $\hat{y}$, and shows the ability of our measurement scheme to reconstruct non-trivial quantum states (Fig. 4c). To quantify the degree of correspondence between the input state ρ_i and reconstructed state ρ_r, Fig. 4 also lists the fidelity [10] $F = [Tr\sqrt{\rho_i^{1/2}\rho_r\rho_i^{1/2}}]^2$ obtained for each test state.

Starting from $|m_F = -4\rangle$ we can also produce a range of spin-coherent states, by applying a magnetic field and letting the state precess for a fraction of a Larmor period. In this case it is more informative to display a Wigner function representation [11] of the measured states, which resemble localized "wavepackets" in the spherical angular coordinates θ and φ. Fig. 6 clearly shows the quasi-classical nature of the spin-coherent states, i. e. the quasi-probability distribution retains its roughly Gaussian shape as it precesses in the spherical phase space. By contrast the state $|m_F = -4\rangle$ along $\hat{y}$ is not quasi-classical, and has a delocalized Wigner function representation spread over much of phase space (Fig. 5).

5 Outlook

In summary, we have observed coherent Rabi oscillations of atomic spinor wavepackets in the optical double-well potentials of a far-off-resonance 1D lin-θ-lin magneto-optical lattice. Strong entanglement between the atomic spin- and center-of-mass degrees of freedom allows us to use the spin as a meter to track the quantum coherent dynamics of the system. In a second experiment we have laid the foundation for a much more comprehensive use of the spin-meter, by developing a method to reconstruct the entire spin density matrix. With better lattice

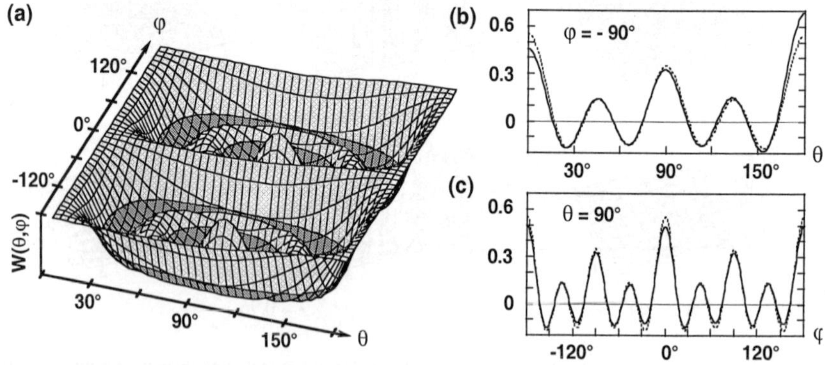

Figure 5. Measured $m_y = 0$ state, represented by its Wigner function $W(\theta, \varphi)$ in spherical phase space. The darker shading indicates negative values. Selected cuts of $W(\theta, \varphi)$ along (b) $\varphi = -90°$ and (c) $\varphi = 90°$ (dotted lines = input state, solid lines = measured state).

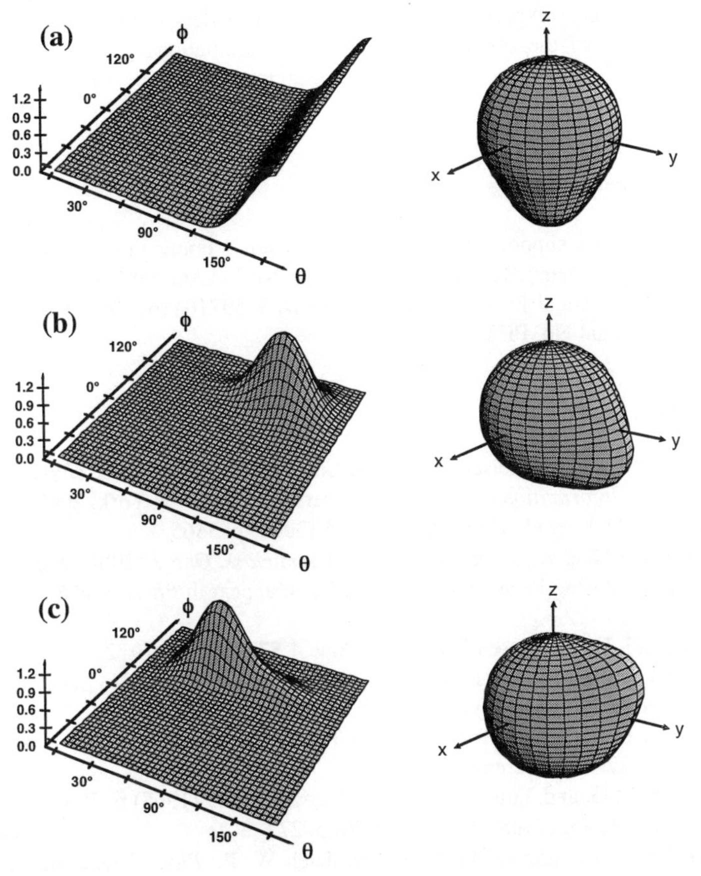

Figure 6. Measured Wigner functions $W(\theta,\varphi)$ for spin coherent states, corresponding to (a) $m_F = -4$ [F $= 0.97$], (b) $60°$ [F $= 0.96$] and (c) $120°$ [F $= 0.92$] rotation of this state around $\hat{x}$.

homogeneity and more powerful measuring techniques we hope to explore coherent dynamics on time scales much longer than the Rabi period. We can then introduce dissipation through laser cooling or photon scattering, and continuous measurement via Faraday rotation spectroscopy [12], to study the fundamental process of decoherence and the transition from quantum coherent to classical dynamics. This last aspect is especially interesting, as the coupled spin-motion Hamiltonian for our system approximates the Tavis-Cummings model without the rotating wave approximation, whose classical counterpart exhibits classical chaos [13]. So far we have carried out a detailed numerical study of the classically chaotic dynamics of our atom-lattice system, including a direct comparison between classical predictions and quantum theory/experiment [4]. This serves to underscore the profoundly non-classical nature of the observed tunneling Rabi oscillations, and points to future work on the effect of decoherence and measurement in the emergence of classical chaos [14].

6 Acknowledgements

PSJ acknowledges support from the National Science Foundation (grant No. PHY-9732612, by the Army Research Office (grant No DAAG 559710165, and by the Joint Services Optical Program (grant No DAAG559710116). IHD was supported by the NSF (Grant No. PHY-9732456).

References

1. See, for example, Nielsen M. A. and Chuang I. L., *Quantum Computation and Quantum Information* (Cambridge University Press, New York, 2000).
2. Haycock D. L. et al., *Phys. Rev. Lett.* **85** (2000) p. 3365.
3. Deutsch I. H. et al., *J. Opt. B: Quantum Semiclass. Opt.* **2** (2000) p. 633.
4. Ghose S., Alsing P. M. and Deutsch I. H., to appear in *Phys. Rev. E.* (Nov. 2001).
5. Deutsch I. H. and Jessen P. S., *Phys. Rev. A* **57** (1998) p. 1972.
6. Klose G., Smith G. and Jessen P. S., *Phys. Rev. Lett.* **86** (2001) p. 4721.
7. Brennen G. K. et al., *Phys. Rev. Lett.* **82** (2000) p. 1060.
8. See, for example, Liebfried D., Pfau T. and Monroe C., *Physics Today* (April 1998) p. 22, and references therein.
9. Newton R. G. and Young B.-L., *Ann. Phys.(N.Y.)* **49** (1968) p. 393,
10. Uhlmann A. *Rep. Math. Phys.* **9** (1976) p. 273.
11. Dowling J. P. Agarwal G. S. and Schleich W. P., *Phys. Rev. A* **49** (1994) p. 4101, and references therein.
12. Kuzmich A., Mandel L. and Bigelow N. P., *Phys. Rev. Lett.* **85** (2000) p. 1594.
13. Milonni P. W., Ackerhalt J. R. and Gaibraith H. W., *Phys. Rev. Lett.* **50** (1983) p. 966.
14. Habib S., Shizume K. and Zurek W. H., *Phys. Rev. Lett.* **80** (1998) p. 4361.

COLD ATOMS IN AN AMPLITUDE MODULATED OPTICAL LATTICE - DYNAMICAL TUNNELLING

W. K. HENSINGER[*†], H. HÄFFNER[*], A. BROWAEYS[*], N.R. HECKENBERG[†], K. HELMERSON[*], C. A. HOLMES[§], C. MCKENZIE[*], G. J. MILBURN[‡], W. D. PHILLIPS[*], S. L. ROLSTON[*], H. RUBINSZTEIN-DUNLOP[†], AND B. UPCROFT[*†]

[*] *National Institute of Standards and Technology, Gaithersburg, MD 20899, USA*

[†] *Centre for Laser Science, Department of Physics, The University of Queensland, Brisbane QLD 4072, Australia*

[‡] *Centre for Quantum Computer Technology, Department of Physics, The University of Queensland, Brisbane QLD 4072, Australia*

[§] *Department of Mathematics, The University of Queensland, Brisbane QLD 4072, Australia*

E-mail: hensinge@physics.uq.edu.au

We give an overview of our experiments on the quantum and classical dynamics of atoms in an amplitude modulated standing wave - the quantum driven pendulum. Atoms are prepared in a magneto-optic trap or as a dilute Bose-Einstein condensate and subjected to a far detuned optical standing wave. We observe the occurrence of a bifurcation sequence in the parameter space of the quantum driven pendulum. We also observe dynamical tunnelling between two classically isolated states of motion. The tunnelling process is coherent and we see more than eight tunnelling periods.

1 Introduction

Graham et al.[1] first proposed the use of atom manipulation experiments to test predictions of "quantum chaos". Laser cooled atoms moving in a far off resonant optical dipole potentials provide an ideal environment for the investigation of the quantum dynamics of non-linear Hamiltonian systems. The ability to achieve non-dissipative dynamics in the quantum domain (i.e. the effective Planck's constant is of the order of unity) has been demonstrated in previous work[2,3,4,5,6,7] by using well controlled time dependent potentials. Dynamical localization was observed using sinusoidal potentials with periodically modulated phase.[1,2] The dynamics of atoms in a phase modulated standing wave have been analyzed[3] and a close approximation to the delta kicked rotor was implemented using a pulsed standing wave.[5] In contrast we consider cold atoms moving in a sinusoidal optical dipole potential with single frequency amplitude modulation which corresponds to the classical driven pendulum, a textbook example for theoretical studies of nonlinear Hamiltonian systems. We present recent results showing the discovery of multiple fixed point bifurcations and the observation of dynamical tunnelling in this system. Both topics are discussed in more detail in two recent papers.[8,9]

We prepare the atoms in either a standard magneto-optic trap or as a dilute Bose-Einstein condensate and subject them to a far detuned optical standing wave. We can control the effective Planck's constant of our system allowing experiments in both quantum and classical regimes. There are two main advantages of quantum chaos experiments utilizing cold atoms. Firstly one can create a good approximation of a one-dimensional potential so that an accurate theoretical description of the quantum evolution is feasible. Secondly the system can be isolated from the environment on time scales long enough to observe quantum effects and to introduce controlled decoherence.

In the limit of large detuning our system may be described by the centre-of-mass Hamiltonian [6,10] given by

$$\mathcal{H} = p^2/2 + 2\kappa(1 - 2\varepsilon \sin\tau)\sin^2(q/2), \tag{1}$$

where $q = 2kx$, $p = (2k/m\omega)p_x$ and $\tau = t\omega$ are the scaled variables. The parameter ε describes the depth of modulation and ω is the angular frequency of the modulation. The time is given by t, p_x and x are respectively the momentum and position variables of the atom along the standing wave, k is the wave number of the light, and m is the mass. The driving amplitude, κ, is given by

$$\kappa = \frac{4U_0\omega_R^2}{\omega^2} \tag{2}$$

where $\omega_R = \hbar k^2/(2m)$ is the atomic recoil frequency and U_0 is the well depth in units of $\hbar\omega_R$. A thorough theoretical study along with the relevant classical and quantum simulation methods can be found in a previous publication.[6]

The phase space of the driven pendulum is two-dimensional, consisting of the position and momentum coordinates along the direction of the standing wave. When the standing wave is modulated a mixed phase space results. In Fig. 1 we show Poincaré sections with the stroboscopic period equal to one modulation period. The two islands of stability (ellipsoid rings) left and right of the centre result from phase space resonances and correspond to atoms which oscillate in phase with the standing wave modulation. The period of a resonance determines how many modulation periods it takes for the atoms to return to their original position in phase space. The presence of ellipses in the Poincaré section implies the existence of Kolmogorov, Arnold and Moser (KAM) surfaces, the crossing of which is forbidden by classical mechanics. The

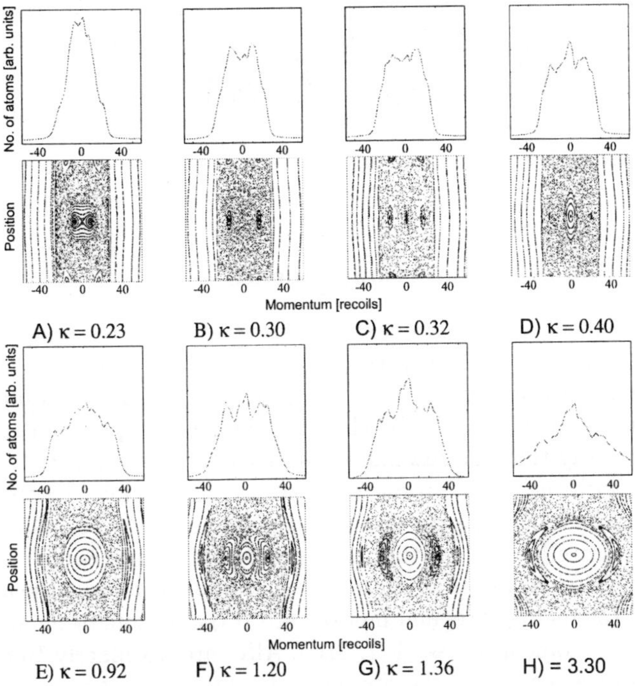

Figure 1: The bifurcations sequence: Poincaré sections along with the corresponding atomic momentum distributions.

inner island corresponds to atoms which are approximately stationary at the bottom of the well. The sea of chaotic motion (dotted region) is bounded in momentum by the region of regular unbound motion which consists of atoms having enough kinetic energy to travel from one well of the standing wave to the next. The islands of regular motion near the region of unbound regular motion are librations: atoms which take multiples of one modulation period to hop from one well to another. Figure 2 shows a graphical representation of period one resonances. The filled circle represents an atom in a period one resonance. As shown for subsequent times the atom oscillates in the modulated well returning to its initial position after one modulation period. The appearance of the second ball at $t = 0.5\ T$ illustrates tunnelling into the initially empty resonance - dynamical tunnelling, this will be explained in section 3.

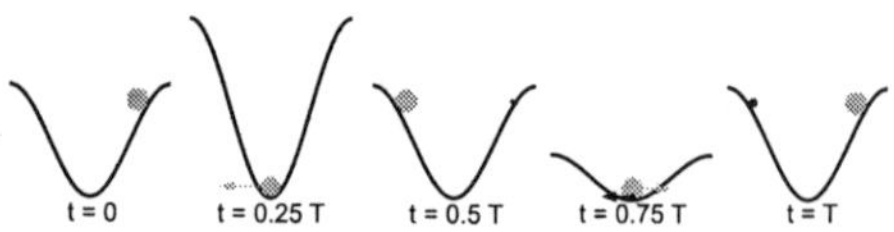

Figure 2: Diagram illustrating period one resonances and dynamical tunneling. T is the modulation period.

2 Observation of a bifurcation sequence

For this experiment we have prepared rubidium atoms in a standard magneto-optic trap (MOT) at a temperature of approximately 8μK. A far detuned optical standing wave is modulated using an acousto-optic modulator. We load resonances experimentally by choosing the starting phase of the modulation in such a way that the resonances are located on the position axis where they overlap with the initial atomic momentum distribution. To observe phase space resonances we choose the end phase of the intensity modulation so that the resonances are located on the momentum axis having equal but opposite momentum. Classically at that phase (n + 0.25 or 0.75 periods) the atoms, contained inside the region of regular motion, are at the bottom of the well, moving with maximum momentum (see Fig. 2B)). After approximately 10 ms ballistic expansion time a picture of the atomic distribution is taken using a freezing molasses method. Resonances can be observed as distinct peaks in the atomic momentum distribution[a] resulting in experimental data as shown in Fig. 1. Details of the experimental setup may be found in previous publications.[6,7]

We observed the occurrence of a bifurcation sequence in the parameter space of the quantum driven pendulum. When one control parameter is varied (the scaled well depth of the standing wave, κ) the phase space undergoes distinct changes where resonances with different period emerge and dissolve as shown in Fig. 1. Figure 1 shows the bifurcation sequence in terms of Poincaré sections along with the corresponding experimental momentum distributions. Classical Hamiltonian perturbation theory (details of this calculation may be found elsewhere[9]) shows that at $\kappa \approx \frac{1}{4}(1 - \epsilon + \epsilon^2 - \epsilon^3)$ the phase space origin destabilizes producing two period two resonances which can be seen in Fig. 1A). As κ is increased the islands move apart and become separated by a sea of chaos as shown in B). Then at $\kappa \approx \frac{1}{4}(1 + \epsilon + \epsilon^2 + \epsilon^3)$ a bifurcation occurs, stabilizing the origin. Between the two curves, the momentum distribution is clearly depressed in the centre as shown in Fig. 1B), whereas for $\kappa >$

[a]The actual measurement is a position distribution, however the position distribution after sufficient free evolution corresponds to the momentum distribution before the evolution.

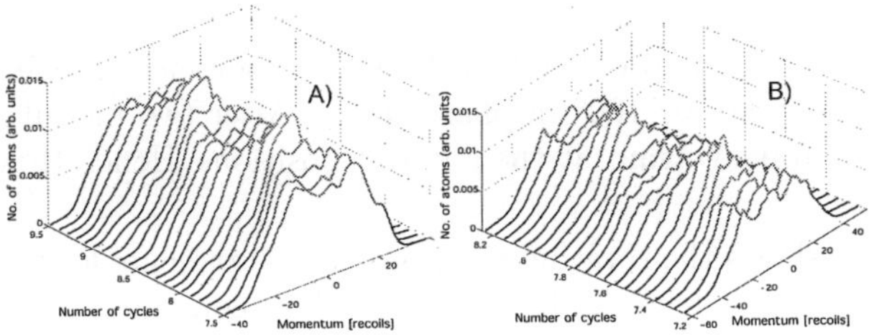

Figure 3: Momentum distributions as function of the interaction time with the standing wave showing period two resonances in A) and period one resonances in B).

$\frac{1}{4}(1 + \epsilon + \epsilon^2 + \epsilon^3)$, at C) the three distinct islands of regular motion appear clearly as three peaks in the momentum distribution. As κ increases further, the period two islands move out, breaking up as they do so and eventually become indistinguishable from the chaotic sea (Figure 1 parts D) and E)).

When κ is increased even further, another set of bifurcations produces period one resonances which are shown in Fig. 1 parts F) and G). The pattern is repeated again at $\kappa \approx \frac{9}{4}$ with two period 2/3 resonances emerging. As one expects the momentum separation between regions of unbound regular motion becomes larger when the value of κ is increased. This is due to the wells becoming effectively deeper. In Fig. 1 many of the Poincaré sections also show librational resonances. As they do not rotate they will not cross the position axis and therefore they are not loaded by the initial atomic momentum distribution in our current experimental setup. Thus they do not appear in the experimentally measured momentum distribution.

To perform an accurate mapping of the experimental results to the bifurcation sequence we measured the rotation frequency of the resonances. This measurement determines the period of the resonances. To conduct these measurements we vary the time the atoms interact with the standing wave (which is equivalent to a variation of the end phase of modulation) measured in cycles and record the resulting atomic momentum distribution. Figure 3A) shows the rotation frequency measurements for the resonances bifurcating at $\kappa \approx 0.25$. We plot atomic momentum distributions as function of interaction time with the standing wave in cycles. Resonances can be observed clearly when they are located on the momentum axis when the standing wave is turned off and will disappear when located on the position axis at that time. At 7.5 modulation periods (cycles) one can see two distinct resonances. These have completely disappeared after 8 modulation periods. The resonances can again be seen as distinct peaks in the momentum distribution after 8.5 periods of the standing wave modulation. Therefore it follows that the resonances needed one modulation period to rotate by 180° in phase space. After two modulation periods the resonances appear again at their initial position. Hence they are period two resonances.

Figure 3B) shows the rotation frequency measurements for the resonances bifurcating at $\kappa \approx 1$ which exhibit different features compared to Fig. 3A). This allows us to distinguish between different bifurcation regimes. After 7.2 cycles one can see two distinct resonances and the centre island of stability. A quarter period later the resonances have rotated on the position axis and therefore no longer visible. Another quarter period later the resonances have rotated by 180° and are now located on the momentum axis. Note that the peaks in the momentum distribution associated with the resonances have become wider and that their intensity has decreased as compared to the distribution at 7.2 periods of the standing wave modulation. Phase space volume is preserved in Hamiltonian systems. However the shape of the resonances has changed significantly under 180° rotation. The momentum width of the resonance after 7.7 periods is larger compared to the one at 7.2 cycles. This results in the experimentally observed broadening of the resonances. These resonances need one modulation period to return to their initial position making them period one resonances.

We have also measured the rotational frequency of the resonances shown in Fig. 2H) and we were able to successfully confirm them to be period 2/3 resonances as predicted by classical theory for this value of κ.

We have compared our experimental results with predictions from quantum trajectory simulations [6] and the classical description. A detailed account may be found elsewhere,[9] here we discuss the main results. We have measured the momentum of the centre of the peak in the atomic momentum distributions corresponding to the resonance both for the experimental data and the quantum simulations. Uncertainty in our measurement results from asymmetries in the experimentally measured momentum distribution. These asymmetries are most likely due to non-uniformities in the initial spatial distribution of the MOT. Other possible sources of uncertainty include an imperfectly zeroed magnetic field in the interaction region. The momenta of resonances in the quantum simulation exhibit some readout uncertainty because the peaks in the momentum distribution do not necessarily have a symmetric Gaussian shape. There is satisfactory agreement between the experimental results and the quantum simulations. We have compared these results with the momentum of the resonances when they are located on the momentum axis, as calculated from numerical solutions of the equations describing the classical system and from classical perturbation theory. The results taken from the Poincaré sections (numerical solution of Hamilton's equations) and the analytical classical results from perturbation theory are in good agreement. However the classical predictions show a larger resonance momentum than the quantum simulations and the findings of our experiment. While the classical theory predicts the momentum of the fixed point (when positioned on the momentum axis), the quantum simulations directly predict the full momentum distribution as observed in the experiment. Part of the above discrepancy could result from the fact that the fixed points are positioned asymmetrically toward the faster side of the region of regular motion which in turn gives rise to the peak in the experimentally measured momentum distribution. However if one assumes the mean of maximum and minimum momenta occupied by the region of regular motion (when positioned on the momentum axis) as a classical momentum approximation for the experimentally observed momentum peak, the classical resonance momenta are still significantly faster than the experimentally measured values. Our results suggest a possible explanation in terms of the theory of quantum slow motion,[11] but a more rigorous and detailed investigation will be needed to confirm this.

3 Dynamical tunnelling

To observe dynamical tunnelling we utilize a sodium Bose-Einstein condensate which is loaded into the bottom of the lowest band of an optical lattice at a quasi-momentum of 0. We selectively load a single resonance by suddenly shifting the position of the standing wave so that the wavepacket of the condensate overlaps the resonance. This is done by inducing a phase shift with an acousto-optical modulator just before turning on the modulation. After the atoms have interacted with the modulated standing wave for a selected number of modulation periods, we turn the standing wave off with the modulation phase chosen so that the resonance lies on the momentum axis. We measure the atomic momentum distribution with absorption imaging after 1.5 ms of free flight. The momentum distribution appears as a set of diffraction peaks at integer multiples of $2\hbar k$ due to the coherence of the atoms between wells. A detailed description of the experimental procedures may be found elsewhere.[8]

Figure 4A) depicts momentum distributions which are taken after 0.25 (front), 2.25 (middle) and 5.25 (back) modulation periods for $\varepsilon = 0.29$, $\kappa = 1.66$ and $\omega/2\pi = 250$ kHz. The momentum distribution after 0.25 modulation periods consists mainly of a pair of diffraction peaks at -4 and -6 $\hbar k$. Classically the atoms should remain in the resonance leaving the stroboscopically measured momentum distribution unchanged. Instead after 2.25 modulation periods about half of the atoms have appeared with opposite momenta, which

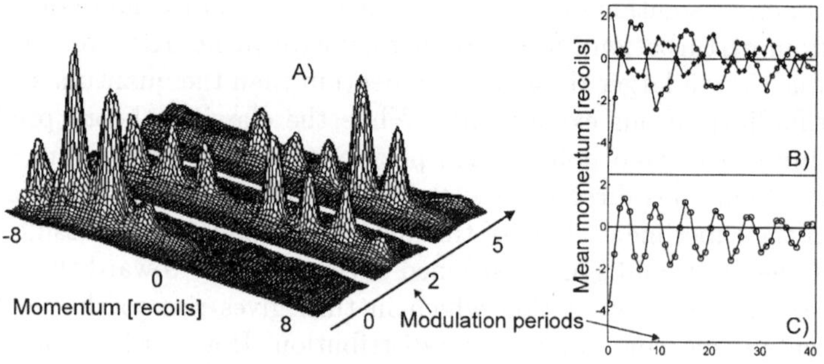

Figure 4: A) Three atomic momentum distributions taken after 0.25, 3.25 and 5.25 modulation periods showing dynamical tunneling. B) and C) Mean momentum as a function of the number of modulation periods for two sets of parameters.

corresponds to the other resonance. By 5.25 modulation periods most of the atoms are in the other resonance. At 9.25 modulation periods the atoms have returned to the original resonance. This transfer of atoms back and forth between the regions of regular motion is coherent dynamical tunnelling.

In Fig. 4B) we plot the mean atomic momentum after multiples of the modulation period for the same parameters used to obtain Fig. 4A)). The circles (diamonds) correspond to turning off the standing wave at the maximum (minimum) of the amplitude modulation (see Fig. 2 at $t = 0.25\ T$ and $t = 0.75\ T$). We observe an oscillation of the mean momentum indicating the occurrence of dynamical tunnelling. The period-one character of the motion is verified by the reversal of the momentum between the two curves, which are separated in time by 0.5 modulation periods. By taking the Fourier transform of the data, we find the tunnelling period to be 10.3(2) modulation periods where the uncertainty is statistical. Quantum theory predicts dynamical tunnelling to occur for various values of the scaled well depth κ, the modulation parameter ε and modulation frequency ω and also predicts a strong sensitivity of the tunnelling period and amplitude on these parameters. An example of this is shown Fig. 4C), for $\varepsilon = 0.30$, $\kappa = 1.82$ and $\omega/2\pi = 222$ kHz we find a tunnelling period of 6 modulation periods with a significantly longer decay time.

We note that there is no significant (above background) zero momentum peak even in the case of approximately zero mean momentum (when the atoms have tunnelled half way). This indicates that at half the tunnel period the system is in a coherent superposition of two distinguishable classical motions: one with positive momenta and one with equal but opposite momenta. We expect this, because Floquet analysis shows that atoms tunnel from one region of regular motion into the other and it is impossible for them to enter the central island of stability at $(p,q) = (0,0)$. Steck et al.[12] recently also reported tunnelling of a different motional state in a similar system.

4 Conclusions

In conclusion we have presented an exploration of the dynamics of cold atoms in an amplitude modulated standing wave. We have observed multiple bifurcations resulting from the classical character of the system and dynamical tunnelling originating in its quantum nature. The observation of dynamical tunnelling marks the beginning of Bose-Einstein condensates as test-bed for quantum non-linear dynamics and quantum chaos. The system can be used to explore decoherence and we can control the effective Planck's constant of the system making it ideal to explore the borderland of quantum and classical physics.

Acknowledgments

The NIST group acknowledges support from ONR, NASA and ARDA, and the University of Queensland group was supported by the ARC. A.B. was partially supported by DGA (France), and H. H. was partially supported by the A. v. Humboldt Foundation. W.K.H. and B.U. thank NIST for the hospitality during the experiments utilizing the Bose-Einstein condensate.

References

1. R. Graham *et al*, *Phys. Rev.* A **45**, R19 (1992).
2. F.L. Moore *et al*, *Phys. Rev. Lett.* **73**, 2974 (1994).
3. J.C. Robinson *et al*, *Phys. Rev. Lett.***74**(20), 3963 (1995).
4. J.C. Robinson *et al*, *Phys. Rev. Lett.* **76**, 3304 (1996).
5. F.L. Moore *et al*, *Phys. Rev. Lett.* **75**, 4598 (1995).
6. W.K. Hensinger *et al*, *Phys. Rev.* A **64** 033407 (2001)
7. W.K. Hensinger *et al*, *J. Opt. B: Quantum Semiclass. Opt.* **2**, 659 (2000).
8. W.K. Hensinger *et al*, *Nature* **412** (5 July), 52 (2001).
9. W.K. Hensinger *et al*, Multiple bifurcations in atom optics, *Phys. Rev.* A (in press)
10. S. Dyrting *et al*, *Phys Rev.* E **48**, 969 (1993).
11. M. Hug and G.J. Milburn, *Phys.Rev.* A **63**, 023413 (2001).
12. D.A. Steck *et al*, *Science Express* 10.1126/science.1061569 (2001)

PHOTONIC INFORMATION STORAGE AND QUANTUM INFORMATION PROCESSING IN ATOMIC ENSEMBLES

J. HAGER, A. FLEISCHHAUER, A. MAIR, D. F. PHILLIPS, R. L. WALSWORTH

Harvard-Smithsonian Center for Astrophysics, Cambridge, MA 02138, USA

M. D. LUKIN

Dept. of Physics and ITAMP, Harvard University, Cambridge, MA 02138, USA

Recently, we proposed the concept for and performed a proof-of-principle demonstration of the technique to store photonic excitations carried by traveling waves in atomic vapor, and showed that this technique is phase coherent and amenable to accurate phase control. Here, we review the experimental details and theoretical underpinnings of our work to date.

1 Introduction

In this paper we describe our recent experimental demonstrations[1,2] of a technique[3,4] to trap, store, and release coherent excitations carried by light pulses. Specifically, a pulse of light which is several kilometers long in free space can be compressed to a length of a few centimeters and then converted into collective spin excitations in a vapor of Rb atoms. After a controllable storage time, the process is reversed and the atomic coherence is converted back into a light pulse. We accomplish this "storage of light" by dynamically reducing the group velocity of the light pulse to zero.

The light-storage technique is based on the recently demonstrated phenomenon of ultra-slow light group velocity[5], which is made possible by Electromagnetically Induced Transparency (EIT)[6]. In "slow-light" experiments an external optical field (the "control field") is used to make an otherwise opaque medium become transparent near an atomic resonance. A weak optical field (the "signal field") at a particular frequency and polarization can then propagate without dissipation and loss but with a substantially reduced group velocity. Associated with slow light is a considerable spatial compression, which allows a signal pulse to be almost completely localized in the atomic medium. In addition, as the signal light propagates the atoms are driven into a collective coherent superposition of (typically) Zeeman or hyperfine states that is strongly coupled to the light via a Raman transition. The coupled light and atomic excitations can be efficiently described as a single form of

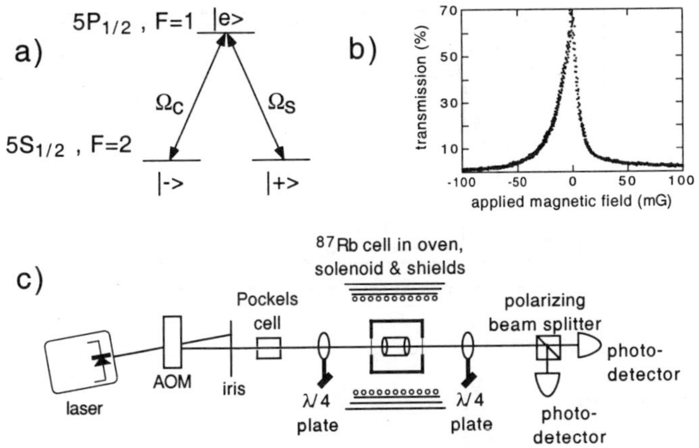

Figure 1: (a) Λ-type configuration of ^{87}Rb atomic states resonantly coupled to a control field (Ω_c) and a signal field (Ω_s). (b) A typical observed Rb Faraday resonance in which the transmission intensity for a cw signal field is shown as a function of magnetic field. The full width of this resonance is 20 mG which corresponds to a 15 kHz shift in the Zeeman levels. (c) Schematic of experimental setup.

dressed-state excitation known as the "dark-state polariton" [4]. In order to store a light pulse, we smoothly turn off the control field, which causes the dark-state polariton to be adiabatically converted into a purely atomic excitation (a collective Rb Zeeman coherence in our experiments) which is confined to the vapor cell. Turning the control field back on reverses the process: the dark-state polariton is adiabatically restored to an optical excitation.

Our recent experiments can be understood qualitatively by considering a "lambda" configuration of three atomic states coupled by a pair of optical fields (see Fig. 1a). Here the control field (Rabi-frequency Ω_c) and signal field (Ω_s) are represented, respectively, by the two helicities of circularly polarized light (σ_+ and σ_-) derived from a single laser beam. These light fields couple pairs of Zeeman sublevels of electronic ground state ($5^2S_{1/2}$) Rb atoms ($|-\rangle, |+\rangle$), with magnetic quantum numbers differing by two, via the excited ($5^2P_{1/2}$) state. In this configuration the effective two-photon detuning can be controlled by applying an external magnetic field, which causes Zeeman level shifts between $|-\rangle$ and $|+\rangle$. The two optical fields drive the atoms into a superposition of Zeeman sublevels which at zero magnetic field is decoupled from the light.

2 Demonstration of Stored Light

We performed light-storage experiments[1] in atomic Rb vapor at temperatures $\sim$ 70-90 °C, which corresponds to atomic densities $\sim 10^{11} - 10^{12}$ cm^{-3}. Under

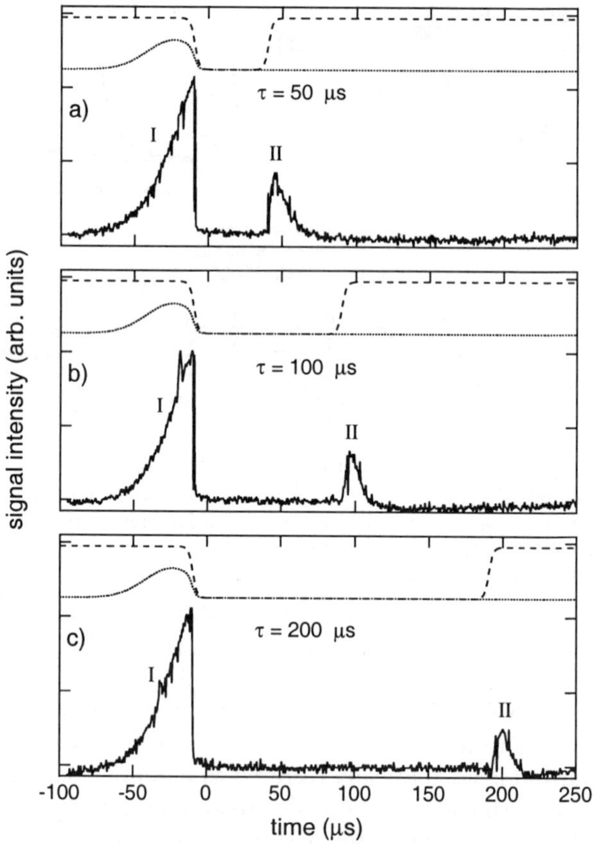

Figure 2: Observed light pulse storage in a ^{87}Rb vapor cell. Examples are shown for storage times of (a) 50 μs, (b) 100 μs, and (c) 200 μs. (Background transmission from the control field, which leaks into the signal field detection optics, has been subtracted from these plots.) Shown above the data in each graph are calculated representations of the applied control field (dashed line) and input signal pulse (dotted line).

these conditions the 4 cm-long sample cell was completely opaque for a weak optical field near the Rb D_1 resonance ($\simeq$ 795 nm). We derived the control and signal beams from an extended cavity diode laser by carefully controlling the light polarization as shown in the experimental schematic in Fig. 1c. For the data presented here we employed the $5^2S_{1/2}, F=2 \rightarrow 5^2P_{1/2}, F=1$ transition in ^{87}Rb. The control field was always much stronger than the signal field ($\Omega_c \gg \Omega_s$); hence most of the relevant atoms were in the $5^2S_{1/2}, F=2, M_F=+2$ magnetic sublevel. In this case the states $|-\rangle, |+\rangle$ of the simplified 3-level model correspond, respectively, to $|F=2, M_F=0\rangle$ and $|F=2, M_F=+2\rangle$. By using a fast Pockels cell we slightly rotated the polarization of the input light to create a weak pulse of σ_- light, which served as the signal field. We monitored transmission of the σ_- light pulse using a $\lambda/4$ waveplate and polarizing beam splitter. In order to ensure long lifetimes of the atomic Zeeman coherences, we magnetically shielded the Rb cell and filled it with about 5 torr of He buffer gas. We used a precision solenoid to control the static magnetic field along the propagation direction of the optical beam.

We first consider the case of cw signal and control fields. Fig. 1b displays a typical transmission spectrum for the signal (σ_-) field obtained by scanning the magnetic field and thereby changing the effective two-photon detuning. Note that due to the induced transparency the signal field transmission is maximal for zero magnetic field, even though most of atoms are in the state $|+\rangle$. Outside of the transparency window (magnetic fields > 20 mG) the Rb vapor is completely opaque to σ_- light.

We next present a demonstration of light storage. Typical input σ_- signal pulses had a temporal length of ~10 to 30 μs. Upon entrance into the Rb cell the signal pulse was spatially compressed due to the reduction in group velocity. In order to trap, store, and release the signal pulse, we used an acousto-optic modulator to turn off the control field smoothly over about 3 μs while much of the signal pulse was contained in the Rb cell. After some time interval, we turned the control field on again, thereby releasing the stored portion of the signal pulse.

An example of the observed light storage is shown in Fig. 2. Typically, two time-resolved σ_- signal pulses were registered by the photodetector. First, a fraction of the signal pulse left the cell before the control field was turned off, which resulted in an observed signal that was not affected by the storage operation (peak I in each plot of Fig. 2). This untrapped light was delayed by about 30 μs as compared to free-space propogation due to the slow group velocity ($v_g \sim 1$ km/s). The second observed signal pulse was light that was stored in atomic excitations for a time interval τ. Note that the released signal pulse was detected only after the control field was turned back on (peak II in each plot of Fig. 2). We observed that the amplitude of the released signal pulse decreased as the τ increased. We could resolve released light pulses without signal averaging for storage intervals up to $\tau \simeq 0.5$ ms.

3 Coherence of Stored Light

The demonstrations of phase coherence and control[2] were similar to our previous work[1]. We applied a pulsed magnetic field during the light storage interval

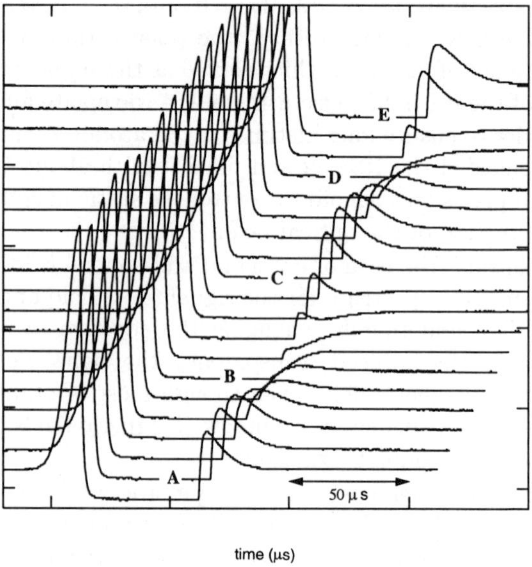

Figure 3: Results of interferometric measurements of released photonic excitations: twenty light storage experiments. The magnetic field was pulsed during the storage interval with increasing strength from trace A to E such that the accumulated phase difference between the output signal pulse and the control field varied from approximately 0 to 4π. Note: there is a small phase offset at zero pulsed magnetic field, caused by the Pockels cell.

to vary controllably the phase of the collective Zeeman coherence in the Rb vapor. We then converted the spin excitations back into light and detected the resultant phase shift in an optical interferometric measurement. A $\lambda/2$ waveplate was placed after the $\lambda/4$ waveplate following the vapor cell (See Fig. 1c). To form an interferometer for the two fields, we adjusted the $\lambda/2$ plate such that a small fraction ($< 10\%$) of the control field was mixed into the signal detection channel.

The phase of an atomic Zeeman coherence can be easily manipulated using an external magnetic field. If a pulsed magnetic field, $B_z(t)$, is applied in a direction parallel to that of the light propagation (the quantization or $\hat{z}$ axis), then the Zeeman sublevels are differentially shifted in energy, producing a phase shift in the Zeeman coherence given by:

$$\Phi = (g_+ - g_-)\frac{\mu_B}{\hbar}\int_0^T B(t')dt'. \tag{1}$$

Here $g_\pm$ are Lande factors corresponding to the Zeeman states $|\pm\rangle$ and T is the time during which the magnetic field is applied. Once the magnetic field is removed, the dark-state polariton can be adiabatically restored to a photonic excitation by turning the control field (Ω_c) back on. The phase-shift Φ of the atomic coherence is thus transfered to a phase difference between the control field and the reconstituted signal pulse emitted from the sample.

To manipulate the phase of the atomic Zeeman coherence, we carefully controlled the magnetic field at the Rb cell. We employed high permeability magnetic shields and a precision solenoid to screen out and cancel external magnetic fields. Additionally, a Helmholtz coil pair generated the pulsed field to manipulate the Rb Zeeman coherence, thereby inducing phase shifts in the stored light. We typically applied current pulses to the Helmholtz pair of 10 to 30 μs duration, with peak magnetic fields as large as 100 mG.

Each trace in Figure 3 is an example of stored light observed with a magnetic field pulse applied during the storage interval which created a phase shift between the signal pulse and the control field. With the σ^+ control field dressing the atoms, we measured a small background signal due to the control field being slightly mixed into the signal detection channel via the slightly-rotated $\lambda/2$ plate. We then applied a σ^- signal pulse of duration ~ 20 μs. With the peak amplitude of the signal pulse inside the cell, we ramped down the control field. The fraction of the signal pulse which left the cell before the control field was turned off was not affected by the phase manipulations and was used as a reference in the experiment. With the remainder of the signal pulse stored in the vapor cell, we applied a pulsed current to the Helmholtz coils to induce a controllable phase shift in the ^{87}Rb ensemble Zeeman coherence. After the pulsed magnetic field, we turned the control field back on and the signal light pulse was reconstituted and observed interferometrically.

Note that for the accurate, coherent control of the phase it is essential to apply the magnetic field only during the storage interval[2]. By adjusting the applied magnetic field pulse during the storage, we could easily modify the phase Φ of the coherent excitation. Fig. 3 shows twenty stored light experiments for which we increased the Zeeman phase shift by approximately $0.2\,\pi$ for each successive run. Trace A in Fig. 3 shows the result for $\Phi \simeq 0$ and hence maximum constructive interference between the output signal light and the control field. As we increased the pulsed magnetic field to change the phase by π, we observed destructive interference (e.g., trace B). As we increased the pulsed magnetic field still further, we alternatively observed constructive and destructive interference as expected at $\Phi \simeq 2\,\pi, 3\,\pi, 4\,\pi$, etc. (traces C-E in Fig. 3). We observed up to 10 periods of phase accumulation (i.e., $\Phi \simeq 20\,\pi$) without loss of coherence, limited by the strength of our pulsed magnetic field power supply rather than non-linear shifts in the Rb Zeeman levels at higher fields or magnetic field inhomogeneity.

4 Discussion and Theoretical Interpretation

We consider the propagation of a signal pulse in an EIT medium (along the z direction) subject to a time-dependent control field. We assume that the signal field is always weaker than control field and that the signal field group velocity is always much smaller than the control field group velocity.

As noted above, the dynamical trapping of signal pulses can be understood in terms of dark-state polaritons. These are coupled superpositions of photonic and spin wave-like excitations, defined by a transformation:

$$\Psi(z,t) = \cos\theta(t)\,\Omega_s(z,t) - \sin\theta(t)\,\sqrt{\kappa}\,\rho_{-+}(z,t), \tag{2}$$
$$\cos\theta(t) = \frac{\Omega_c(t)}{\sqrt{\Omega_c^2(t)+\kappa}}, \quad \sin\theta(t) = \frac{\sqrt{\kappa}}{\sqrt{\Omega_c^2(t)+\kappa}}.$$

Here ρ_{-+} is the atomic coherence between states $|-\rangle$ and $|+\rangle$. Also, $\kappa = 3n\lambda^2\gamma_r c/8\pi$, where n is the ^{87}Rb density, λ is the wavelength and γ_r is the natural linewidth of the D_1 transition, and c is the free-space speed of light.

In the ideal limit, corresponding to vanishing dephasing of the atomic coherence and perfect adiabatic following, the polariton propagation is described by [4]

$$\Psi(z,t) = \Psi\left[z - \int_{t_0}^{t} \mathrm{d}t' v_g(t'), t = t_0\right], \tag{3}$$

where the time-dependent group velocity is

$$v_g(t) = \cos^2(\theta) = \frac{\Omega_c^2(t)}{\Omega_c^2(t)+\kappa}. \tag{4}$$

In the stationary case, $v_g(t) = v_g^0$, and after initial pulse compression at the entrance of the cell, the polariton describes the well-known EIT-like propagation of coupled light and atomic coherence. Remarkably, however, when the intensity of the control field is changed during the pulse's propagation through the atomic medium, the polariton can preserve its shape, amplitude, and spatial length, while its group velocity and the ratio of the light and matter components are altered. In particular, when the group velocity is reduced to zero by turning off the control beam, the polariton becomes purely atomic ($\cos\theta = 0$) and its propogation is stopped. The state of the input light pulse is thereby mapped into the atomic coherence ρ_{-+}. The coherence stored within the cell ($0 < z < L_{cell}$) after switching off the control field over the time interval $[t_0, t_1]$ is given by:

$$\rho_{-+}(z,t_1) = -\sqrt{\frac{c}{v_g^0\kappa}}\,\Omega_s\Big(z=0, \frac{\int_{t_0}^{t_1} v_g(t')dt' - z}{v_g^0}\Big). \tag{5}$$

If the control beam is turned back on after a storage interval τ, the polariton is accelerated and the atomic coherence ρ_{-+} is mapped back into light. The released light pulse has a shape, amplitude, and spatial length proportional to the coherence *after* the storage interval:

$$\Omega_s(z,t) = -\cos\theta(t)\sqrt{\kappa}\rho_{-+}(L_{cell} - \int_{t_2}^{t} v_g(t')dt', t_2), \tag{6}$$

where $t_2 = t_1 + \tau$. In principle, complete storage and retrieval of the input light pulse is possible.

Under realistic conditions, the light storage time is always limited by loss of atomic coherence. In our recent experiments, for example, the Rb atoms diffuse through the buffer gas and escape from the region of interaction with the light beam, leading to a coherence time of $\sim 150\ \mu$s.

It is also important to consider the assumption of adiabatic following and its effect on the *dynamic* method for group velocity reduction, as compared to the conventional, stationary approach based on EIT. Naively, Eq. (4) indicates that long light pulse delays can be obtained by simply using a cw control field of sufficiently low intensity. However, for pulses similar to those used in the dynamic trapping method we failed to observe long delays ($\sim 100\ \mu$s) using the stationary EIT technique. This failure is due to *the breakdown of adiabatic following*.

The essence of adiabatic following is that the light pulse spectrum should be contained within a relatively narrow transparency window $\Delta\omega$ (Fig. 1b) to avoid loss and dissipation. The magnitude of $\Delta\omega$ is determined by both the control field intensity and the opacity of the atomic medium [7]. In conventional EIT propagation, a weaker control field induces a narrower transmission spectrum. For a fixed bandwidth of the propagating signal pulse, such spectral narrowing causes absorption of certain pulse spectral components and inevitably destroys propagating light.

The observed light pulses stored and released by *dynamic* reduction of the group velocity are not destroyed in spite of the narrowing of the transparency window. This important result is in agreement with theoretical predictions[3,4] which pointed out that adiabatic following occurs as long as the product of the propagation distance and the normal opacity of the medium is smaller than the square of the spatial light pulse length in the medium. In other words, adiabaticity can be preserved with the dynamic light-storage method as long as the input pulse bandwidth is within the *initial* transparency window. A remarkable feature of the dark-state polariton is that its spatial length remains unchanged in the process of deceleration. Hence, a dynamic reduction in group velocity is accompanied by a *narrowing* of the polariton frequency spectrum (bandwidth). In this case, adiabatic following occurs even when the group velocity is reduced to zero, which is in good agreement with the experimental results presented here.

5 Conclusions

We have demonstrated that it is possible to control the propagation of light pulses in optically thick media by dynamically changing the group velocity. In particular, a light pulse can be trapped and stored in an atomic coherence; after a controllable delay this coherence can be converted back into a light pulse. We have also demonstrated that this "light storage" technique is phase coherent. In so doing we have performed accurate, coherent manipulation of information that is stored in an atomic spin coherence and then transferred back into light and released.

Acknowledgments

It is a pleasure to thank M. Fleischhauer and S. Yelin for many fruitful ideas and collaboration on theoretical aspects of this work. We also thank S. E. Harris and M. O. Scully for many stimulation discussions. This work was partially supported by the Nathional Science Foundation through the grant to ITAMP and by the ITR program, the Office of Naval Research, and NASA.

References

1. D. F. Phillips *et al.*, Phys. Rev. Lett. **86**, 783 (2001). C. Liu *et al.*, Nature **409**, 490 (2001).
2. A. Mair, J. Hager, D. F. Phillips, R. L. Walsworth and M. D. Lukin, submitted to Phys. Rev. Lett. (2001).
3. M. D. Lukin, S. F. Yelin, and M. Fleischhauer, Phys. Rev. Lett. **84**, 4232 (2000); L. M. Duan, J. I. Cirac, and P. Zoller (unpublished).
4. M. Fleischhauer and M. D. Lukin, Phys. Rev. Lett. **84**, 5094 (2000);
5. L. V. Hau, S. E. Harris, Z. Dutton, and C. H. Behroozi, Nature **397**, 594 (1999); M. Kash *et al.* Phys. Rev. Lett. **82**, 5229 (1999); D. Budker *et al. ibid* **83**, 1767 (1999).
6. See e.g. M.O.Scully and M.S. Zubairy, Quantum Optics (Cambridge University Press, Cambridge, UK, 1997); S. E. Harris, Physics Today **50**, 36 (1997).
7. S. E. Harris, Phys. Rev. Lett. **70**, 552 (1993); M. D. Lukin *et al.*, Phys. Rev. Lett. **79**, 2959 (1997).

EXPERIMENTAL EVIDENCE OF BOSONIC STATISTICS AND DYNAMIC BEC OF EXCITON POLARITONS IN GaAs AND CdTe QUANTUM WELL MICROCAVITIES

Y. YAMAMOTO[a], R. HUANG, H. DENG, F. TASSONE
Quantum Entanglement Project, ICORP, JST
Ginzton Laboratory, Stanford University, Stanford, CA 94305, USA

J. BLEUSE, H. ULMER, R. ANDRÉ
Dept. de Recherche Fondamentale sur la Materie Condensee
CEA Centre d'Etudes de Grenoble, France

We present experimental evidence for the final state stimulation for the exciton-exciton scattering rate into the upper and lower exciton-polaritons in a GaAs quantum well microcavity. We have studied the reservoir exciton density, lower polariton population, and time delay dependence of the stimulated scattering rate, and have found the behavior to be in agreement with a theoretical prediction. We then describe a dynamic BEC of exciton-polaritons in a CdTe quantum well microcavity. By performing a pulsed pump and probe experiment, we observe unambiguous evidence of real excited exciton-exciton scattering gain in the form of $\exp(const \cdot N_{exc}^2)$, where N_{exc} is the exciton reservoir population.

1 Introduction

According to the spin-statistics theorem of quantum mechanics, a composite particle with integer spin obeys the symmetrization postulate and Bose-Einstein statistics. One of the distinct features of Bose particles is final state stimulation: the presence of a particle in the final state causes an enhancement of the scattering rate into that final state, at a rate proportional to $1+N$, where N is the population in the final state. Final state stimulation is the consequence of constructive interference between the direct and exchange terms in a symmetrized multi-particle wavefunction and the origins of lasers and Bose-Einstein condensation (BEC). It is of great interest to demonstrate final state stimulation for bosonic particles other than a photon, especially for massive composite particles [1]. The exciton, which is composed of an electron and a hole in a semiconductor, is expected to behave as a composite boson at low density, i.e., when $n_{exc}a_B^d \ll 1$, where n_{exc} is the exciton density, a_B is the Bohr radius of the exciton, and d is the dimensionality of the system [2,3]. Because of its fermionic constituents, at higher densities, $n_{exc}a_B^d \approx 1$, the bosonic approximation fails [4] and a new ordered state should emerge [5].

[a] also affiliated with NTT Basic Research Laboratories.

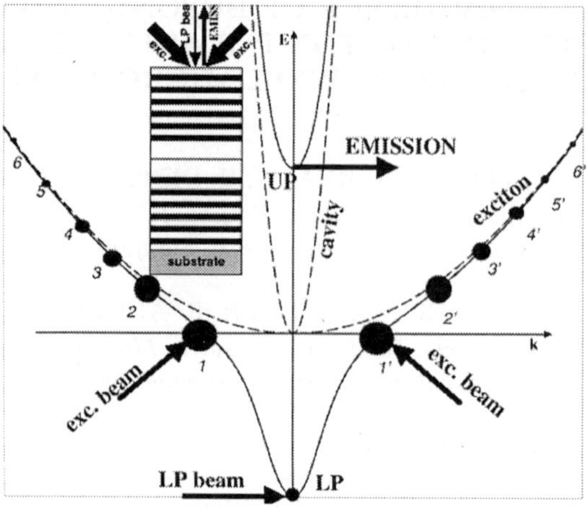

Figure 1: Dispersion of the MEP (solid lines), QW exciton and cavity photon (dashed lines). The exciton population is depicted by solid circles of proportional size. Inset: the cross section of the microcavity structure. Quarter wavelength dielectric stacks (DBR's) confine the photon inside the cavity, into which a GaAs quantum well, confining the exciton, is embedded. The experimental excitation scheme is also shown.

Microcavity exciton-polaritons (MEP) are hybrid states between a photon and an exciton in a semiconductor quantum well (QW) microcavity [6,7,8]. In the energy dispersion of the MEP (Fig. 1), strong coupling between the QW exciton and the cavity photon occurs only for small in-plane momenta, i.e., for $k_{\parallel} < 0.1k_0 \equiv 0.1n\omega_0/c$, where n is the index of refraction, ω_0 is the cavity frequency, and c is the speed of light. The radiative lifetime of the excitons at $k_{\parallel} \geq 0.1k_0$ is at least two orders of magnitude longer than for the MEP with $k_{\parallel} \approx 0$. Due to the reduced density of states at $k_{\parallel} < 0.1k_0$, the thermalization of the excitons with larger $k_{\parallel}$ to the MEP with $k_{\parallel} \approx 0$ by acoustic phonon emission is suppressed [9]. The excitons accumulate at $k_{\parallel} \simeq 0.1k_0$ and in the larger $k_{\parallel}$ region (bottleneck effect).

2 Stimulated Scattering of Excitons into MEP

We experimentally test the final state stimulation in binary exciton-exciton scattering, in which two excitons are created by two pump pulses at large $k_{\parallel}$ and are scattered into two final states, the upper polariton (UP) and lower polariton (LP) states at $k_{\parallel} = 0$ [10]. To artificially control a final state population, another probe pulse excites the LP at $k_{\parallel} = 0$. This LP population is expected to stimulate the exciton-exciton scattering rate by the bosonic enhancement factor. Because of the short radiative lifetime (approximately 2 ps) and the long thermalization time of the LP at low temperatures [9], and the long radiative lifetime and short thermalization time of the exciton, we can create the incoherent quasi-thermal exciton population and the coherent quasi-monochromatic LP population.

The quantum well microcavity sample consists of a 20 nm thick, single GaAs quantum well in a $\lambda/2$ $Al_{0.30}Ga_{0.70}As$ cavity layer. The cavity is sandwiched between two Bragg mirrors, consisting of 15.5 and 30 pairs of alternating layers of $Al_{0.13}Ga_{0.87}As$ and AlAs. The cavity mode is on resonance with the heavy-hole exciton, where the MEP splitting is 3.6 meV at 4.8 K. A mode-locked Ti:Al_2O_3 laser with pulse duration of 150 fs and spectral bandwidth of 16 meV is spectrally filtered via two gratings. We obtain two pulses of bandwidth 0.4 meV and duration 9 ps at different wavelengths. Their linewidth is 4.5 times narrower than the separation between the pump and probe energies. Two pump beams, resonant with the exciton energy, are incident on the sample at angles of $\theta = \pm 45°$. This incidence angle was chosen in order not to satisfy the energy and momentum conservation for parametric polariton amplification [11]. A probe beam is incident on the sample at an angle of 3° in order to excite the LP near $k_{\parallel} = 0$. The luminescence at the UP energy is detected in the normal direction. We have calibrated the exciton density n_{exc} and LP density n_{LP} from measurements of the average power, the spot size of the laser when focused on the sample, and an estimate of the absorption in the quantum well. We note that there is a negligible shift of the peak position of the UP and LP for the entire range of pump and probe intensities. We have confirmed that the initial exciton density is well below its saturation density, $N_s = 4 \times 10^{10}$ cm^{-2}. [12]

The UP population is described by a rate equation

$$\frac{dN_{UP,k}}{dt} = -\frac{N_{UP,k}}{\tau_{UP,k}} + a_{UP,k} n_{exc} + [b_{UP,k} + b'_k(1 + N_{LP})] n_{exc}^2. \qquad (1)$$

Here $\tau_{UP,k}$ is the radiative lifetime of the UP and N_{LP} is the LP population. The subscript k in each term indicates the in-plane momentum $k_{\parallel}$. The coefficient $a_{UP,k}$ represents the scattering rate of the exciton to the UP due to acoustic phonon absorption and emission, and was calculated by Fermi's golden rule applied to the Hamiltonian for deformation potential scattering (e.g., 6 $\rightarrow$UP$\pm$phonon in Fig. 1) [13]. The coefficient $b_{UP,k}$ represents the rate of exciton-exciton scattering, where the initial states are two excitons in the reservoir and the final states are the UP and another exciton in the reservoir (e.g., 1+6′ $\rightarrow$UP+4′ in Fig. 1). The coefficient b'_k represents the rate of exciton-exciton scattering, where the initial states are the two excitons at opposite

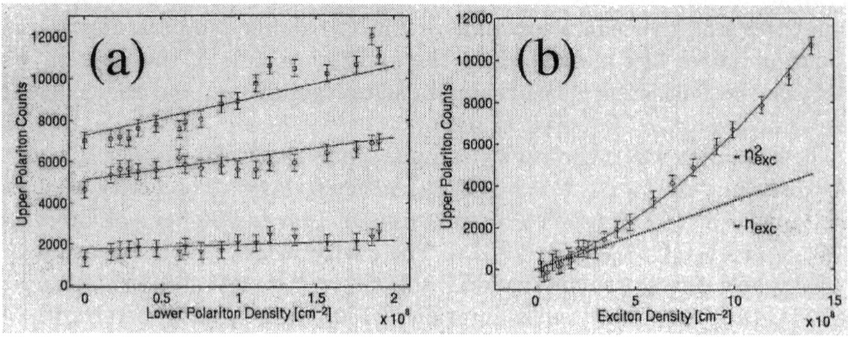

Figure 2: (a) The integrated UP emission intensity as a function of n_{LP}, for varying n_{exc}. The experimental points are indicated by the points with error bars, and the theoretical prediction is shown by the solid lines. The initial exciton densities are 1.5×10^9 cm^{-2} (squares), 1.2×10^9 cm^{-2} (circles), and 5.4×10^8 cm^{-2} (triangles). (b) The integrated UP emission intensity as a function of n_{exc}, for a fixed $n_{LP} = 1.1 \times 10^{-9}$ cm^{-2}. The dotted line indicates the acoustic phonon scattering contribution to the UP intensity, in the absence of exciton-exciton scattering.

in-plane momentum $\pm k_{\|}$ and the final states are UP and LP at $k_{\|} \approx 0$ (e.g., $1+1' \rightarrow$UP+LP in Fig. 1). The b coefficients have been calculated via Fermi's golden rule applied to the Hamiltonian for exciton-exciton scattering [10].

The rate equation model predicts that the UP luminescence has a quadratic dependence on n_{exc} and a linear dependence on n_{LP}. The quadratic dependence on n_{exc} has its physical origin in the two-body nature of the exciton-exciton collision. The linear dependence on n_{LP} is a consequence of the $1+N_{LP}$ dependence of the final state stimulation. Both dependencies are clearly shown in the experiment. The linear dependence on n_{LP} is shown in Fig. 2 (a): the positive slope of each line is a signature of the final state stimulation. The quadratic dependence on n_{exc} is shown in Fig. 2 (b).

The solid lines in Fig. 2 represent the numerical solution of Eq. (1), with subsequent time-integration of the numerical solution, in order to model the experimental time-integration of pulsed excitation. The only parameter varied to match the theory and experiment was the total quantum efficiency of detection. We measured the quantum efficiency of detection and used a value differing from the measured value by 10% in order to improve the fit.

To confirm that the observed final state stimulation is not a coherent optical four-wave mixing but is based on real collision of incoherent excitons, we changed the time delay τ between the pump and the probe pulses. On the positive τ side (the pump pulse arrives before the probe pulse), the UP intensity due to stimulated scattering should decrease exponentially with a time

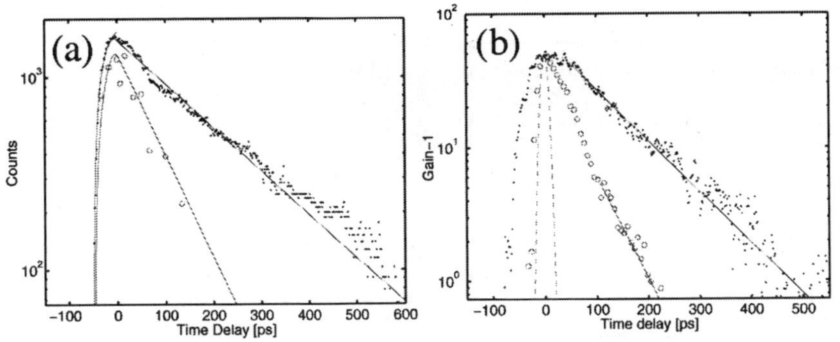

Figure 3: (a) Time dependence of the stimulated scattering rate and the bottleneck exciton population. The solid dots represent the exciton luminescence intensity measured by a streak camera; the open circles are the stimulated scattering counts into the UP state. The solid lines indicate exponential curves with time constants of 96 ps and 190 ps. (b) Gain-1 (open circles) and bottleneck exciton emission (solid dots) as a function of time delay. Solid lines indicate exponential decay curves of 114 ps and 57 ps. The dashed line indicates the convolution of the pump and the probe pulse envelopes for comparison.

constant which is exactly one-half the exciton population lifetime. The factor of one-half originates from the quadratic dependence of the scattering rate on exciton density, n_{exc}^2. In Fig. 3 (a), we show the time delay τ dependence of the stimulated scattering intensity. The pump and probe intensities are held constant with $n_{exc} = 1.0 \times 10^9$ cm^{-2} and $n_{LP} = 2.5 \times 10^9$ cm^{-2}. The stimulated scattering intensity (open circles) decreases with a time constant of ~96 ps. We have independently measured the lifetime of the exciton and obtained ~190 ps (solid dots). In the parametric polariton amplifier (or coherent optical four-wave mixing mediated by MEP), we would expect a decay time of about a few ps instead of the measured 96 ps [11].

We have also noted that if one of the two pump beams is blocked, the stimulated scattering is still observed. This indicates that the excitons quickly relax in the momentum space, creating $\pm k_{\parallel}$ bottleneck exciton populations which then scatter into the UP and LP at $k_{\parallel} \simeq 0$.

3 Dynamic BEC of Exciton-Polaritons

Recently, a proposal for the dynamic BEC of MEP was made [14]. In the proposed system, a coherent population of LP is formed spontaneously by the enhancement of the phonon-assisted exciton scattering rate into a LP ground state. Experimentally, the effect was not observed for the case of zero detuning GaAs MEP systems [15], due to the slow thermalization into the LP ground

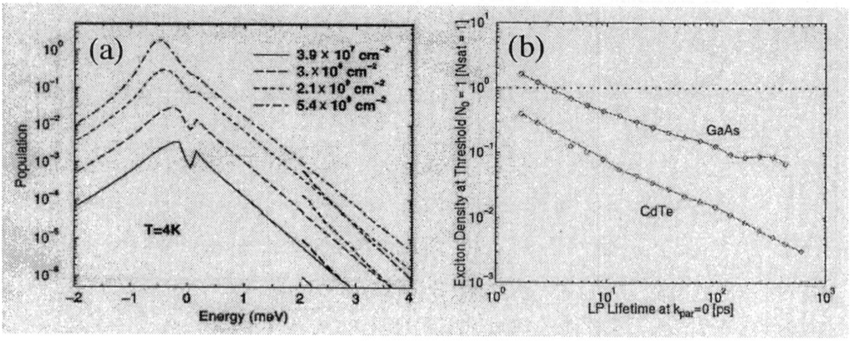

Figure 4: (a) The LP and UP populations per state vs. energy (measured from a bare exciton energy) at different exciton densities for a GaAs quantum well microcavity at 4K. The LP population starts at $E = -2meV$ and features a bottle-neck effect, while the UP population starts at $E = 2meV$ and features a conventional thermal distribution. (b) The threshold exciton density n_{exc}^{th} for the dynamic BEC vs. cavity photon lifetime for GaAs and CdTe systems.

state. More recently, several groups have reported positive results of final state stimulation in high-Q GaAs and CdTe microcavity systems [16].

Theoretical prediction for spontaneous build-up of bottleneck excitons in a GaAs quantum well microcavity is shown in Fig. 4 (a), in which the multi-mode rate equations [10] are numerically solved to obtain the steady state populations of the LP and UP as a function of their energies. At the exciton density of $n_{exc} = 5.4 \times 10^9$ cm^{-2}, the bottle-neck exciton population with $k_\parallel \simeq 0.1k_0$ and $E \simeq 0$ (measured from a bare exciton energy) exceeds the threshold of dynamic BEC: one exciton per state. In Fig. 4 (b), the threshold exciton density n_{exc}^{th} for the dynamic BEC, normalized by the exciton saturation density $n_{exc}^{s} \sim 0.1 \times (\pi a_B^{*2})^{-1}$, is plotted as a function of a cavity photon lifetime for GaAs and CdTe systems. We note that a cavity photon lifetime of longer than $\sim 10ps$ is required for observing the dynamic BEC in a GaAs system. On the other hand, a CdTe system should feature a dynamic BEC even for a relatively short cavity photon lifetime of $\sim 1ps$. The difference stems from ten-times larger saturation density of CdTe excitons compared to GaAs excitons.

The sample has two 80 $\mathring{A}$ CdTe quantum wells at the center, anti-node position of the $\lambda/2$ $Cd_{0.6}Mg_{0.4}Te$ cavity region. The distributed Bragg reflector mirrors are made of alternating index $\lambda/4$ layers of $Cd_{0.75}Mg_{0.25}Te$ and $Cd_{0.4}Mg_{0.6}Te$, with 16.5 (20) pairs on the top (bottom). The MEP splitting at zero detuning is 8.4 meV at T=4.5 K. The incident angle of the pump beam is about 25° (in-plane momentum $k_\parallel = 3.5 \times 10^4$ cm^{-1}), so that the

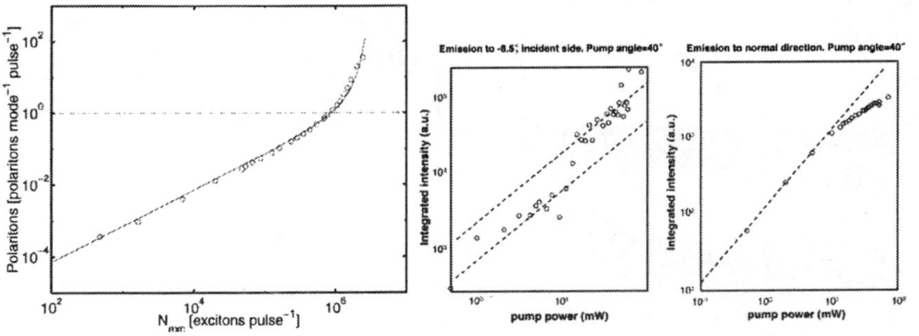

Figure 5: (a) The bottleneck exciton emission intensity, that is normalized to the population per mode, as a function of pump power. The solid line represents the theoretical result. (b) and (c): The emission intensities of the bottleneck exciton with $k_{\parallel} = 1.8 \times 10^4$ cm^{-1} and LP with $k_{\parallel} = 0$ vs. pump power.

polariton parametric amplification process, for which the maximum incident angle of the pump beam is $\theta_M \simeq 13.5°$,[17] disappears and the injected excitons only thermalize to the bottleneck excitons by elastic and inelastic scattering processes.

Fig. 5 (a) shows the bottleneck exciton emission intensity vs. the Ti:Al_2O_3 laser power. We monitor the bottleneck exciton emission at $k_{\parallel} = 1.8 \times 10^4$ cm^{-1} (emission angle of 13° in air). For a small pump power, the observed emission is created via acoustic phonon emission, leading to a linear dependence on the pump power. As the pump power is increased above about 10 mW, a dynamic BEC threshold is observed. The corresponding exciton density is approximately 3.5×10^{10} cm^{-2}, which is an order of magnitude smaller than the saturation density of 3×10^{11} cm^{-2} for CdTe QW excitons. We also confirmed the observed exciton emission intensity at the threshold corresponds to an exciton population per transverse mode equal to $\sim$1, which suggests the nonlinear increase of the emission intensity is indeed the onset of bosonic final state stimulation. To model the results shown in Fig. 5 (a), we use the coupled discrete rate equations for the LP with different in-plane momenta[10], with the CdTe parameters for the exciton-exciton and exciton-acoustic phonon scattering coefficients. The theoretical result is shown by the solid line in Fig. 5 (a). Fig. 5 (b) and (c) show the bottle-neck exciton emission at $k_{\parallel} \simeq 1.8 \times 10^4$ cm^{-1} and the LP emission at $k_{\parallel} \simeq 0$ vs. pump power over a wider range. At the threshold, the bottleneck exciton emission efficiency increases while the LP emission efficiency drops. The behavior is similar to the lasing and non-lasing modes in a photon laser threshold.

Using angular-resolved detection, the emission intensity vs. in-plane mo-

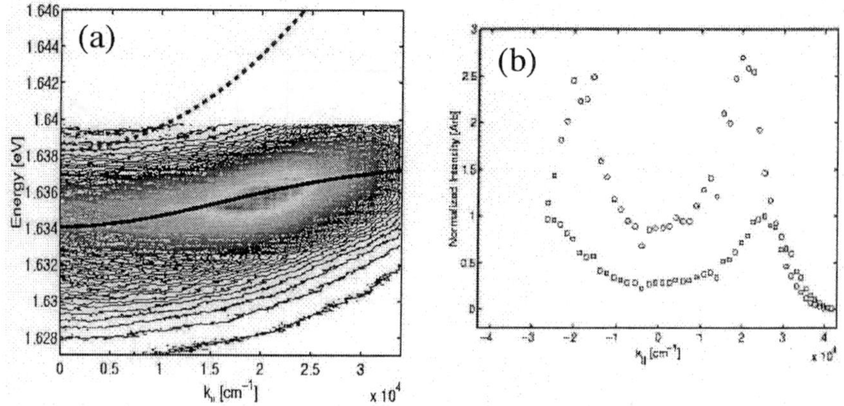

Figure 6: (a) The polariton spectra taken as a function of in-plane wavevector. The system is above the BEC threshold, and the solid line indicates the LP dispersion relation, while the dashed line indicates the cavity photon dispersion. (b) The LP emission intensity vs. in-plane wavevector for the pump rates below (squares) and above (circles) BEC threshold. LP emission intensities are normalized to the input pump power.

mentum of the LP is mapped out (Fig. 6 (a)). The LP population is peaked in the bottleneck regime, with $k_{\|} = 1.8 \times 10^4$ cm^{-1}. As the pump power is increased past the BEC threshold, the maximum of the emission slightly shifts toward smaller $k_{\|}$ and a transverse mode structure shows up due to the decreased spectral linewidth. The theoretical dispersion curves for the cavity photon and LP are plotted in Fig. 6 (a). It is clear from the effective mass of the condensed mode $m = \hbar^2 k_{\|}(dE/dk_{\|})^{-1} \gg m_{ph}$ (cavity photon effective mass), that the observed nonlinear emission is not a normal photon laser but a dynamic BEC of excitons. The LP emission intensity vs. emission angle is plotted in Fig. 6 (b) for the pump rates below and above threshold. The fact that the emission is symmetric (conical emission) is a unique and unambiguous signature of the dynamic BEC of bottleneck excitons. In the parametric polariton amplifier, the emission is strongly asymmetric due to the momentum conservation [11,17].

4 Matter-Wave Amplification of Excitons

We performed experiments probing the gain as a matter-wave amplifier below the BEC threshold, using a pump and probe set up. The pump bandwidth is about 3 meV, and selectively excites the excitons with $k_{\|} = 3.5 \times 10^4$ cm^{-1} (θ_p=25°). The probe bandwidth is about 0.6 meV, and selectively excites the bottleneck exciton with $k_{\parallel} = 1.8 \times 10^4$ cm^{-1} (θ=13°). The probe pulse produces a narrow bandwidth bottleneck exciton population, which is amplified by the surrounding exciton reservoir. Both pump and probe beams are horizontally polarized, so that there is no selective exciton spin excitation.

In Fig. 7 (a), we plot the reflected probe emission intensity for three initial exciton populations close to BEC threshold. The pump and probe pulses have zero time delay between them. The linearity of the reflected probe intensity as a function of input probe intensity is a signature of linear amplification. The dynamics of the bottleneck exciton population is modeled by [10]

$$\frac{dN_{LP}}{dt} = P_{LP} - \frac{N_{LP}}{\tau_{LP}} + a_{LP}N_{exc}(1 + N_{LP}) + b_{LP}N_{exc}^2(1 + N_{LP}), \quad (2)$$

where τ_{LP} is the bottleneck exciton lifetime, P_{LP} is the external injection rate, $a_{LP} = 47s^{-1}$ is the acoustic phonon scattering rate, and $b_{LP} = 0.45s^{-1}$ is the exciton-exciton scattering rate. The solid lines in Fig. 7 (a) represent the numerical integration of Eq. (2).

To probe the physical origin of the gain, we study the gain dependence on the pump power. In Fig. 7 (b) (c), the gain is plotted as a function of pump power. The gain follows an exponential-type dependence which is well reproduced by the rate equation model (solid line). The gain g is proportional to $\exp(const \cdot N_{exc}^2)$ in the high-gain regime, and g - 1 is proportional to $const \cdot N_{exc}^2$ in the low-gain regime. These experimental results directly indicate that the gain is provided by the real collision of reservoir excitons. In the polariton parametric amplification [11,17], the parametric gain is proportional to the pump intensity, that is, $g \propto \exp(const \cdot N_{exc})$.

The gain persists after the pump pulse has turned off, up to a time constant determined by the reservoir exciton lifetime. In Fig. 3 (b), we plot the measured gain as a function of time delay between pump and probe. The solid dots represent the measured bottleneck exciton emission intensity as a function of time. The characteristic decay time is ~114 ps. The gain, however, has a characteristic decay time of ~57 ps in the small gain regime. The difference of a factor of 2 in the decay times is due to the N_{exc}^2 quadratic dependence of the gain.

5 Conclusion

The dynamic BEC of bottleneck excitons is observed at an exciton density well below the exciton saturation density. In this regime, the quantum well is not inverted by electron-hole pairs, so the standard stimulated emission of photons does not take place. In this sense the dynamic BEC of bottleneck excitons

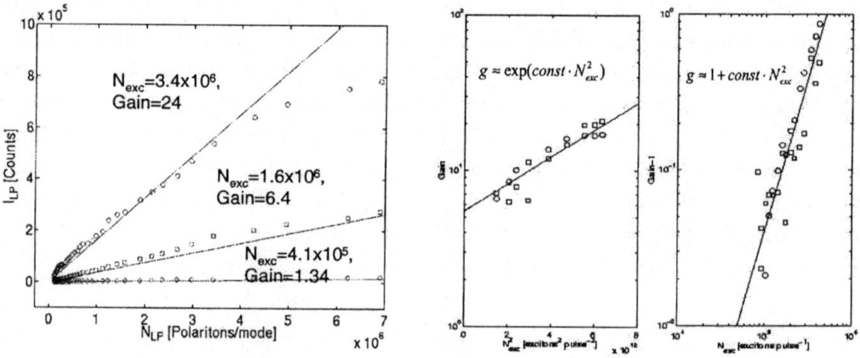

Figure 7: (a) The reflected probe intensity as a function of input probe power for three different total exciton population (gain). The solid lines represent the numerical integration of the rate equation 2. (b) Semi-log plot for gain as a function of the squared initial exciton population, restricted to the high gain regime. Solid line is linear, which indicates that $g \approx \exp(const \cdot N_{exc}^2)$ in the high gain regime. (c) Log-log plot for gain-1 as a function of initial exciton population. This plot is restricted to the low gain regime. Solid line has a slope of 2, which indicates that gain $g \approx 1 + const \cdot N_{exc}^2$ in the small gain region.

provides a much lower threshold for coherent light emission than a standard photon laser based on population inversion. The electrical injection of QW excitons, in particular the stimulated tunneling into a QW exciton-polariton state [18], is interesting from this viewpoint but yet to be demonstrated. The difference between parametric polariton amplifier and dynamic BEC of excitons has an optical analogue: an optical parametric amplifier vs. a laser amplifier [19]. This difference manifests in the $\exp(const \cdot |E_p|)$ gain dependence of the optical parametric amplifier, in contrast to the $\exp(const \cdot |E_p|^2)$ gain dependence of the laser amplifier, where $|E_p|$ is the pump laser amplitude. In the laser amplifier, real absorption of the pump photon occurs by a phase breaking transition. In the optical parametric amplifier, the pump photon is only virtually absorbed and split into signal and idler photons.

References

1. *Bose-Einstein Condensation*, ed. A. Griffin, D.W. Snoke and S. Stringari (Cambridge, New York, 1995).
2. J.L. Lin and J.P. Wolfe, *Phys. Rev. Lett.* **71**, 1222 (1993).
3. A. Mysyrowics *et al*, *Phys. Rev. Lett.* **77**, 896 (1996).
4. P. Kner *et al*, *Phys. Rev. Lett.* **81**, 5386 (1998).
5. C.D. Jeffries and L.V. Keldysh, *Electron-Hole Droplets in Semiconduc-*

tors (North Holland, Amsterdam, 1983); M. Nagai *et al*, *Phys. Rev. Lett.* **86**, 5795 (2001).
6. C. Weisbuch *et al*, *Phys. Rev. Lett.* **69**, 3314 (1992).
7. G. Khitrova *et al*, *Rev. Mod. Phys.* **71**, 1591 (1999).
8. Y. Yamamoto, F. Tassone and H. Cao, *Semiconductor Cavity Quantum Electrodynamics* (Springer, Berlin, 2000).
9. S. Pau *et al*, *Phys. Rev.* B **51**, 7090 (1995).
10. F. Tassone and Y. Yamamoto, *Phys. Rev.* B **59**, 10830 (1999); R. Huang *et al*, *Microelectron. Eng.* **47**, 325 (1999); R. Huang *et al*, *Phys. Rev.* B **61**, R7854 (2000).
11. P.G. Savvidis *et al*, *Phys. Rev. Lett.* **84**, 1547 (2000).
12. S. Schmitt-Ring *et al*, *Adv. Phys.* **38**, 89 (1989).
13. F. Tassone *et al*, *Phys. Rev.* B **53**, R7642 (1996).
14. A. Imamoḡlu *et al*, *Phys. Rev.* A **53**, 4250 (1996).
15. H. Cao *et al*, *Phys. Rev.* A **55**, 4632 (1997); M. Kira *et al*, *Phys. Rev. Lett.* **79**, 5170 (1997).
16. P.V. Kelkar *et al*, *Phys. Rev.* B **56**, 7564 (1997); L.S. Dang *et al*, *Phys. Rev. Lett.* **81**, 3920 (1998); X. Fan *et al*, *Phys. Rev.* B **56**, 15256 (1997); P. Senellart and J. Bloch, *Phys. Rev. Lett.* **82**, 1233 (1999).
17. C. Ciuti *et al*, *Phys. Rev.* B **62**, R4825 (2000).
18. H. Cao *et al*, *Phys. Rev. Lett.* **75**, 1146 (1995); G. Klimovitch *et al*, *Phys. Lett.* A **267**, 281 (2000).
19. Y. Yamamoto, *Nature* **405**, 629 (2000).

THEORETICAL ASPECTS OF PRACTICAL QUANTUM KEY DISTRIBUTION

NORBERT LÜTKENHAUS

Center for Modern Optics (ZEMO), Universität Erlangen-Nürnberg, Staudtstr. 7/B2, D-91054 Erlangen, Germany

E-mail: luetkenhaus@kerr.physik.uni-erlangen.de

Quantum key distribution (QKD) is the leading example in the field of quantum information that shows, how the explicit use of quantum mechanics allows to solve tasks that cannot be solved in the classical regime. Despite noisy and lossy transmission lines and detectors, we can make these systems work with todays technology. In doing so, we are able to maintain the rigorous notion of unconditional security that characterizes the idealized protocol by Bennett and Brassard [1]. In this article I introduce the concept of QKD and show the problems that occur in the realizations. These problems impose limitations on what we can achieve with QKD which we review together with a recent positive security proof.

1 Introduction

In a modern society, privacy of communication is important, e.g. to protect valuable commercial, military or diplomatic information. Protecting information against eavesdropping has a long history that is characterized by development of new schemes that are secure against known eavesdropping attacks, followed by new eavesdropping attacks aimed at cracking these new schemes. Since Shannon's information theory, we are able to investigate the security of cryptographic schemes in abstract terms independent of specific eavesdropping attacks.

It turns out, that we know only one encryption scheme that is secure against all eavesdropping attacks: the Vernam cipher (or one-time-pad). With this method, both parties need to share a secret key, expressed e.g. as a binary sequence, before the secure transmission can start. The message is expressed as binary sequence as well and a cryptogram is formed by a bitwise XOR operation between the message and the key of same length. The cryptogram can be sent over publicly accessible channels to the receiver, who recovers the message by another bitwise XOR connection between the cryptogram and his copy of the secret key.

This method can be proven to be unconditionally secure against any eavesdropping attack on the cryptogram given that the key is a) as long as the message, b) is never reused, and c) is chosen at random with flat distribution from all keys of same length.

All other currently used encryption schemes draw their security claim from assumptions that certain mathematical functions are hard to invert (one-way or trap-door function). An example is the task of factorization of large numbers. So far it has not been possible to prove that these functions are indeed hard to invert, in the sense of computational complexity, although a long history of failure to develop efficient algorithms is a strong indication in this direction. Therefore, these schemes are not unconditionally secure, nor can they be proven to be computational secure.

Let us come back to the Vernam cipher. We see that we can achieve secure communication, but at a prize: we need to distribute the shared secret key. Many application today require key distribution, and this problem is either solved by couriers (at the high end of the market), with the assumption that they can be trusted, or by key distribution systems like Diffie-Hellman algorithms that rely again on complexity assumptions.[2] On first sight, this shift of the problem from secure communication to the secure key distribution does not help, but this is not true. Indeed, it is an important conceptual step from the distribution of secure messages to the distribution of secure keys. There are two differences that are important here. The first is that in key distribution one is not forced to establish a specific key between sender and receiver; the key may emerge during a key distribution protocol without being determined beforehand. The second difference is that during or at the end of a key distribution protocol sender and receiver may decide whether to proceed to use the emerging key, or whether to abort the protocol. In the latter case no secret information has been exchanged and, therefore, can not have leaked to an eavesdropper. Only if the emerging key is approved, the secret information is encoded with it. At this point quantum mechanics comes into the game and allows to test the signals that lead to the key for eavesdropping activity. This step is based on the principle, that any measurement will change the signals. Only in absence of signs of such activity, the key will be approved for use.

These two elements, emergence of the key during the protocol and testing for eavesdropping, are essential parts of the quantum key distribution protocol presented by Bennett and Brassard. It will be described in the next section. In section 3 I will list the obstacles for a realization of the ideal protocol, e.g. noise, loss, non-ideal sources etc., and show how these obstacles can be overcome. In Sec. 4 I will show the limitations imposed by non-ideal realizations. Moreover, in that section we will show that we can nevertheless implement QKD with non-ideal components to obtain unconditional security.

2 The Bennett-Brassard Protocol (BB84)

Why should it be of advantage to use quantum mechanical states for key distribution? The reason is that quantum states can be made susceptible against eavesdropping attacks. Note that there exists no measurement that distinguishes two non-orthogonal signal states perfectly. Moreover, any attempt to get a hint which state out of the two has been prepared, will necessarily lead on average to disturbed states. This can be summarized in the following theorem:

THEOREM: If a measurement device leaves two non-orthogonal signal states $|u\rangle$ and $|v\rangle$ undisturbed, e.g. the output density matrices of the measurement ρ_u and ρ_v satisfy $\langle u|\rho_u|u\rangle = \langle v|\rho_v|v\rangle = 1$, then this means that the measurement device performs the trivial measurement on the signal subspaces spanned by $|u\rangle$ and $|v\rangle$.

PROOF: Any two signals can be written as

$$|u\rangle = a|0\rangle + b|1\rangle \qquad (1)$$
$$|v\rangle = a|0\rangle - b|1\rangle$$

where the states $\{|0\rangle, |1\rangle\}$ is an orthonormal basis and a, b are real numbers satisfying $a^2 + b^2 = 1$. The general ansatz for a measurement by Eve is represented by a unitary evolution U on the Hilbert space of the signal and the state space of an auxiliary system initially prepared in some state $|0\rangle_E$. This can be described by

$$\begin{aligned} &U\left(a|0\rangle \pm b|1\rangle\right)|0\rangle_E \qquad (2)\\ &= a\left(|0\rangle|U_{00}\rangle_E + |1\rangle|U_{01}\rangle_E\right)\\ &\quad \pm b\left(|0\rangle|U_{10}\rangle_E + |1\rangle|U_{11}\rangle_E\right)\end{aligned}$$

with $|U_{nm}\rangle_E$ as a set of four unnormalized and not necessarily orthogonal states. The condition that the signal states are unchanged implies that the projection of the final state onto the respective orthogonal complement $b|0\rangle \mp a|1\rangle$ vanishes. In two lines it follows that

$$|U_{00}\rangle_E = |U_{11}\rangle_E \qquad (3)$$
$$|U_{01}\rangle_E = |U_{10}\rangle_E\,.$$

Similarly, we require that the conditional state of the environment after projection of the state of Eqn. (2) onto the respective input state have unit norm. By making use of the partial results (3), this leads to the condition

$$\langle U_{00}|U_{00}\rangle = 1 \qquad (4)$$

$$|U_{01}\rangle_E = 0\,.$$

In consequence, the whole process is described by

$$U\left(\tilde{a}|0\rangle \pm \tilde{b}|1\rangle\right)|0\rangle_E = \left(\tilde{a}|0\rangle \pm \tilde{b}|1\rangle\right)|U_{00}\rangle_E\,, \tag{5}$$

and can therefore, at most, describe a unitary operation on the auxiliary system. In this way, no information at all about the complete Hilbert-space spanned by $\{|0\rangle, |1\rangle\}$ can be gained by Eve, since this equation holds for all values of $\tilde{a}$ and $\tilde{b}$.

The Bennett-Brassard protocol (BB84) [1] is the first complete protocol for quantum key distribution. Earlier ideas in that direction were brought up by Wiesner[3]. Sender and receiver, traditionally called Alice and Bob, are connected by a quantum channel that is under control by Eve, the eavesdropper. As signals pass through the channel, Eve can perform measurements on the signals to extract information. Additionally, Alice and Bob are connected by a public channel to exchange classical information. Eve can listen into this channel, but she is assumed not to be able to change or forge signals.

Alice and Bob need to test the quantum channel by sending non-orthogonal quantum states so that Eve's activity will be revealed. In our presentation of the protocol, we will use the language of the polarization Hilbert space of photons. So Alice will prepare at random single photons in the basis states of either the horizontal-vertical linear polarization basis, or the right-left circular polarization basis. States chosen from different bases are non-orthogonal. Bob measures the polarization at random and independently from Alice in the linear or circular polarization basis. They use the public channel to exchange the basis information of the signals and of the measurement and discard all those events where the bases do not match. What is left is the *sifted key* [4]. All we need to do, we can do with this sifted key.

In the absence of an eavesdropper, these signals are perfectly correlated and chosen at random, so they are suitable for the one-time pad. Moreover, by testing that the correlation is perfect, e.g. by publicly comparing some bits, we check that the theorem above applies and Eve cannot have any information on this key. In case that we find errors, we can discard the key and try a new key distribution session. It is clear that Eve can therefore block Alice and Bob from building up a secure key, but she cannot force a situation where Alice and Bob use a corrupted key. Once Alice and Bob declare a key safe, it is safe.

3 Problems in realistic implementations of the BB84 protocol

The above protocol has been presented in an idealized manner. We will now summarize the problems encountered in a practical realization of QKD and show how those problems can be solved.

3.1 Public Channel

The first problem addressed here is that of the public channel itself. It is assumed to be faithful, that means, Eve cannot change signals sent over this channel. The only thing she can do is to listen to the communications exchanged between Alice and Bob. We might think of a kind of radio broadcast with two stations for sender and receiver. However, there are two problems. First, there is no guarantee that any such implementation really cannot be influenced by Eve, and second, any implementation trying to approximate this scenario by physical means will become unpractical. What we really would like to do is to use simple connections such as telephone lines or digital data links.

At the latest with these implementations, we run into the problem of the separate-world attack (or man-in-the-middle attack). Here Eve isolates Alice from Bob both in the quantum and in the classical channel. Then she takes on Bob's role in the protocol at the end of the channels coming from Alice, and she takes on Alice's role at the channels leading to Bob. This means, that she establishes a key with Alice, and a different key with Bob. In this scenario there is no way for Alice and Bob to find out that they are sharing their keys not with each other, but with Eve respectively. Therefore they will use their respective keys to encrypt their secret messages. For example, Alice sends a message encrypted with her key. Eve will decrypt the message with the key she shares with Alice, read it, and then encrypt it with Bob's key and send it on to Bob. We see, that secure communication via QKD and Vernam cipher becomes impossible if we cannot guarantee that the public channel is faithful such that the communication between Alice and Bob cannot be altered.

We can solve this problem by adding the well established tool of *message authentication codes (MAC)* [5,6] on top of any physical communication channel. Precondition is that Alice and Bob do share some short initial key at the outset of the protocol. We use this initial key to authenticate the messages on the public channel. This will allow us to generate a new secret key that will be much longer than the initial shared key. However, it also implies an important change of the goal of QKD away from key generation between any parties towards a key growing protocol between parties sharing already some initial secret key. This change limits the use of quantum cryptography. For example, it cannot replace public key cryptography where sender and receiver need only a trusted distribution of public keys, but no shared secret between sender and receiver. Nevertheless, even in the for of quantum key growing we find that QKD is useful and fulfills a task that we cannot do within classical information theory.

It is important to note that MAC's do not rely on computations assumptions. They are proven to be secure with a clear security statement bounding the probability that Eve could forge an authentication tag. To authenticate a message of length m one needs a shared secret of approximate length $n_{auth} \approx 4t \log_2 m$ to assure that the probability P_{forge} that Eve can forge the authentication is bounded by $P_{forge} < 2^{-t+1}$. Note that for long messages we find $\lim_{m \to \infty} \frac{n_{auth}}{m} = 0$. We need to take the cost of authentication therefore only into account when doing an analysis for finite key sizes. However, the principle, that a shared secret key is required is independent of the key size.

3.2 Occurrence of errors

Clearly, in any real implementation we will find errors in the sifted key. This is due to noise that affects the quantum channel and the detection devices. If follows, that we can no longer abort the protocol whenever we find errors in the sifted key. Instead, we have to extend the protocol to cope with the existence of errors. The occurrence of errors implies two things: Alice and Bob do not share their key and Eve can have some partial information on the key.

For cryptographic purposes it is necessary for Alice and Bob to reconciliate their key. This can be done using error correction methods via the public channel. In general, those techniques involve Alice and Bob exchanging some more information about the key. It is necessary to protect this information from falling into Eve's hands. One way to achieve that is to encode the crucial information with secret bits via the one-time pad. This makes sense since Alice and Bob need to share anyway some initial secret key. Again, one has to be careful to assure that the protocol will generate more secure key bits than it uses up in the process. For authentication, the required number of secret bits per bit of the sifted key goes against zero in the limit of long keys, meaning that the cost of authentication is negligible if we only deal with large blocks of key material. This is different in the case of error correction. Here the number of required secret bits scales linearly with the number of sifted bits. The minimum number n_{ec} of those secret bits needed to correct n bits that are affected by an error rate e is given by the Shannon limit [7]

$$n_{ec} \geq -n\left[e\log_2(e) + (1-e)\log_2(1-e)\right] . \tag{6}$$

In principle, we cannot distinguish between errors caused by noise and errors caused by eavesdropping activity. Therefore, we have to assume that all errors are due to eavesdropping with the consequence, that Eve is now in possession of partial information on the key. To estimate the magnitude of this effect, we investigate a specific eavesdropping strategy: the intercept/resend attack. Here Eve intercepts the signals and measures it in one of the signal bases. Then she prepares a new photon corresponding to her measurement result and sends it to Bob. The analysis shows, that this attack gives an error rate of 25% and leaks 50% of the bits to Eve. Now, the typical error rate in an experiment might be only 5%, so that this can be explained as an eavesdropping strategy that uses this intercept/resend attack only on every fifth signal. Nevertheless, this attack provides Eve with 10% of the signals. This is too much for a direct use of the key for cryptographic purposes.

Fortunately, under many circumstances it is possible for Alice and Bob to obtain an unconditionally secure key from the sifted key. To see this, look at the following situation: Assume that Eve is in possession of partial information on all bits such that she guesses each bit correctly with probability $p_e = \frac{1}{2}(1+\epsilon)$. Alice and Bob can now divide their key into blocks of length m and go over to a new key formed by the parity bits of each block. A short calculation shows that Eve will guess each bit of the new key correctly with probability $p'_e = \frac{1}{2}(1+\epsilon^m)$. Therefore it is asymptotically possible to cut out Eve's knowledge from the key. This happens at the costs of the length of the key. The method proposed above is not efficient: more efficient methods exists and can be found either through hash functions [5], as was promoted in [8] and used in [9,10], or in a new interpretation of those results, in the form implied by quantum error correction codes [11]. The later way is another excellent way to understand how it is possible to create a secure key despite noise and loss. The idea is that one can protect the signal states against noise and loss on the quantum channel by encoding them with a quantum error correction code [12,13]. In this encoding, the nonorthogonal signal states can be transferred asymptotically error free from Alice to Bob, therefore guaranteeing security. Fortunately, the required operations can be mapped into the simple prepare-and-measure scheme of the BB84 protocol with some classical privacy amplification algorithm that involves taking parity bits of random substrings of the sifted and corrected key as the final key.

The important message of these methods is that it is possible to obtain a secure key as long as the raw material, here the sifted and corrected key, contains correlations of quantum mechanical nature.

3.3 Imperfect sources

Another problem for realizations is that we do not have a single-photon source at our hand so far. Several groups are making advances into that direction[14,15], but those sources are not ready for use in QKD yet. Other sources use parametric down-conversion together with coincidence measurements. [16,17] More widely used, however, are signal sources preparing weak coherent pulses (WCP), that is dimmed laser pulses with average photon numbers $\mu \approx 0.1$ or less.[18,19,20,21,22,23] The problem is that those sources have a Poissonian photon number statistics, that means that they emit with some probability not one but two or more photons. This gives a new handle for Eve to attack the signals. In these multi-photon signals, all photons are prepared in the same signal polarization. This means that Eve can split off one photon, store it until the basis of the signal is revealed in the public discussion between Alice and Bob, and then measure this photon in the signal basis. This Photon Number Splitting (PNS) attack [24,25] differs strongly from attacks on single photons: it reveals the complete information to Eve while it does not introduce errors, since the polarization of the photons remaining in the signal pulse is undisturbed. In that case the back-action of the measurement is noticed in the reduction of the photon number. This change, however, is compatible with a lossy channel. Therefore, this attack becomes more important as the losses in the channel grow, as we will see below. Still, it is clear that we can expect to be able to cope with this situation with the help of privacy amplification, within some limits that are discussed below.

4 Performance of practical QKD

Let us first describe the limitation of implementations of the BB84 protocol using weak coherent pulses over a lossy channel in the presence of transmission and detection errors. Then we will show that we can still extract unconditionally secure bits in the scenario developed in the previous section.

In all our considerations we will take the conservative point of view that *all* errors and *all* loss is due to eavesdropping activity although we know that some errors and some loss will always arise in Bob's detection apparatus, notably through detector dark counts and detection inefficiencies. In principle, it should be possible to take a less restricted viewpoint acknowledging the mechanism in a real detection device. However, it turns out that it is not only an open problem to derive security statements in that case, it is even a

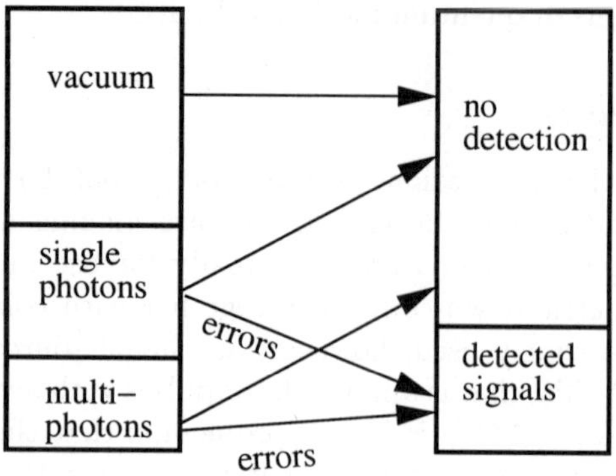

Figure 1. Transfer process in a noisy and lossy quantum channel. All processes leading to detection contribute to the error rate.

problem to draw a clear limit between what an eavesdropper can do and what he cannot do. For example, the dark count rate and detection efficiency of a gated detector might be influenced by a light pulse arriving between the active periods. It is, therefore, a problem to define to what extend those mechanism can change the dark count rate and detection efficiency from those values measured in the absence of the eavesdropper. Certainly, to go beyond the here adopted conservative point of view, one would need to make additional physical assumptions about Bob's detection devices. We would like to avoid such assumptions in our rigerous approach here

4.1 Bounds on the achievable secret bit rate

The use of weak coherent pulses in the presence of loss leads to a direct limitation of the achievable rate of secret bits that we can create. In a lossy channel (see Fig. 1) the various signals consisting to zero, one or more photons contribute to the detected and non-detected signals according to a simple pattern. All detected signals are affected by errors caused by the noisy quantum channel. The number of detected signals is less than the number of non-vacuum signals emitted by the source due to the loss in the channel.

Eve can replace this imperfect channel by a perfect one. Moreover, instead of having loss affecting the signal at random, she can determine which signals should reach Bob. Especially, she will make sure that as many multi-photon signals as possible will contribute to the detected signals, since those signals leak the complete contained information to Eve without the creation of errors.

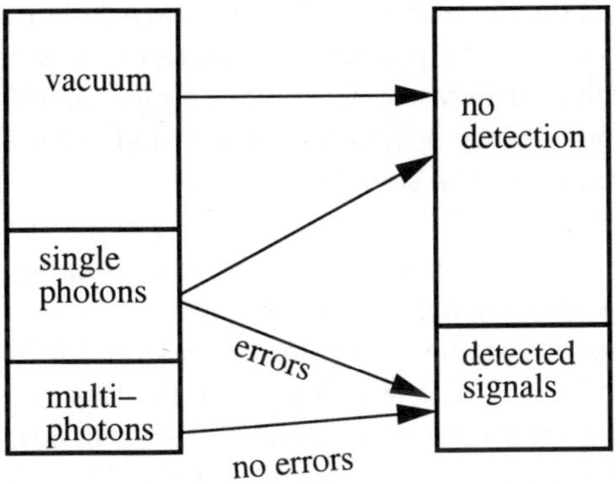

Figure 2. Transfer process in a quantum channel organized by the eavesdropper. All multi-photon signals lead to detection events. Only single-photon processes contribute to the error rate.

Her optimal pattern of transfer is shown in Fig. 2. Note that in this scenarios errors can stem only from eavesdropping on single-photon signals that enter the set of detected signals. As pointed out before, eavesdropping on multi-photon signals is possible without the occurrence of errors.

It is clear that only those signals can lead to secure keys that stem from single-photon signals. We denote the multi-photon probability of the source as p_{multi} and the fraction of detected signals as p_{exp}. From the argument made above, we can find immediately an upper bound on the number of secret key bits we can generate in our BB84 scheme with imperfect sources, namely by bounding the gain of secret bits per signal, G as [25]

$$G < \frac{1}{2}\left(p_{exp} - p_{multi}\right) . \tag{7}$$

Here the leading factor 1/2 stems from the sifting procedure of the BB84 protocol. It is instructive to evaluate this expression for a source with Poissonian photon number distribution with mean photon number μ and a quantum channel with a single-photon tranmission efficiency η. We then find

$$G < \frac{1}{2}\left\{(1-\mu)\exp(-\mu) - \exp(-\eta\mu)\right\} \tag{8}$$

We can optimize G by varying μ. In the limit of small η we obtain $\mu_{opt} \approx \eta$ and we find [25]

$$G < \frac{1}{4}\eta^2 . \tag{9}$$

This seems to indicate that even for arbitrarily high losses we can have a positive secure bit rate. In practical realizations, however, this is not the case, since single-photon detectors exhibit a signal independent dark-count rate. The above optimization scenario shows that with increasing loss the average photon number of the signals has to be reduced. That leads to even weaker signals arriving at Bob's detector. Meanwhile the dark count rate remains constant. At some point it will be the dominating effect and the quantum cryptographic scheme will break down.

We are able to determine an upper bound on the tolerable loss for a given dark count rate. [24] The starting point for these consideration is the question how many errors we can tolerate. We will argue that this error threshold is 25% within the single photon signals. Consider for this purpose again the intercept/resend attack of Eve where Eve measures the signals in one of the signal bases. Then she prepares a single photon corresponding to her measurement outcome and sends it to Bob. This procedure will result in an error rate of 25%. Moreover, it turns out that the correlations between Alice, Bob and Eve resulting from this procedure can be reproduced by public discussion. To see this note that whenever Eve happened to measure in the signal basis, all three parties share the same bit information. If Eve measured in the wrong basis, the bits in Alice's, Bob's and Eve's hand are random and independent of each other. Clearly, the three parties could have fixed such a distribution by first selecting at random half of the time slots and filling them with publicly announced bit values. Then each party would fill the remaining time slots independent of each other with random bits. It is clear that from such a public discussion no secret key can be distilled.

In order to apply this result to our scheme we have to rescale the observed error rate in the sifted key ϵ to that within the single photon signals. For this we go again back to the scenario that all errors result from eavesdropping on single-photon signals only. Then we have

$$\epsilon \frac{p_{exp}}{p_{exp} - p_{multi}} < \frac{1}{4} \tag{10}$$

as criterion for the possibility of creation of secure bits. Again, we can apply this criterion to a source with Poissonian photon number distribution (where we optimize over the mean photon number), and we can insert a lossy channel with single photon transmission efficiency η and a detection unit build up with two realistic photon detectors, each with a dark count rate d_B and a single photon detection efficiency η_B. This leads to the assignments (up to neglected higher order terms not relevant for our application)

$$\begin{aligned} \epsilon &= \frac{d_B}{p_{exp}} \\ p_{exp} &= 1 - \exp\left(-\mu\,\eta\,\eta_B\right) + 2d_B \\ p_{multi} &= 1 - (1+\mu)\exp\left(-\mu\right) \ . \end{aligned} \tag{11}$$

As a result, we find after optimization over μ for the transmission efficiency η has to be higher than the lower bound η_{WCP} [24]

$$\eta > \eta_{WCP} \approx 2\sqrt{d_B}/\eta_B \ . \tag{12}$$

The bounds given above can be combined into an estimate on the achievable rate for a source with multi-photon probability p_{multi} and an observed error rate ϵ within the detected and sifted key. This bound is given by

$$G < \frac{1}{2}\left(p_{exp} - p_{multi} - 4\,\epsilon\, p_{exp}\right) \tag{13}$$

and stems from the fact that at most this fraction of the records of Alice and Bob can not be explained by a process that can be replaced by classical communication between Alice, Eve and Bob. Note again, that this is the upper bound on the key generation rate given our conservative view that all loss and all errors are due to eavesdropping. The upper bound is evaluated for one example in figure 3.

Again, we can evaluate this bound for a Poissonian photon number distribution and a detection unit using two detectors described by d_B and η_B as above. Note that it is possible to determine the three quantities from Eqns. 11 in an experiment in order to obtain a direct upper bound on the number of secret bits that can be extracted by a error correction and privacy amplification protocol. We find for higher losses the bound

$$G < \frac{1}{4}\eta^2\eta_B^2 - 2d_B \ . \tag{14}$$

4.2 Provable unconditionally secure bit rate

In the previous section we derived clear bounds on the achievable rate and the tolerable loss, there evaluated for Poissonian photon sources. After working on bounds to show when secure quantum key distribution can no longer be secure, we finally present a result showing that it is indeed possible to create an unconditionally secure secret key even with an imperfect signal source that emits multi-photon signals with some probability p_{multi}. As we show in [26]

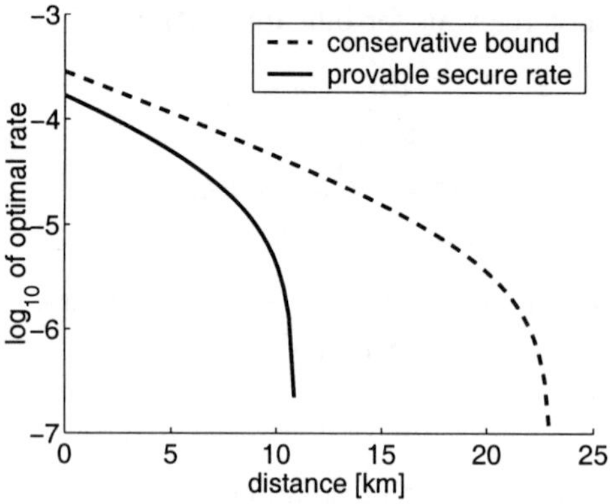

Figure 3. We show the upper bound of Eqn. 13 and the provable secure rate of Eqn. 15 for experimental parameters used and reported in Marand and Townsend[20]. These parameters describe an experiment at 1300μm with a channel loss of 0.38 dB/km, a receiver loss of 5 dB, an intrinsic error rate of 0.8%, and detectors with a dark count rate of 10^{-5} per slot and a detection efficiency of 11%.

the corresponding secure bit rate, taking into account privacy amplification and error correction, is given by

$$G_{sec} = \frac{1}{2}\left(p_{exp} - p_{multi}\right)\left[1 - H_1\left(\frac{2\epsilon}{1 - \frac{p_{multi}}{p_{exp}}}\right)\right] - p_{exp} H_1(\epsilon)\,. \tag{15}$$

Here we used the binary entropy function $H_1(x) = -x\log_2 x - (1-x)\log_2(1-x)$. We present here only the result for the limit of long keys. For the effects of finite key size see [26]. The leading factor $1/2$ is explained by the match of bases by Alice and Bob. The second factor $p_{exp} - p_{multi}$ takes care of the multi-photon signals, while the third factor $1 - H_1\left(\frac{2\epsilon}{1-\frac{p_{multi}}{p_{exp}}}\right)$ corresponds to the privacy amplification necessary due to the eavesdropping on the single-photon process (see Fig. 2). Note that therefore in this term the rescaled error rate $\frac{\epsilon}{1-\frac{p_{multi}}{p_{exp}}}$ appears. The last term takes care of the flow of information during error correction.

Note that for $\epsilon = 0$ and $p_{exp} > p_{multi}$ the key creation is positive. Since the rate depends continuously on ϵ, this implies a whole range of tolerable error rates. Actually, this formula is evaluated for an experimental set-up in Fig. 3. As we see, provably unconditionally secure QKD is possible over several kilometers even in our conservative picture. Here unconditional means that we do not put restrictions on Eve's ability to attack the signals, she is allowed to do the most general attack on all signals simultaneously that quantum mechanics allows. However, we do make assumptions about Alice's and Bob's apparatus. For Alice, we assume that the signals are prepared ideally according to our description (e.g. perfect polarization preparation of the weak coherent pulses). For Bob's apparatus we require that the detection probability for a signal is independent of the detection basis Bob chooses. Moreover, for both sides we assumed that Eve cannot intrude into the respective set-ups to read off polarization settings. Although this is a natural assumption for cryptographic devices (for classical or for quantum strategies) we have to be very careful in our situation. The reason is that we provide an optical link from the outside to the central pieces of our devices that might be used by Eve for intrusion attempts. Careful engineering is here required to assure that our assumption holds for real devices.

5 Conclusion

In this article we reviewed some of the problems we are faced with when implementing QKD with realistic means. We found, that the presence of noise and loss requires additions to the original protocol, namely error correction and privacy amplification, in order to make the scheme practical. It is interesting to note that it is not necessary to develop single-photon sources in order to make QKD a reality. Nevertheless, the Poissonian photon number distribution is at the moment the limiting factor in the realization of QKD.

Acknowledgment

Part of this research has been undertaken while the author was with MagiQ Technologies, Inc., in New York. The author is supported by the Deutsche Forschungsgemeinschaft via the Emmy-Noether-Programme.

References

1. C. H. Bennett and G. Brassard. Quantum cryptography: Public key distribution and coin tossing. In *Proceedings of IEEE International Conference on Computers, Systems, and Signal Processing, Bangalore, India*, pages 175–179, New York, December 1984. IEEE.
2. A. J. Menezes, P. C. van Oorschot, and S. A. Vanstone. *Handbook of applied cryptography.* CRC Press, Boca Raton, 1997.
3. S. Wiesner. Conjugate coding. *Sigact News*, 15:78, 1983.

4. B. Huttner and A. K. Ekert. Information gain in quantum eavesdropping. *J. Mod. Opt.*, 41(12):2455–2466, 1994.
5. J. L. Carter and M. N. Wegman. Universal classes of hash functions. *J. Comp. Syst. Sci.*, 18:143–154, 1979.
6. M. N. Wegman and J. L. Carter. New hash functions and their use in authentication and set equality. *J. Comp. Syst. Sci.*, 22:265–279, 1981.
7. T. C. Cover and J. A. Thomas. *Elements of information theory.* Wiley, 1991.
8. C. H. Bennett, G. Brassard, C. Crépeau, and U. M. Maurer. Generalized privacy amplification. *IEEE Trans. Inf. Theory*, 41:1915–1923, 1995.
9. D. Mayers. Quantum key distribution and string oblivious transfer in noisy channels. In *Advances in Cryptology — Proceedings of Crypto '96*, pages 343–357, Berlin, 1996. Springer. available as quant-ph/9606003.
10. D. Mayers. Unconditional security in quantum cryptography. Report quant-ph/9802025v4, (1998).
11. P. W. Shor and J. Preskill. Simple proof of security of the BB84 quantum key distribution protocol. *Phys. Rev. Lett.*, 85:441–444, 2000.
12. A. R. Calderbank and P. W. Shor. Good quantum error-correcting codes exist. *Phys. Rev. A*, 54:1098, 1996.
13. A. M. Steane. Error correcting codes in quantum theory. *Phys. Rev. Lett.*, 77:793, 1996.
14. P. Michler, A. Kiraz, C. Becher, W. V. Schoenfeld, P. M. Petroff, L. Zhang, E. Hu, and A. Imamoglu. A quantum dot single-photon turnstile device. *Science*, 290:2282–2285, 2000.
15. C. Santori, M. Pelton, G. Solomon, Y. Dale, and Y. Yamamoto. Triggered single photons from a quantum dot. *Phys. Rev. Lett.*, 86:1502–1505, 2001.
16. T. Jennewein, C. Simon, G. Wiehs, H. Weinfurter, and A. Zeilinger. Quantum cryptography with entangled photons. *Phys. Rev. Lett.*, 84:4729–4732, 2000.
17. G. Ribordy, J. Brendel, J.-D. Gautier, N. Gisin, and H. Zbinden. Long-distance entanglement-based quantum key distribution. *Phys. Rev. A*, 63:012309, 2001.
18. S. Chiangga, P. Zarda, T. Jennewein, and H. Weinfurter. Towards practical quantum cryptography. *Appl. Phys. B*, 69:389–393, 1999.
19. M. Bourennane, F. Gibson, A. Karlsson, A. Hening, P. Jonsson, T. Tsegaye, D. Ljunggren, and E. Sundberg. Experiments on long wavelength (1550nm) "plug and play" quantum cryptography systems. *Opt. Express*, 4:383–387, 1999.
20. C. Marand and P. T. Townsend. Quantum key distribution over distances

as long as 30 km. *Opt. Lett.*, 20(16):1695–1697, 1995.

21. P. D. Townsend. Experimental investigation of the performance limits for first telecommunications-window quantum cryptography systems. *IEEE Photonics Technology Letters*, 10:1048–1050, 1998.
22. R. J. Hughes, G. L. Morgan, and C. G. Peterson. Practical quantum key distribution over a 48-km optical fiber network. *J. Mod. Opt.*, 47:533–547, 2000.
23. G. Ribordy, J.-D. Gautier, N. Gisin, O. Guinnard, and H. Zbinden. Fast and user-friendly quantum key distribution. *J. Mod. Opt.*, 47:517–531, 2000.
24. G. Brassard, N. Lütkenhaus, T. Mor, and B. Sanders. Limitations on practical quantum cryptograpy. *Phys. Rev. Lett.*, 85(6):1330–1333, 2000.
25. N. Lütkenhaus. Security against individual attacks for realistic quantum key distribution. *Phys. Rev. A*, 61:052304, 2000.
26. H. Inamori, N. Lütkenhaus, and D. Mayers. Unconditional security of practical quantum key distribution. quant-ph/0107017, 2001.

DIAGNOSING INVISIBLE CANCER WITH TRI-MODAL SPECTROSCOPY

MARKUS G. MÜLLER, IRENE GEORGAKOUDI AND MICHAEL S. FELD
Massachusetts Institute of Technology, George R. Harrison Spectroscopy Laboratory, 77 Massachusetts Avenue, Room 6-014, Cambridge, MA, USA

Abstract

Dysplasia, which is often invisible, is the precursor to cancer. This study evaluated the use of three spectroscopic techniques: intrinsic fluorescence, diffuse reflectance and light scattering spectroscopy (LSS), to detect precancerous lesions. We evaluated low and high-grade dysplasia in Barrett's esophagus, and squamous intraepithelial lesions in the cervix. Fluorescence spectra at 11 excitation wavelengths and a white light diffuse reflectance spectrum were obtained in less than one second from a tissue site prior to biopsy, using an optical fiber probe. The intrinsic fluorescence was extracted by combining the measured fluorescence spectra with the diffuse reflectance spectrum at the same site, from which biochemical information was obtained. Analysis of the diffuse reflectance provided morphological information about tissue structure and biochemical information about tissue absorbers. LSS was used to obtain information about epithelial cell nuclear enlargement and crowding. The combination of the three techniques, tri-modal spectroscopy (TMS), gave more accurate results than any one technique alone. The excellent results indicate the potential of using TMS for diagnosis without tissue removal.

1 Introduction

Approximately 90% of all cancers develop in the epithelium, a thin layer of cells lining the body's surfaces. Changes in the epithelium and the underlying connective tissue (stroma) are often referred to as *neoplasia*, and the pre-invasive state proceeding most cancers as *dysplasia*. Dysplastic changes, mostly invisible, include increased metabolic activity and cell proliferation, reduced structural organization and increased size and density of cell nuclei [1].

Spectroscopic techniques can be implemented clinically to provide information about microstructural changes (morphology) and biochemistry. This can provide a rich source of diagnostic information and further our understanding of the basic processes involved. Numerous studies have used fluorescence as a tool for diagnosing disease. Its success in detecting neoplastic change is due to the fact that the development of neoplasia is accompanied by modifications in the biochemical composition of tissue which reflect themselves in the fluorescence spectrum. Promising results have been obtained in a number of tissues, demonstrating the potential of this technique as a clinical tool to guide or even replace biopsy [2-4].

However, quantitative analysis of tissue fluorescence has been difficult, as the spectra can be highly distorted by tissue scattering and absorption. Elimination of these distortions, mostly due to hemoglobin absorption, and recovery of the intrinsic tissue fluorescence, is not trivial. In the following, we present a method, intrinsic fluorescence spectroscopy (IFS), that overcomes these difficulties and enables the intrinsic fluorescence to be extracted by combining information from the measured fluorescence and the diffuse reflectance.

Diffuse reflectance spectroscopy (DRS) has also shown promise as a diagnostic tool. Diffusely reflected white light enters the tissue and scatters multiple times before it is collected at the tissue surface. Spectral analysis of the re-emerging light, based on diffusion theory, provides biochemical information due to absorption and structural information due to scattering [5]. Such information has been used to differentiate between normal and dysplastic colon tissue [5].

Light scattering spectroscopy (LSS) studies the spectral features of photons singly backscattered from cell nuclei. Using Mie theory analysis, the size distribution of nuclei contained in the epithelial cell layer can be obtained [6]. The potential of LSS as a diagnostic tool has been demonstrated in the esophagus, the bladder, the oral cavity, the colon [7] and the cervix [8]. Changes in nuclear morphology are well-established biomarkers of pre-cancerous and cancerous change. Nuclear enlargement and crowding are key parameters used by pathologists to diagnose dysplasia.

IFS, DRS and LSS can be used individually to diagnose disease, or in combination, a method we call tri-modal spectroscopy (TMS) [8,9]. Acquiring fluorescence and diffuse reflectance from the same spot at the same time enables analysis of each tissue site with all three techniques. This provides a combination of biochemical and structural information about early neoplastic change, without the need for tissue excision and processing. Early detection of dysplasia is very important for successful treatment [10].

In this paper we apply TMS to two conditions for which patients are at high risk for cancer, Barrett's esophagus (BE) and dysplasia in the uterine cervix. BE develops due to acid reflux from the stomach, which causes an inflammatory response in which the normally pink tissue (squamous epithelium) is replaced by red (columnar) epithelium. Patients with BE have a high risk of developing dysplasia and cancer. Since dysplasia is invisible to the eye, patients with BE undergo periodic GI endoscopic surveillance in which random biopsies are taken. Unfortunately, random sampling often misses sites of dysplasia. In any case, if high grade dysplasia is found, removal of the esophagus is recommended.

The first step in detecting cervical dysplasia is Papanicolaou smear screening, a simple but inaccurate technique. Abnormal Pap smears are followed by examination of the cervix with a colposcope, a low power microscope with a long working distance. Application of acetic acid creates whitened areas which may indicate the presence of dysplasia. Such regions are biopsied and examined by a pathologist. This process is expensive and time consuming, and not always accurate.

2 Procedures

Fluorescence and reflectance spectra were taken in patients undergoing colposcopy and GI endoscopy using a "FastEEM" [11]. This instrument, built in our laboratory, provides 11 laser excitation wavelengths between 337 and 610 nm used to acquire a fluorescence excitation-emission matrix (EEM), and broadband white light excitation (350-750 nm) for obtaining diffuse reflectance. (An EEM is a 3-D or contour map in which fluorescence intensity is plotted as a function of excitation and emission wavelength to fully characterize the fluorescence properties of a sample.) The excitation light is coupled into a 1 mm diameter optical fiber probe, which consists of a single light delivery fiber surrounded by six collection fibers (fused silica, 200 μm core diameter, 0.22 NA). Light emitted from tissue is collected by the probe and coupled to a spectrograph and optical multichannel analyzer detector. Data acquisition and storage are computer-controlled.

The optical fiber probe is brought into gentle contact with the tissue, and fluorescence spectra at 11 excitation wavelengths and white light reflectance spectra are acquired in less than five seconds. The probe is then removed, leaving a temporary mark at the site at which spectra were taken, facilitating accurate biopsy sampling of the same site. The BE spectra were obtained from 7 patients during routine GI endoscopy. Cervical spectra were collected from 44 patients undergoing colposcopy. Biopsies from the cervix and BE were examined and classified by experienced pathologists, and then the classification was correlated with the results of spectroscopy.

We used a statistical technique called logistic regression [12] to correlate the results of spectroscopy with pathology, used as the standard. To establish the ability of spectroscopy to accurately predict the presence of disease (sensitivity) and non-disease (specificity), we employed a cross-validation scheme in which one data point was removed, the remaining data was used to develop a classification scheme, and the result was applied to the omitted data point [13]. This process was repeated until all the data was evaluated.

3 Methods and Results

The measured fluorescence and reflectance spectra were analyzed with three spectroscopic techniques: IFS, DRS and LSS.

A) **IFS**: To extract the intrinsic tissue fluorescence, fluorescence and reflectance spectra were combined using a photon migration-based model [14]. This model is based on the fact that fluorescent and reflected photons undergo similar scattering and absorption events as they traverse the tissue. Thus, information from the distortions introduced in tissue reflectance spectra by scattering and absorption can be used to remove such distortions from the measured tissue fluorescence spectra. Unlike the measured tissue fluorescence spectra, which consist of a non-

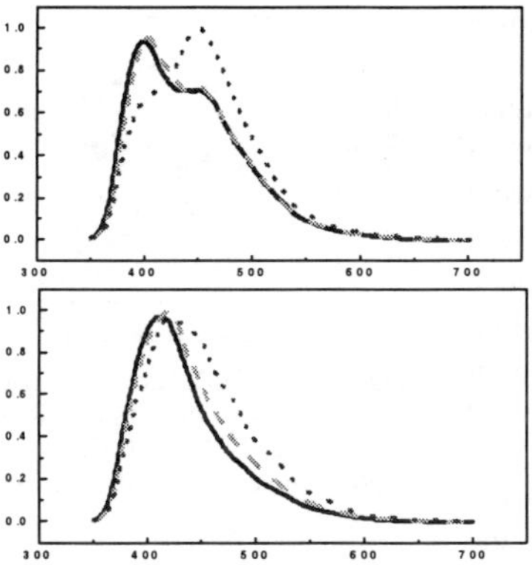

Figure 1: Non-dysplastic BE (solid line), low-grade (dashed line) and high-grade dysplasia (dotted line) sites at 337 nm excitation. Top: Measured fluorescence. Bottom: Corresponding intrinsic fluorescence. All spectra are normalized.

linear combination of contributions from tissue fluorescence, scattering and absorption, the intrinsic tissue fluorescence spectra are a linear combination of the fluorescence spectra of individual tissue biomolecules. Measured and modeled intrinsic fluorescence spectra for normal BE (solid line), low-grade dysplasia (dashed line), and high-grade dysplasia (dotted line) sites at 337 nm excitation are shown in Figure 1. Note the difference in line shape between the measured and intrinsic fluorescence, especially in the 420 nm region of the spectrum, where hemoglobin distortions are prominent. As seen in the lower panel, as the extent of dysplasia increases there is a shift to the red. In addition, we observe a higher intensity in the intrinsic fluorescence from normal sites than from dysplastic sites (not shown).

Similar behavior is observed in the cervix (Fig. 2). We compared tissue sites with cervical dysplasia (squamous intraepithelial lesions, SIL); squamous metaplasia (SQM), a normal condition in which tissue is transformed from columnar to squamous; and normal squamous epithelium (NSE). The intensities of SQM and SIL sites are smaller than those of NSE sites, and their line shapes are broadened and shifted to the red region of the spectrum. For both tissue types, changes in the fluorescence line shape and intensity are most prominent at 337 nm excitation, and more pronounced in the intrinsic fluorescence than in the measured fluorescence. The dominant fluorophores in the visible range of the spectrum are NADH and collagen (Sec. 4) [15]. The intrinsic fluorescence tissue spectra could be accurately fit to a linear combination of NADH and collagen EEMs, from which the NADH and collagen contributions could be extracted. The

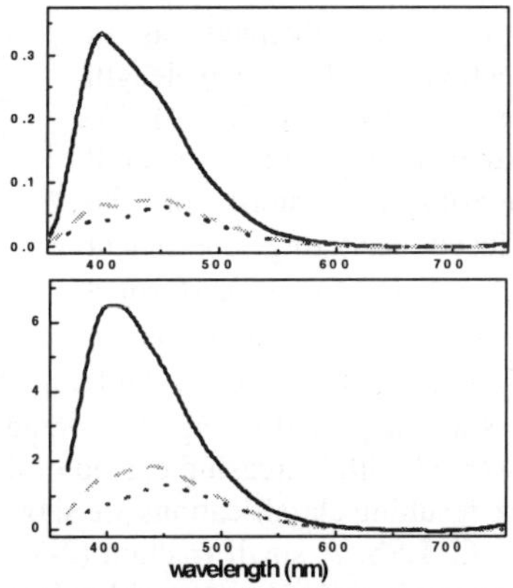

Figure 2: NSE (solid line), SQM (dashed line) and SIL (dotted line) sites at 337 nm excitation. Top: Measured fluorescence. Bottom: Corresponding intrinsic fluorescence.

resulting classifications for the data set are shown in Fig. 3 and the first line of Table 1.

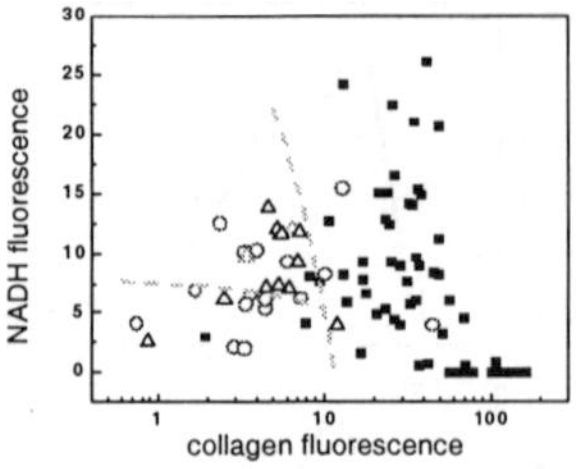

Figure 3: Collagen and NADH contributions to the intrinsic fluorescence EEMs of NSE (squares), SQM (circles) and SIL (triangles) sites.

B) **DRS:** An example of a reflectance spectrum, from a non-dysplastic BE site, is shown in Fig. 4. Diffuse reflectance spectra were analyzed using a model based on light diffusion theory to extract the absorption and reduced scattering coefficients of tissue [5]. Hemoglobin, which is present only in the connective tissue layer underlying the epithelium, is the main absorber in the visible range of the spectrum for these tissues. The scattering properties of diffusely reflected light are determined mainly by the collagen fibers of the underlying stromal layer. Thus, DRS mainly provides information about the morphology and biochemistry of the stroma. The diffusion model was fit to the experimental data, from which the absorption, (μ_a) and reduced scattering coefficient spectra (μ_s'), were obtained as a function of wavelength (smooth curve of Fig. 4). To characterize the overall intensity and slope of the reduced scattering coefficient (μ_s'), a linear fit was performed to $\mu_s'(\lambda)$. The intercepts at λ=0 nm and the slopes of the lines fitted to the extracted $\mu_s'(\lambda)$ for the cervical data sites are displayed in Fig. 5. A trend towards smaller and flatter values of $\mu_s'(\lambda)$ is observed with increasing extend of dysplasia. Similar results were obtained for BE. The resulting classifications are shown in line 2 Table 1.

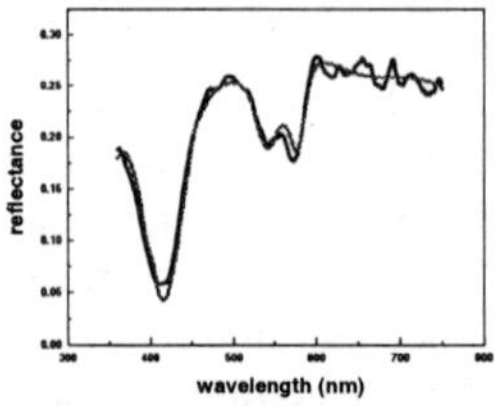

Figure 4: Reflectance spectrum of a non-dysplastic BE site. The smooth curve is the fit is the model of Ref. [5].

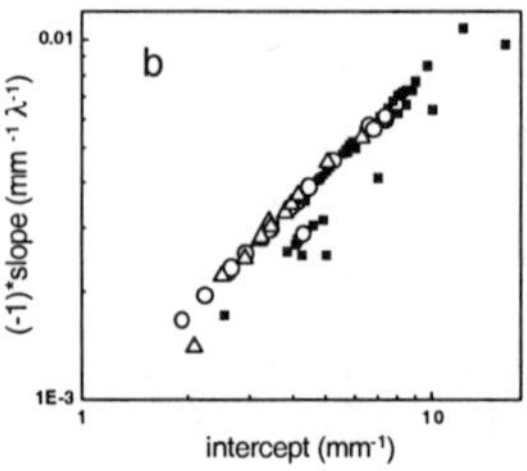

Figure 5: Slopes and intercepts of the wavelength dependent reduced scattering coefficients, extracted from the model. Squares: NSE; circles: SQM; triangles: SIL.

C) **LSS**: A small fraction (2-5%) of the back reflected light is due to photons that are *singly* backscattered by the cell nuclei of the epithelium The nucleus is the major contributor to this type of light scattering because it has a slightly (6%) higher index of refraction than the cytoplasm. The LSS spectrum in the backward direction can be analyzed by Mie scattering theory to extract nuclear size and density (crowding) [6]. This spectrum is extracted from the experimentally obtained reflectance spectrum by subtracting the diffuse component, obtained using the model of Zonios *et al* [5]. The intensity of the backscattered LSS spectrum varies in an oscillatory manner in inverse wavelength space. The frequency of these oscillations is proportional to the size of the scatterers (cell nuclei), and their amplitude is proportional to the density of scatterers, which in this case indicates nuclear crowding. The size distribution of the epithelial cell nuclei can be obtained by fitting this data to Mie theory or the van de Hulst approximation [6].

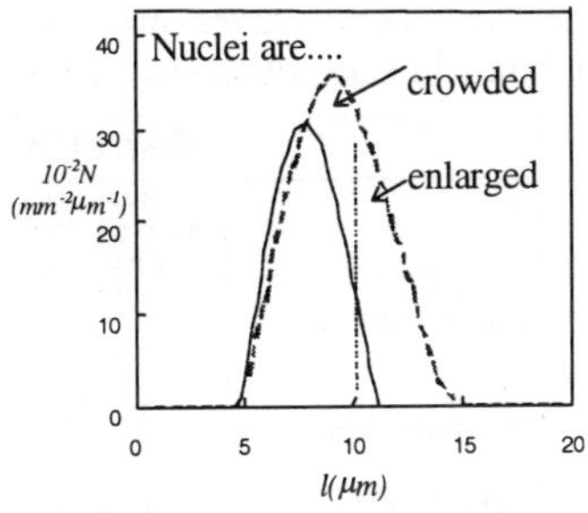

Figure 6: LSS analysis of a non-dysplastic BE site (solid line) and a dysplastic site (dashed line). The enlargement and increased nuclear number density associated with dysplasia can be seen.

Figure 6 shows nuclear size distributions of non-dysplastic and dysplastic BE tissue sites obtained using LSS analysis. As can be seen, the mean diameter of cell nuclei is increased for the dysplastic site, and the area under this curve is larger, indicating crowding. Distribution curves of this type were used to classify the data. For the BE data two parameters, percentage of enlarged nuclei (defined as nuclear diameters greater than 10 μm) and number density of nuclei (crowding) were used as diagnostic parameters. For the cervix data, crowding and width of size distribution curves were the two parameters used. A binary decision plot for the cervical tissue data is shown in Fig. 7. For both BE and cervix data, good agreement with the results of pathology is obtained (Table 1, line 3).

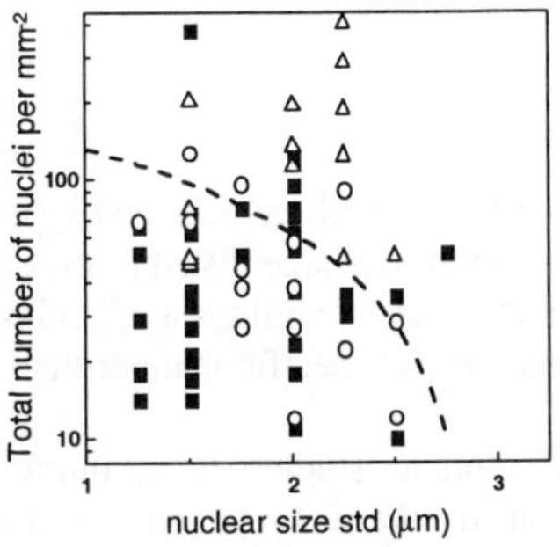

Figure 7: Binary decision plot of cervix data. Nuclear number density is plotted as a function of the standard deviation of the nuclear size distribution for NSE (squares), SQM (circles) and SIL (triangles) sites.

D) **TMS** Table 1 lists the results of spectral diagnosis for each of the techniques separately and the three techniques combined (TMS), using pathology as the standard. The combined diagnosis of each site was obtained as the diagnosis for which the results of at least two of the three techniques agreed. For the cervix data, LSS exhibits the highest sensitivity of the three spectroscopic techniques. However, the highest specificity was achieved by IFS. As can be seen, prediction accuracy increases significantly for the combined TMS analysis (last row of Table 1). In BE, the small data set (7 patients) gives perfect agreement with pathology for each technique. To show the advantage of having several diagnostic parameters, and therefore the value of TMS, we show a larger data set in Table 1, where diagnostic variability becomes apparent. In the latter data set a red rise in fluorescence at 397 and 410 nm excitation was also used as diagnostic indicator in IFS. Here, it can also be seen that the combination of the three techniques gives an almost perfect prediction of disease, with the best individual results obtained for LSS.

	Barrett's esophagus		Uterine cervix	
	dysplastic spectroscopy / pathology	*non-dysplastic* spectroscopy / pathology	*dysplastic* spectroscopy / pathology	*non-dysplastic* spectroscopy / pathology
Intrinsic fluorescence (IFS)	**11/14**	**23/26**	**8/13**	**64/70**
Diffuse Reflectance (DRS)	**11/14**	**23/26**	**8/13**	**58/70**
Light scattering (LSS)	**13/14**	**25/26**	**9/13**	**58/70**
Combination of IFS, DRS and LSS	**13/14**	**26/26**	**12/13**	**63/70**

Table 1: Statistical analysis of the cervical and BE data for each technique and the three techniques combined (last row).

4 Discussion

TMS analysis of fluorescence and reflectance provides three different types of information: biochemical, stromal scattering/absorption, and size distribution of the epithelial nuclei. Combining this information enables us to noninvasively classify cervical and esophageal epithelium in a more sensitive and specific manner than any one technique alone.

A number of researchers have conducted clinical studies to examine the potential of fluorescence to diagnose dysplasia in the cervix and esophagus [4,16,17]. Most of these studies simply attempt to correlate spectral features with development of disease, and none of them have related the presence of specific tissue fluorophores to diagnosis. In contrast, the current study uses modeling to extract quantitative biochemical and morphological features of the type employed qualitatively by pathologists in making diagnoses, and thus to create a tissue classification scheme that is based on the underlying causal features. We acquire data from a macroscopic sample approximately 1 mm^3 in volume, from which we extract average information about tissue structure and chemistry at the microscopic level. In both the cervix and esophagus spectra, the undistorted (intrinsic) fluorescence exhibited broadening and a shift to longer wavelengths with development of disease (especially with 337 nm excitation), which suggests that the underlying biochemical changes are similar in the two tissue types. The intrinsic fluorescence spectra were shown to be a linear combination of two biochemical spectral components, NADH and collagen, and these components were used to make diagnostic decisions. In the cervix, we found that the collagen contribution could distinguish well between NSE and SQM/SIL, and the NADH contribution between SQM and SIL.

Decomposition of the intrinsic tissue fluorescence spectra into its biochemical components indicates the occurrence of interesting biochemical changes. The collagen content was found to be significantly lower in abnormal sites than in normal sites. This decrease could result from an increase in epithelial thickness of the abnormal sites, since in that case a reduced amount of light reaches the stromal layer, where collagen is found. It could also result from a reduced number of collagen cross-links within the abnormal tissue sites. It has been shown that collagenases, enzymes responsible for the degradation of collagen cross-links, are usually present in tissue areas undergoing significant architectural changes, as in the case of wound healing and tissue regeneration [18]. In addition, small differences can be found in the levels and/or patterns of expression of matrix metalloproteinases (MMPs), a family of collagenases, in normal squamous epithelium, squamous metaplastic, and SIL sites of the cervix [19].

Changes were also observed in the scattering properties of dysplastic and non-dysplastic tissue sites, as determined by analysis of the diffuse reflectance. The scattering properties of bulk tissue, measured with our system, are affected mainly by the collagen network of the stromal layer [20]. Thus, the observed decrease in the reduced scattering coefficient associated with cancer could be the result of changes in the density of the collagen network. However, an increase in the thickness of the epithelial layer could also contribute to this decrease.

As discussed in Sec. 3.C, we can also detect changes in epithelial cell nuclei associated with processes which occur in the development of cancer. Dysplastic nuclei contain more genetic material due to extra, often faulty copies of chromosomes, associated with genetic alterations in tissue. LSS is a powerful tool to measure this. Furthermore, because of the interferometric nature of Mie scattering processes, LSS can characterize structures two orders of magnitude smaller than the optical wavelength. In recent work we have also started to characterize this structure, and find it to be fractal; this will be discussed elsewhere.

The final step of our analysis combines the results of all three spectroscopic techniques (TMS) to obtain a diagnosis that classifies a particular tissue site. This results in an enhanced sensitivity and specificity for separating dysplastic from non-dysplastic tissue, as compared to the results of each one of the techniques used individually. This enhancement is expected, since the three spectroscopic techniques provide complementary information about different aspects of tissue biochemistry and morphology. These results suggest that TMS can serve as a guide to biopsy by enhancing the physicians ability to detect lesions at an early stage.

In summary, TMS analysis provides a rich source of quantitative information that allows us to further our understanding of the changes that take place during tissue transformation. We are currently in the process of developing software that will allow data analysis to take place in real time while spectroscopic data is acquired. In addition, we shall continue to acquire data to further develop and refine our diagnostic algorithms. We plan to perform further basic research studies to understand the origin of the biochemical and morphological changes we detect, which, in turn, could lead to refinements in the manner with which we collect and analyze spectra. Ultimately, we want to incorporate TMS into an LSS imaging modality, in order to be able to first screen a large area, and then zoom in on suspicious sites that require more thorough analysis by TMS.

5 Acknowledgement

This work was carried out at the MIT Laser Biomedical Research Center, supported by NIH grants P41RR02594 and CA53717. Irene Georgakoudi acknowledges support from a NIH NRSA fellowship. We also would like to thank Vadim Backman, Ellen Sheets, Christopher Crum, Brian Jacobson and Jacques Van Dam for helpful discussions.

Reference

1. Epithelial cells are typically 20 μm in dimension and the nuclear diameters can range from 5 μm (normal) - 15 μm (dysplastic). The epithelial layer can be one or more cells thick
2. Wagnières G.M., Star W.M., and Wilson B.C., *J Photochem Photobio*, 1998. 68: p. 603-632.
3. Ramanujam N., *Neoplasia*, 2000. 2: p. 89-117.
4. Vo-Dinh T., Panjehpour M., and Overholt B.F., *Ann NY Acad Sci* 1998; 838: p. 116-122.
5. Zonios G., et al., *Appl Opt*, 1999. 38: p. 6628-6637.
6. Perelman L., et al., *Phys Rev Let*, 1998. 80: p. 627-630.
7. Backman V., et al., *Nature*, 2000. 406: p. 35-36.
8. Georgakoudi I., et al., *Am J of Obstetrics and Gynecology*, in press 2001.
9. Georgakoudi I., et al. *Gastroenterology*, 2001. 120, p. 1620-1629.
10. Van Sandick J.W., et al., *Gut*, 1998; 43: 216-222.
11. Zângaro R.A., et al., *Appl Opt*, 1996. 35: p. 5211-5219.
12. Fisher L. and Van Belle G., *Biostatistics: A methodology for the health sciences* (John Wiley & Sons, New York, 1993).
13. Schumacher M., Hollander N., and Sauerbrei W., *Statistics in Medicine*, 1997. 16: p. 2813-2827.
14. Zhang, Q., et al., Opt Lett, 2000. 25: p. 1451-1453.
15. Richards-Kortum R. and Sevick-Muraca E., *Annu Rev Phys Chem*, 1996. 47: p. 555-606.
16. Ramanujam N., et al., *Proc Natl Acad Sci*, 1994. 91: p. 10193-10197.
17. Ramanujam, N., et al., *J Photochem Photobio*, 1996. 64: p. 720-735.
18. Ellis D.L. and Yannas I.V., *Biomaterials*, 1996. 17: p. 291-299.
19. Talvensaari A., et al., *Gynecol Oncol*, 1999. 72: p. 306-311.
20. Saidi, I., Jacques S., and Tittel F., *Appl Opt*, 1995. 34: p. 7410-7418.

NEW ADVANCES IN COHERENT ANTI-STOKES RAMAN SCATTERING (CARS) MICROSCOPY

JI-XIN CHENG, ANDREAS VOLKMER, LEWIS BOOK AND X. SUNNEY XIE

Harvard University, Department of Chemistry and Chemical Biology,
12 Oxford Street, Cambridge, Massachusetts, 02138
Email: xie@chemistry.harvard.edu

We present three approaches for improving the sensitivity and spectral selectivity of coherent anti-Stokes Raman scattering (CARS) microscopy by reducing the non-resonant background. First, epi-detection (or backward-detection) avoids the forward scattered background signal from a bulk solvent. Second, excitation with two synchronized near IR picosecond laser beams increases the signal to background ratio. Third, polarization-sensitive detection suppresses the non-resonant background from the sample.

Multiphoton microscopy has become a powerful tool for biology. The nonlinear power dependence of a multiphoton process reduces the excitation volume and provides the three-dimensional sectioning capability. A multiphoton process can be incoherent (*e.g.*, fluorescence) or coherent (*e.g.*, multi-wave mixing). The coherent signals include second-harmonic generation, third-harmonic generation, and CARS, among which CARS is particularly attractive because of its inherent chemical selectivity. CARS is a four wave mixing process that involves a pump beam at frequency of ω_p, a Stokes beam at frequency of ω_s, a probe beam usually at the same frequency as the pump beam, and a signal beam at the anti-Stokes frequency of $2\omega_p-\omega_s$. The CARS signal is enhanced when the frequency difference between the pump and the Stokes beam (referred as the anti-Stokes shift below) coincides with a Raman-active vibrational frequency. Thus, CARS microscopy can be used for vibrational imaging of specific species without any fluorescent labels.

Duncan *et al.* constructed the first CARS microscope in 1982, with two visible dye lasers and non-collinear beam geometry.[1] However, the spectral selectivity of their setup was limited by the large non-resonant background that arises from the two-photon enhanced electronic contributions. In 1999, Zumbush et al. revived CARS microscopy using two near IR femtosecond laser beams.[2] The use of near IR excitation avoids two-photon enhancement of the nonresonant background and thus improves the vibrational contrast. Another example of progress in the same work is the use of tightly focused collinear beams, which simplifies the alignment and allows high spatial resolution in three dimensions. The phase matching condition for a collinear geometry can be easily satisfied because of the large cone angles of the k vectors associated with the tight focusing condition.

In the 1982 and 1999 work, the signal was detected in the forward direction, the same direction as the excitation beams. In CARS spectroscopy, the constructive interference of the CARS radiation generates a large and directional signal from a bulk homogeneous sample in the forward direction. However, this advantage introduces a drawback in CARS microscopy: when the objects of interest are embedded in a bulk medium such as water, the forward-detected signal from a small scatterer can be easily buried in a large forward-going solvent background. On the other hand, for a scatterer with size much smaller than the excitation wavelength, the CARS radiation pattern is nearly identical in both forward and backward directions, similar to a Hertzian dipole.[3,4] Therefore, epi-detection provides a way of imaging tiny features in a nonlinear medium.[3,5] The epi-detected signal can also arise from an interface of two sizable media with different $\chi^{(3)}$ or $\chi^{(1)}$ ($\chi^{(3)}$ is the third order nonlinear susceptibility and Re $\chi^{(1)}$ is the refractive index).

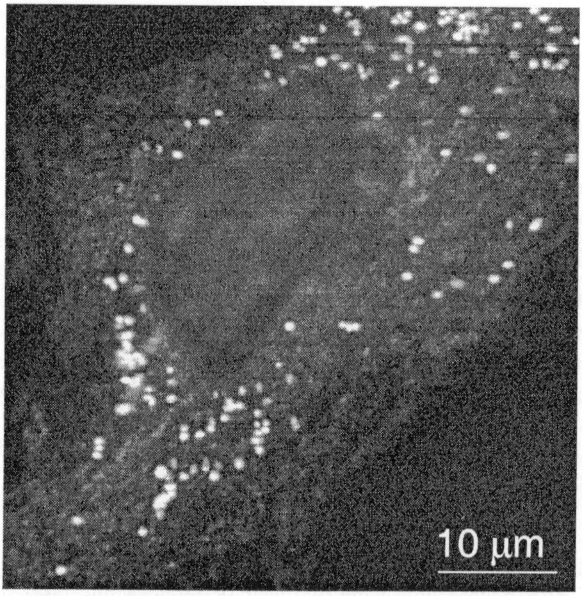

Figure 1. An epi-detected CARS image of a live and unstained bovine capillary endothelial cell. The anti-Stokes shift was tuned to the aliphatic C-H vibration at 2860 cm^{-1}. The average pump and Stokes powers were 1.8 mW and 1.5 mW at a repetition rate of 800 KHz. The acquisition time was 12 minutes. The CARS signal at each pixel is square rooted so that the image intensity is linearly proportional to the density of C-H oscillators. The details of the experimental setup were described in ref [5].

Vibrational imaging of live cells based on aliphatic C-H vibration was demonstrated with forward-detected CARS microscopy [2]. The imaging sensitivity can be greatly improved with epi-detection. An epi-detected CARS image of live bovine endothelial cells is illustrated in Figure 1. The signal to background ratio for the bright spots is about 140:1 with the anti-Stokes shift tuned to the aliphatic C-H vibration at 2860 cm^{-1}. The signal from these scatterers became 10 times weaker when the anti-Stokes shift was tuned away from the C-H vibration to 2648 cm^{-1}. This proves that the contrast in the E-CARS image is mainly contributed by the resonant CARS signal from the C-H vibration. The bright features are expected to be mitochondria that have an outer membrane and folded inner membranes.

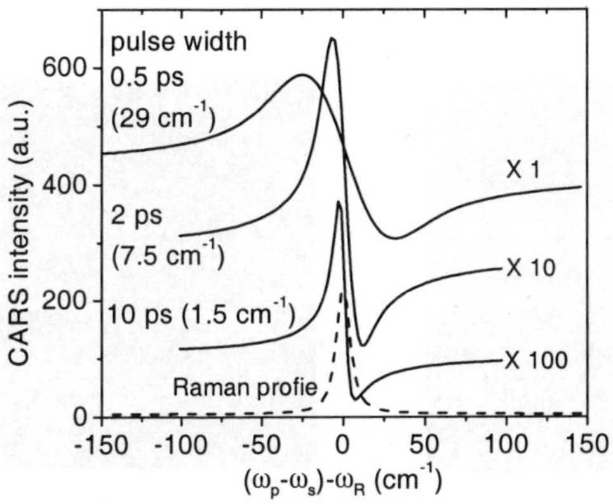

Figure 2. Calculated CARS spectra as a function of the pulse width of the pump and Stokes beams. The third-order susceptibility, $\chi^{(3)}$, has the form of $A/(\omega_R-(\omega_p-\omega_s)-i\Gamma)+\chi_{NR}$. ω_R is the vibrational frequency. The Raman line width, 2Γ, is set as 10 cm^{-1}. The ratio of A/χ_{NR} is assumed to be 5 cm^{-1}. The model was described in ref [5].

In multiphoton microscopy, the high peak power of femtosecond excitation pulses is usually advantageous. However, this conventional wisdom does not hold for vibrational imaging with CARS microscopy [5]. Figure 2 depicts the calculation of the dependence of CARS spectral shape on the width of the excitation pulse. With two 0.5-ps pump and Stokes beams, one can see a broad CARS band superimposed on a large non-resonant background. The peak is negatively shifted from the center of the spontaneous Raman band by 25 cm^{-1}. With two 2-ps beams, the signal to background ratio is obviously increased. Further increase of the pulse width to 10 ps does not improve the signal to background ratio but significantly reduces the signal level. Our calculation indicates that the optimal pulse width for

CARS imaging is around 2 ps, of which the spectral width (7.5 cm^{-1}) is comparable to the Raman line width (around 10 cm^{-1}) in condensed phase. The use of shorter (femtosecond) pulses only increases the nonresonant electronic contributions and thus reduces the signal to background ratio.

In Figure 2, one notes that the CARS spectrum with 2-ps pulse excitation still has a significant non-resonant background. Recently, we have demonstrated that use of polarization-sensitive detection in CARS microscopy allows effective suppression of this nonresonant signal [6]. This method makes use of the polarization difference between the resonant CARS signal and the non-resonant background. An analyzer polarizer is used before the detector to cross the linearly polarized non-resonant background. A maximal signal to background ratio can be obtained when the angle between the polarization directions of the two linearly polarized beams is 72°.

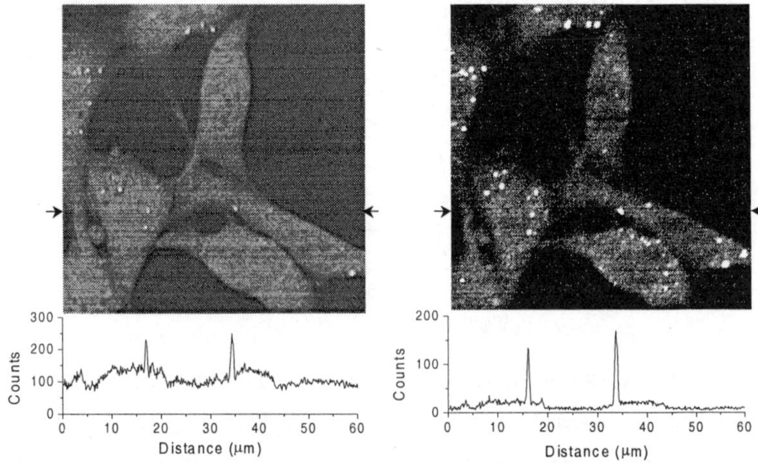

Figure 3. The left panel is a CARS image of live NIH3T3 cells without polarization-sensitive detection. The average pump and Stokes power were 0.5 and 0.25 mW at a repetition rate of 400 KHz. The right panel is a polarization CARS image of the same NIH3T3 cells. The average pump and Stokes power were 2.0 and 1.0 mW at a repetition rate of 400 KHz. For both images, the acquisition time was 12 min. and the anti-Stokes shift was tuned to 2860 cm^{-1}. The details of a polarization CARS microscope were described in ref [6].

A demonstration of suppression of the nonresonant background by polarization CARS microscopy is shown in Figure 3. The anti-Stokes shift was tuned to the aliphatic C-H stretching vibration at 2860 cm^{-1}. The CARS image in the left panel was taken with two parallel-polarized pump and Stokes beams. The signal from the cellular components is superimposed on the non-resonant water background. In the polarization CARS image of the same cells, the water background is effectively suppressed and some small features that cannot be seen in the left CARS image are clearly detected. It has been shown that polarization CARS microscopy provides a way of imaging the global distribution of protein in live and unstained cells [6].

Acknowledgements This work was supported by a NIH grant (GM62536-01). The bovine endothelial cell culture was kindly provided by Justin Jiang.

References

1. Duncan M. D., Reintjes J., and Manuccia T. J., Scanning Coherent Anti-Stokes Raman Microscope, Opt. Lett. **7**, (1982) pp. 350-352.

2. Zumbusch A., Holtom G. R., and Xie X. S., Three-Dimensional Vibrational Imaging by Coherent Anti-Stokes Raman Scattering, Phys. Rev. Lett. **82**, (1999) pp. 4142-4145.

3. Volkmer A., Cheng J. X., and Xie X. S., Vibrational imaging with high sensitivity via epi-detected coherent anti-Stokes Raman scattering (E-CARS) microscopy, Phys. Rev. Lett. **87**, (2001) pp. 023901(1-4).

4. Cheng J. X., Volkmer A., and Xie X. S., Theory and experimental implementation of CARS microscopy, J. Opt. Soc. Am. B, submitted.

5. Cheng J. X., Volkmer A., Book L. D., Xie X. S., An epi-detected coherent anti-Stokes Raman scattering (E-CARS) microscope with high spectral resolution and high sensitivity, J. Phys. Chem. B **105** (7), (2001) pp. 1277-1280.

6. Cheng J. X., Book L. D., and Xie X. S., Polarization coherent anti-Stokes Raman scattering microscopy, Opt. Lett. **26** (17), (2001) pp. 1341-1343.

IN VIVO DIFFUSE OPTICAL SPECTROSCOPY AND IMAGING OF BLOOD DYNAMICS IN BRAIN

A. G. YODH, C. CHEUNG, J. P. CULVER, AND T. DURDURAN

Department of Physics and Astronomy, University of Pennsylvania, Philadelphia, PA 19104-6396, USA

E-mail: yodh@physics.upenn.edu

J. H. GREENBERG, K. TAKAHASHI, AND D. FURUYA

Department of Neurology, University of Pennsylvania, Philadelphia, PA 19104, USA

In this paper we provide a cursory review on diffuse optical probes and their relevance to the investigation of human tissues. Then we describe a recent instrument we have built and used to study variations of blood flow, hemoglobin concentration, and blood oxygen saturation in the functioning rat brain. The instrument combines two novel optical methodologies. Diffuse correlation spectroscopy monitors changes in the cerebral blood flow by measuring the optical phase-shifts caused by moving blood cells. Diffuse near-infrared absorption spectroscopy concurrently measures tissue absorption at two wavelengths to determine hemoglobin concentration and blood oxygen saturation in the same tissue volume. The optical probe is non-invasive and was employed through the intact skull. The utility of the technique is demonstrated *in vivo* by measuring the temporal changes in the regional vascular dynamics of rat brain during hypercapnia and hypoxia.

1 Introduction

Light spectroscopies are particularly well suited for the study of simple, homogenous, optically thin samples. Unfortunately many practical materials are not so simple. Human tissues, for example, are highly scattering media. Light penetration in tissues is limited, and the effects of tissue absorption and internal motion must be separated from the effects of tissue scattering. Nevertheless, the use of light to investigate the human body interior has grown enormously in recent years, in part as a result in advances in our fundamental understanding about light transport in highly scattering materials, and in part as a result of technological innovations in optics [1].

The application of near-infrared (NIR) optical methods for tissue characterization is attractive for several reasons. The techniques utilize non-ionizing radiation, are non-invasive, and are often technologically simple and fast. These features have lead to compact, portable, and inexpensive clinical instruments in medicine, e.g. the pulse oximeter. The optical method also has several unique measurable parameters. Blood dynamics, blood volume, blood oxygen saturation, and water content of a rapidly growing tumor are often substantially different than normal tissue, and will alter tissue optical absorption coefficients. An increase in organelle population (e.g. mitochondria) often accompanies the higher metabolic activity of rapidly growing tissues, and leads to an increased scattering coefficient. Similarly the optical absorption, fluorescence, and scattering of exogenous contrast agents such as Indocyanine green (ICG) that occupy vascular and extravascular space provide useful forms of sensitization.

Arguably the most critical advance in the field has been the recognition and widespread acceptance that light transport over long distances in tissues is well approximated as a diffusive process. Using this physical model it is possible to quantitatively separate tissue scattering from tissue absorption, and to accurately incorporate the influence of boundaries, such as the air-tissue interface, into the transport theory. Waves of diffuse light energy density or their time-domain analogs propagate deeply in tissues (i.e. ~10 cm) and exhibit simple "optical" phenomena such as refraction, diffraction, interference, and dispersion as they encounter variations in tissue optical properties [2–6]. The diffusion approximation is also critical because it provides a tractable basis for tomographic approaches to image reconstruction of tissue optical properties using highly scattered light [7–14].

Physiological applications of diffuse light imaging and spectroscopy include functional imaging and detection of brain bleeds, quantitation of brain physiology and function, study of mitochondrial diseases, study of muscle function and physiology, optical mammography, optical dosimetry in photodynamic therapy of cancer, and early detection of cancer. This paper focuses on some recent advances made in our lab to monitor oxygen metabolism in rat brain [15].

The vascular changes that accompany neuronal activity represent a critical and incompletely understood component of neurophysiology. Energy is required for cerebral functions, and its conversion into a useful form is facilitated by oxygen delivered to the brain through the cerebral vasculature. A range of non-invasive techniques have been used to probe activation of the cerebral cortex. Electrical signals originating from the firing of neurons can be measured with electroencephalography (EEG), changes in local cerebral glucose utilization and local cerebral blood flow can be measured with positron emission tomography (PET) [16], and changes in the concentration of deoxyhemoglobin and adenosine triphosphate can be measured with magnetic resonance imaging (MRI) [17,18]. Some relatively more invasive methods for studying the brain include laser Doppler flowmetry (LDF) and confocal microscopy. Both of these methods are used to study near-surface (or nearprobe) tissue volumes. LDF measures the blood flow speed and the number of moving blood cells [19]. However, LDF is primarily a surface technique, and in order to make these measurements in brain tissue the skull must be thinned to achieve sufficient signal. Confocal microscopy has been used to image the cortical capillary bed and erythrocytes through a cranial window. In this case, blood flow is estimated by counting the number of blood cells flowing through a capillary per unit time [20]. Finally, a number of other techniques have been applied to measure cerebral tissue oxygen. These include microelectrodes [21], phosphorescence quenching by oxygen in the microvasculature [22], and near-infrared spectroscopy (NIRS). NIRS relies on the difference between the absorption spectra of oxyhemoglobin and deoxyhemoglobin to determine blood oxygenation [23]. Generally these methods provide complementary information about brain tissue, and no technique has emerged as a universal choice for experiments in neurophysiology.

In our study we combine diffuse photon correlation methods, originally used to investigate highly scattering complex fluids [24–26], with a near-infrared diffuse wave regional imager [23] for hemoglobin spectroscopy. The instrument enables us to simultaneously measure regional blood perfusion, hemoglobin concentration and blood oxygen saturation in the same, deep tissue volume. In particular, we present *in-vivo* rat brain data derived from models of hypercapnia (excess CO_2) and hypoxia (O_2 deficit); changes in cerebral blood flow, hemoglobin concentration and oxygenation are measured and discussed. This combination of tools has not been previously employed, and because of the increased information it provides about vascular response, we anticipate that the probe will be valuable for future research on the physiology of deep, highly scattering tissue.

2 Background on Theory and Instrumentation

A turbid medium such as tissue is characterized by a reduced scattering coefficient μ_s', and an absorption coefficient μ_a. In general these quantities can depend on position and time. Physically, the reciprocal of μ_s' is the photon random walk step, corresponding to the length traveled by a photon in the medium before its initial direction becomes randomized; the reciprocal of μ_a is the absorption length, corresponding to the length scale over which a photon traveling in the medium will be absorbed. For many conditions of practical interest, e.g. in tissues, the photon fluence rate satisfies a diffusion equation [1].

In our experiments light is collected in remission, i.e. light sources and detectors are arranged in a plane parallel to the air/tissue surface, and are oriented to receive or emit light normal to the air/tissue interface. We thus approximate the tissue samples as semi-infinite. The solution for diffuse photon density waves in this geometry is readily obtained using an image source approach [27]. Multidistance measurements of the diffuse photon fluence rate enable one to deduce the optical properties (μ_s' and μ_a) of the medium by fitting the known analytical solution to the measurements.

In the near-infrared region, the dominant chromophore in tissue is hemoglobin, and the absorption spectra of oxyhemoglobin and deoxyhemoglobin are different [1,28]. Thus, determination of μ_a at two different optical wavelengths (e.g. 786 nm and 830 nm) provides information sufficient for calculation of the concentrations of oxyhemoglobin and deoxyhemoglobin if one assumes a water/lipids ratio.

We determine the motional dynamics of the medium by measuring the time dependence of detected diffuse light intensity and then computing the (normalized) intensity autocorrelation function. We have recently shown (for continuous light sources) that this autocorrelation function also satisfies a diffusion equation, though the equation differs slightly from that obeyed by the diffuse light fluence rate; in particular there is an additional absorption term that depends on the mean-square displacement of the scatterers (e.g. blood cells) during the autocorrelation time τ [26,29,30].

Two models can be used to describe the dynamics of cerebral blood flow. The first model was used originally in colloidal suspensions where the dynamics is Brownian [25,31]. The second model argues that in a capillary network, blood flow resembles a random flow, i.e. both the speed and direction of the flow at any point in space are random. In such cases [32,33] the decay rate of the correlation function depends on the mean squared velocity of the scatterers. We have analyzed our data with both models, but parameterize our data using the Brownian diffusion model. Thus the effective blood flow speed is parameterized by the Brownian diffusion coefficient, which is proportional to the exponential decay rate of the measured temporal correlation functions. The relative change of this decay rate equals the relative change of the cerebral blood flow (CBF). Justification and discussion of our assumptions is discussed more fully in [15]. We introduce the term 'diffuse correlation flowmetry' to represent the correlation method described above.

Figure 1 shows a schematic diagram of the instrument. The sources of near-infrared (NIR) light were: (a) a single mode laser source with a coherence length (>1 m) much longer than a typical photon path length, for the diffuse correlation flowmetry measurements and (b) two intensity modulated laser diodes for the diffuse photon density wave measurements. The wavelength of the single-mode laser was tunable from 770 nm to 802 nm (model TC40, SDL Inc., San Jose, CA). The laser diodes operated at 786 nm and 830 nm (manufactured by Sharp and Hitachi respectively). The diffuse photon density wave modulation frequency was 70 MHz. All three lasers were multiplexed into nine optical source fibers in the probe.

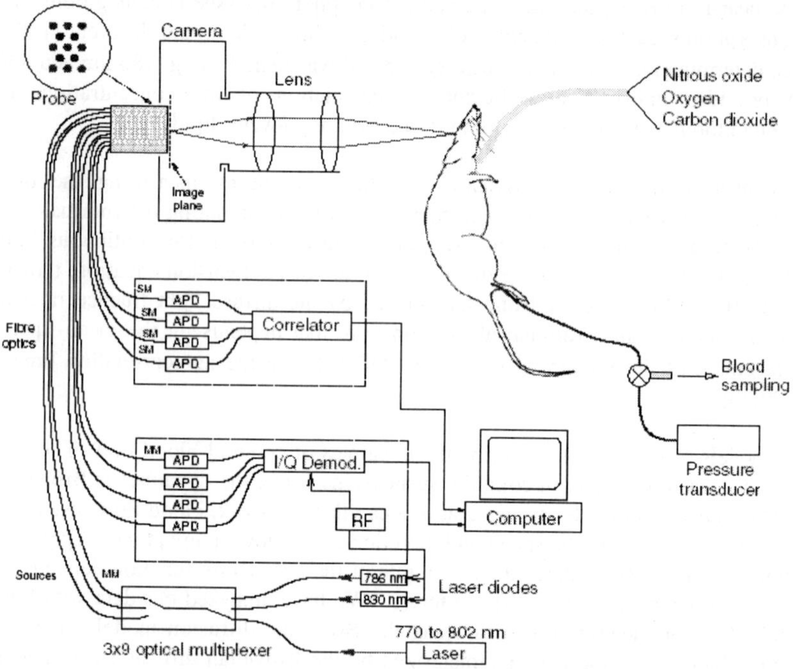

Figure 1. A schematic diagram of the experiment. Key: avalanche photodiode (APD), radiofrequency generator (RF), multimode fiber (MM), single-mode fiber (SM), *I/Q* demodulation (I/Q Demod.).

Two sets of four avalanche photodiodes were employed for detection of diffuse light at four positions, one set for the diffuse photon density waves and another set for the diffuse correlation measurement. The amplitudes and phases of the DPDW were measured for each source–detector pair. For diffuse correlation flowmetry, signals from the photodiodes were directed to a correlator chip (Correlator.com, Bridgewater, NJ). The correlator chip computed the temporal intensity autocorrelation functions. A relay lens projecting the probe onto the measuring sample permitted non-contact measurements. The source and detector optical fibers were arranged in a two-dimensional pattern as shown in the upper left corner of the figure. With this design, we could reconstruct low-resolution images of the dynamics and the optical properties from the measurements. The optical multiplexer was controlled remotely by a computer.

Both the DPDW and diffuse correlation measurements take about the same amount of time, and are operated in a time-shared manner with the DPDW measurements interlaced between successive diffuse correlation measurements. The data presented in this paper were made at about 2.5 min per combined measurement.

3 Experimental

Adult male Sprague-Dawley rats weighing 300-325 g were fasted overnight prior to measurements. The animals were anesthetized (Halothane 1-1.5%, N_2O 70%, O_2 30%) and catheters were placed into a femoral artery to monitor the arterial blood pressure and into a femoral vein for drug delivery. The body temperature was maintained at 37±2C. The animal was tracheotomized, mechanically ventilated and fixed on a stereotaxic frame. The probe source-detector plane was placed relay imaged and located symmetrically about midline; it covered a region from 2mm anterior to 6mm posterior of the rhinal fissure. The hypercapnic challenge consisted of elevating blood p_aCO_2 by adding carbon dioxide (8%) into the breathing mixture. The hypoxic modulation involved decreasing the arterial p_aO_2 by reducing the inspired O_2 to 10%.

4 Results and Discussion

Figure 2 (*left side*) shows the percent increase from the baseline level of the cerebral blood flow (i.e. the correlation function temporal decay rate) in one experiment where the animal breathed 9% CO_2 for 10 minutes. The numbers shown in the figure were averaged over the area of the imaged brain, which comprised an 8x10 mm ^{2} area of cortex. During the CO_2 inhalation, the p_aCO_2 increased from 38 mmHg to 75 mmHg, and the average relative increase of cerebral blood flow was about 0.77 (77%) from the baseline. Figure 2 also shows the concurrent changes in the hemoglobin concentration and cerebral blood flow as measured by NIR absorption spectroscopy averaged over the same cortical area during hypercapnia. A 0.17 (17%) increase in the hemoglobin concentration was observed. The hemoglobin concentration appears to increase at a slower rate than the cerebral blood flow and also decays slower after the CO_2 was switched off. The average relative increase of the hemoglobin oxygenation in the cortex during the same CO_2 stimulation is shown in the bottom graph of Figure 2. The increase was about 16%, with the oxygenation dynamics following the flow changes more closely than the total hemoglobin concentration.

We also obtained time courses for hypoxia (not shown). In Figure 2 (*right side*), the relative changes averaged over the last 10 minutes of the hypercapnia are plotted along with the relative changes during hypoxia. The values are obtained averaging the signal from 5 to 15 minutes following the perturbation. For both hypercapnia and hypoxia, vasodilatation increased blood flow (CBF) and blood volume (CBV).

Since oxygen metabolism remains roughly constant during hypercapnia and hypoxia, the flow increases are countered by a reduced oxygen extraction fraction. The measured tissue blood oxygen saturation (Y_t) is a combination of arterial, capillary, and venous tissue compartments. During hypercapnia the arterial saturation is constant so that the reduced oxygen extraction fraction results in increased tissue saturation, while during hypoxia, the drop in arterial saturation combined with a decreased oxygen extraction fraction results in decreased tissue saturation. From these two modulations we have measured two different trends in the hemoglobin saturation and blood flow.

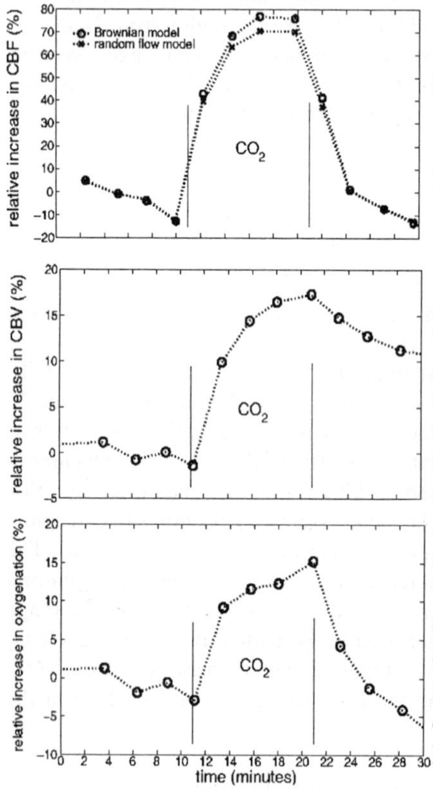

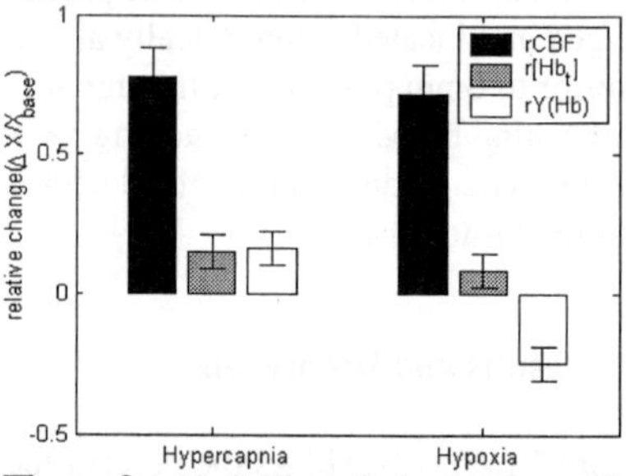

Figure 2. . (*Left*) The relative increase of various quantities in a rat brain during carbon dioxide stimulation. Carbon dioxide was inhaled during the time interval between the two vertical lines in the figures. *Top left*: Diffusion coefficient (circles) and root-mean-squared velocity (crosses), extracted from the correlation function decay. Both are effective measures of relative cerebral blood flow (CBF). *Middle left*: Tissue hemoglobin concentration. *Bottom left*: Hemoglobin oxygen saturation. (*Right*) Relative changes in CBF, total hemoglobin (Hb_t) and blood oxygenation saturation (Y), for hypoxia and hypercapnia, averaged over several rats.

Our combined measurements also offer us the possibility to quantitatively verify that CBF and blood oxygenation changes are self-consistent. We can use a simplified model for oxygen metabolism to relate the two measurements, wherein we assume that the product of the difference in oxyhemoglobin concentration between the artery perfusing the tissue and the vein draining the tissue and blood perfusion rate equals the oxygen consumption rate. The two measurements are consistent within our signal-to-noise. Ultimately this type of analysis will allow us to make low resolution maps of oxygen metabolism, which in turn will provide crucial information about tissue viability and function.

To conclude, we have demonstrated the ability to concurrently measure relative changes in cerebral blood flow, hemoglobin concentration and hemoglobin oxygenation with a single non-contact, non-invasive instrument. Although our pilot measurements are preliminary, results from a rat hypercapnia and hypoxia models are in reasonable agreement with the data taken with other techniques in the literature, and offer the possibility for further growth and quantitation. The optical techniques used in this study are attractive, also, because they enable us to measure the vascular response of deep tissues. We anticipate that the instrument will facilitate other protocols such as hypoxia and hemodilution that affect whole-brain hemodynamics. Finally, the new instrument and concept may be applicable to human studies especially in infants and neonates.

5 Acknowledgements

This work is supported by the National Institutes of Health under grant number HL57835-01. We thank B Chance, A Villringer and R Cheung for helpful comments and discussions. Technical assistance from R Choe and L Zubkov is also gratefully acknowledged.

References

1. Yodh A. G. and Chance B., *Phys. Today* **48** (1995) pp. 34–40.
2. O'Leary M. A., Boas D. A., Chance, B. and Yodh, A. G., *Phys. Rev. Lett.* **69** (1992) pp. 2658–2661.
3. Boas D. A., O'Leary M. A., Chance B. and Yodh, A. G., *Phys. Rev. E* **47** (1993) pp. R2999–3003.
4. Fishkin J. and Gratton E., *J. Opt. Soc. Am. A* **10** (1993) pp. 127–140.
5. Schmitt J. M., Knuttel A. and Knudsen J. R., *J. Opt. Soc. Am. A* **9** (1992) p. 1832.
6. Tromberg B., Svaasand L. O., Tsay T. and Haskell R. C., *Appl. Optics* **32** (1993) pp. 607–616.

7. Walker S. A., Fantini S. and Gratton E., *Appl. Optics.* **36** (1997) pp. 170–179.
8. Schotland J. C., *J. Opt. Soc. Am. A* **14** (1997) pp. 275–279.
9. O'Leary M. A., Boas D. A., Chance B. and Yodh, A. G., *Opt. Lett.* **20** (1995) pp. 426–428.
10. Arridge S. R., *Appl. Optics* **34** (1995) pp. 7395–7409.
11. Arridge S. R. and Schweiger M., *Appl. Optics* **34** (1995) pp. 8026–8037.
12. Yao Y. Q., Wang Y., Pei Y. L., Zhu W. W. and Barbour, R. L., *J. Opt. Soc. Am. A* **14** (1997) pp. 325–342.
13. Jiang H. B., Paulsen K. D., Osterberg U. L. and Patterson M. S., *Appl. Optics* **36** (1997) pp. 52–63.
14. O'Leary M. A., Boas D. A., Li X. D., Chance B. and Yodh A. G., *Opt. Lett.* **21** (1996) pp. 158–160.
15. Cheung C., Culver J. P., Takahashi K., Greenberg J. H. and Yodh A. G., *Phys. Med. Biol.* **46** (2001) pp. 2053–2065.
16. Phelps M. E. and Mazziotta J. C., Positron emission tomography: human brain function and biochemistry. *Science* **228** (1985) pp. 799–809.
17. Kwong K. K. *et al., Proc. Natl Acad. Sci. USA* **89** (1992) pp. 5675–5679.
18. Chance B., *NMR Biomed.* **2** (1989) pp.179–87.
19. Shepherd A. P., and Oberg P. Å., *Laser-Doppler Blood Flowmetry* (Boston: Kluwer) (1990).
20. Villringer A., Them A., Lindauer U., Einhaupl K., and Dirnagl U., *Circ. Res.* **75** (1994) pp. 55–62.
21. Nair P., Whalen W. J., and Buerk D., *Microvasc.Res.* **9** (1975) pp. 158–165.
22. Wilson D. F., Gomi S., Pastuszko A., and Greenberg J. H., *J. Appl. Physiol.* **74** (1993) pp. 580–589.
23. Danen R. M., Wang Y., Li X. D., Thayer W. S., and Yodh A. G., *Photochem. Photobiol.* **67** (1998) pp. 33–40.
24. Maret G., and Wolf P. E., *Z. Phys.* B **65** (1987) pp. 409–413.
25. Pine D. J., Weitz D. A., Chaikin P. M., and Herbolzheimer E., *Phys. Rev. Lett.* **60** (1988) pp. 1134–1137.
26. Boas D. A., Campbell L. E., and Yodh A. G., *Phys. Rev. Lett.* **75** (1995) pp. 1855–1858.
27. Kienle A., and Patterson M. S., *J. Opt. Soc. Am.* A **14** (1997) pp. 246–254.
28. Zijlstra W. G., Buursma A., and Meeuwsen van der Roest W. P., *Clin. Chem.* **37** (1991) pp. 1633–1638
29. Boas D. A., and Yodh A. G., *J. Opt. Soc. Am.* A **14** (1997) pp. 192–215.
30. Hackmeier M., Skipetrov S. E., Maret G., and Maynard R., *J. Opt. Soc. Am.* A **14** (1997) pp. 185–191.
31. Maret G., and Wolf P. E., *Z. Phys.* B **65** (1987) pp. 409–413.
32. Nossal R., Chen S. H., and Lai C. C., *Opt. Commun.* **4** (1971) p. 35.
33. Bonner R., and Nossal R., *Appl. Opt.* **20** (1981) p. 2097.

SELECTED POSTERS

SPEEDY BEC IN A TINY TRAP: COHERENT MATTER WAVES ON A MICROCHIP

J. REICHEL, W. HÄNSEL, P. HOMMELHOFF, R. LONG, T. ROM, T. STEINMETZ, AND T. W. HÄNSCH

Max-Planck-Institut für Quantenoptik and Sektion Physik der Ludwig-Maximilians-Universität, Schellingstrasse 4, D-80799 München, Germany
phone +49 89 2180-2055, http://www.mpq.mpg.de/~jar

A Bose-Einstein condensate of ^{87}Rb has been produced in a microscopic magnetic trap on a chip. The evaporative cooling time is on the order of 2 s. The experimental setup is simple, with much less stringent vacuum requirements than in previous BEC experiments. The integrated approach gives easy access to complex magnetic manipulation. One example is an integrated atom interferometer working with trapped atoms.

1 Integrated Magnetic Traps

Magnetic chip traps for neutral atoms employ currents through lithographic conductors on a microchip to create the trapping field[1]. This approach leads to an unprecedented liberty in magnetic potential design, which is especially useful for complex atom manipulation tasks, such as in proposed schemes for quantum computing[2] and atom interferometry. Moreover, modest currents can produce large magnetic field gradients and curvatures in close proximity to a planar arrangement of wires[3], making chip traps an ideal system for evaporative cooling with collision times in the millisecond range. A crucial issue is how to transfer cold atoms into such a chip trap. We use a mirror-MOT to prepare a laser-cooled atom cloud and transfer it directly into the chip trap. This leads to Bose-Einstein condensation (BEC) in record time and greatly relaxes the vacuum conditions for BEC experiments[4], as first reported on the ICOLS 2001 conference and summarized below. Simultaneously with our result, the group of C. Zimmermann at Tübingen achieved BEC in a hybrid chip-and-wire trap using a multistep loading scheme with evaporative pre-cooling in a coil-based trap. Due to this additional step, their experiment requires the same ultra-high vacuum level as conventional BEC experiments, but leads to much larger condensates[5] (see the contribution by J. Fortagh et al. in this volume).

2 Evaporative cooling to BEC in record time

Fig. 1 shows the conductor layout that is used in our experiments. The trapping potential is produced by superposing the field generated by currents through the lithographic conductors with a homogeneous, external bias field $\mathbf{B}_0$. A Ioffe-Pritchard potential can be created using the Z-shaped conductor (I_0) in fig. 1. Fig. 1(b) shows the magnetic potentials created by this wire with a current of $I_0 = 2$ A for two different values of the external bias field, corresponding to the two extreme values used in the experiment. With increasing bias field, the trap center moves closer to the surface, from $z_0 = 445\,\mu$m to $z_0 = 70\,\mu$m.

The chip which creates these potentials is mounted face down in a small glass cell, which is part of a simple, single-chamber system pumped by a 25 l/s ion pump and a small titanium sublimator. The magnetic trap lifetime is up to 5 seconds, which indicates a background pressure in the 10^{-9} mbar range. Thermal rubidium vapour is produced from a rubidium dispenser[6]. We operate it at low, constant current, so that the rubidium partial pressure remains low and reduces the magnetic trap lifetime by less than 20 %. During the magneto-optical trap (MOT) loading phase, a halogen light bulb is switched on to temporarily increase the rubidium pressure by light-induced desorption[7]. The MOT loaded in this way contains about $5 \cdot 10^6$ ^{87}Rb atoms after a loading time of 3 to 7 seconds.

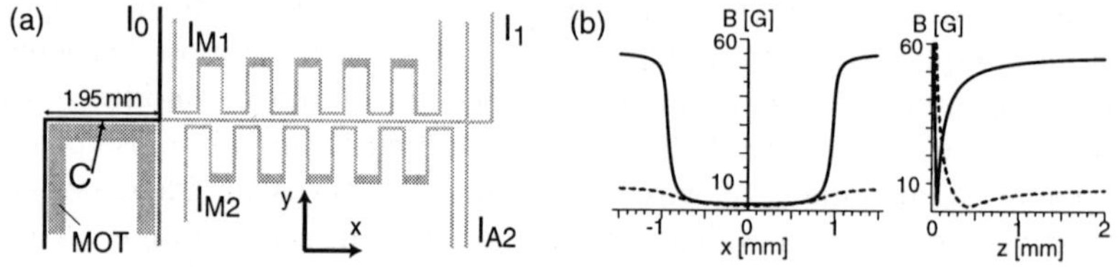

Figure 1. (a) Conductor layout on the chip. Only the conductor labeled "I_0" is needed to produce the magnetic trap for BEC. "C" marks the position of the trap center when this conductor is used . The "MOT" conductor creates the quadrupole field for an intermediate step of magneto-optical trapping. All other conductors are available for subsequent manipulation of the condensate, such as transport with the "magnetic conveyor belt", or splitting and merging. (b) Potentials created by a wire current $I_0 = 2$ A for two different values of the external bias field, ($B_x = 0, B_y = 8$ G) (dashed lines) and ($B_x = 1.9$ G$, B_y = 55$ G) (solid lines).

The initial trap (dashed lines in fig. 1(b)) has frequencies $\nu_x = 28$ Hz and $\nu_{y,z} = 220$ Hz. The temperature and peak density, measured 200 ms after transfer from the MOT, are $T \sim 45\,\mu K$ and $\rho \sim 5 {\cdot} 10^{10}$ cm^{-3}; the trap contains

$\sim 3 \cdot 10^6$ atoms. Immediately after the transfer, we ramp up the bias field B_y in 300 ms to a final value of 55 G, leading to the potential shown in solid lines in fig. 1(b). This very anisotropic potential has a transverse curvature of $2.4 \cdot 10^7\,\mathrm{G/cm^2}$ near the center, leading to a transverse oscillation frequency of $\nu_{y,z} = 6.2\,\mathrm{kHz}$, while the longitudinal frequency is only $\nu_x = 17\,\mathrm{Hz}$.

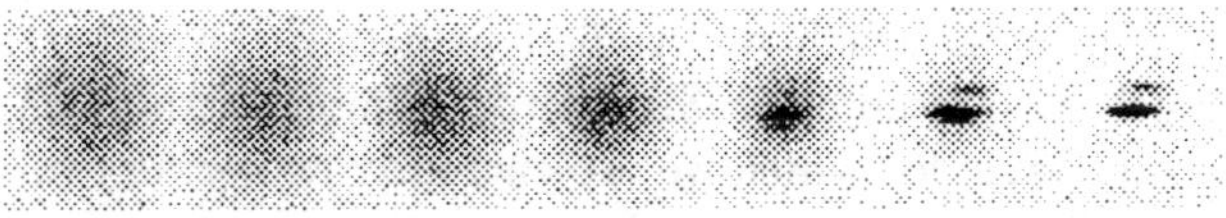

Figure 2. Time-of-flight absorption images of the atom cloud, 21 ms after release from the magnetic trap. The final radio frequency decreases from left to right. The appearence of a bimodal distribution signalizes the formation of a condensate. (The "shadow" appearing in the last images is an artefact caused by atoms transiting to the $m = 1$ state during the detection process.)

Immediately after compression, we initiate forced evaporative cooling with a sequence of one linear and two exponential radiofrequency (rf) sweeps and a total duration of 2 s. At the end of the second ramp, $\sim 7 \cdot 10^4$ atoms are left at a temperature of $\sim 6\,\mu\mathrm{K}$, a density of $5 \cdot 10^{13}\,\mathrm{cm^{-3}}$ and a phase space density in the lower 10^{-2} range. The external bias field is now reduced to ($B_x = 1.2\,\mathrm{G}, B_y = 40\,\mathrm{G}$) within 150 ms to decompress the trap. This is done to avoid an excessively high density (which would lead to trap loss by three-body collisions), and to reduce the heating rate, which was measured to be $2.7\,\mu\mathrm{K/s}$ in the compressed trap at $B_y = 55\,G$, and reduced to $1.1\,\mu\mathrm{K/s}$ at $B_y = 40\,G$. (We are currently investigating the source of this heating rate; possible sources include current noise and surface effects.) The decompressed trap has frequencies $\nu_{x,y,z} = (20, 3900, 3900)$ Hz. After a final rf sweep to $\sim 1.6\,\mathrm{MHz}$, a condensate appears. Fig. 2 shows time-of-flight absorption images which were taken for descending values of the final rf frequency. These images show the appearance of the sharp, nonisotropic peak in the momentum distribution which is a key signature of Bose-Einstein condensation. A bimodal distribution is first observed for a temperature of $T \sim 630\,\mathrm{nK}$, in agreement with the theoretical value of the transition temperature $T_c = 670\,\mathrm{nK}$ (for 11000 atoms). The number of condensate atoms in a distribution well below the transition temperature is typically around 3000. The condensate lifetime is on the order of 500 ms and can be prolonged to $\sim 1.3\,\mathrm{s}$ when the radiofrequency is left on at a frequency just above resonance with the condensed atoms.

3 Applications

A decisive advantage of the chip trap lies in the versatility of the lithographic wire structures. With relatively simple wire layouts such as the one of fig. 1, many complex potentials can be realized. A particularly interesting example is a potential which adiabatically splits and merges a magnetically trapped atom cloud. This mecanism can be used to constructed a trapped-atom interferometer, a process which we studied in detail[8]. Another extension of this work, which we presented in Snowbird, is "nondestructive" single-atom detection in chip traps, for which we are currently exploring two different approaches, one utilizing a fiber resonator, the other, a microsphere resonator. These projects had to be omitted from the present article due to limited space.

4 Conclusion

The short evaporation time entails a very fast cycle time of 10 seconds or less, including MOT loading and detection. Moreover, it relaxes the ultrahigh-vacuum requirements which have been one of major restrictions of "traditional" BEC experiments. On-chip magnetic manipulation of condensates opens the door to a large variety of new applications in the new field of integrated, coherent atom optics.

References

1. J. Reichel, W. Hänsel, and T. W. Hänsch, Phys. Rev. Lett. **83**, 3398 (1999).
2. T. Calarco, E. A. Hinds, D. Jaksch, J. Schmiedmayer, J. I. Cirac, and P. Zoller, Phys. Rev. A **61**, 022304 (2000).
3. J. D. Weinstein and K. G. Libbrecht, Phys. Rev. A **52**, 4004 (1995).
4. W. Hänsel, P. Hommelhoff, T. W. Hänsch, and J. Reichel, Accepted for publication in Nature (2001).
5. H. Ott, J. Fortagh, G. Schlotterbeck, A. Grossmann, and C. Zimmermann, submitted to Phys. Rev. Lett. (2001).
6. J. Fortagh, H. Ott, A. Grossmann, and C. Zimmermann, Appl. Phys. B **70**, 701 (2000).
7. B. P. Anderson and M. A. Kasevich, Phys. Rev. A **63**, 023404 (2001).
8. W. Hänsel, J. Reichel, P. Hommelhoff, and T. W. Hänsch, accepted for publication in Phys. Rev. A (2001).

BOSE-EINSTEIN CONDENSATE IN A SURFACE MICRO TRAP

J. FORTAGH, H. OTT, G. SCHLOTTERBECK, A. GROSSMANN,
AND C. ZIMMERMANN

Physikalisches Institut der Universität Tübingen,
Auf der Morgenstelle 14, D-72076 Tübingen, Germany
http://www.pit.physik.uni-tuebingen.de/zimmermann

We report on Bose-Einstein condensation in a surface micro trap. The strongly anisotropic trapping potential is generated by a microstructure which consists of micro-fabricated linear copper conductors at a width ranging from 3 to 30 μm. The ratio of the trap frequencies for the motion in longitudinal and transverse direction can be varied continuously in a wide range up to values of 100.000. After capturing of 5×10^8 ^{87}Rb atoms from a pulsed thermal dispenser source directly into a magneto-optical trap (MOT) the magnetically stored atoms are transfered into the micro trap by adiabatic transformation of the trapping potential. In the micro trap with high collision rate the atoms are cooled by forced evaporation into the degenerate regime. The pure condensate contains of up to 4×10^5 atoms. The complete *in vacuo* trap design is compatible with ultrahigh vacuum conditions below 2×10^{-11} mbar.

1 Micro traps in atom optics

If ultracold atoms are loaded into a trap with a level spacing that is comparable with the thermal energy of the atoms and their chemical potential, it is possible to enter the regime of quantum degeneracy. It is an appealing goal to combine both, degenerate quantum gases and magnetic micro-potentials which may allow for coherent atom optics on a surface of a micro-structured "atom chip" [1]. The use of integrated current conductors as potential generating elements provides an easy and straightforward way to build and structure strongly confining potentials in a 'micro trap'. Intriguing applications are conceivable in atom interferometry, in quantum computation, for waveguides where the atomic matter waves propagate in the transversal ground state, and for the study of one-dimensional quantum phenomena. However, so far all experiments using micro-fabricated structures have been carried out with thermal, non-degenerate atomic ensembles and therefore small coherence lengths. A break-through has been recently achieved in Tuebingen where we succeeded in loading a Bose-Einstein condensate into a linear micro trap.

Our approach to combine a Bose-Einstein condensate with a micro trap is conceptionally simple. The idea is to generate a magnetic field that can be continuously changed from a rather shallow spherical geometry into the tightly confining field of the micro trap. The atoms are initially collected in a MOT and subsequently loaded into the shallow magnetic quadrupole trap employing standard techniques of polarisation gradient cooling and optical pumping. Then the atoms are adiabatically compressed into the micro trap by a gradual transformation of the magnetic field. The compression enhances the collision rate and the thermalization is accelerated. When the collision rate is several 10 per second the atomic gas can be efficiently cooled by forced evaporation into the degenerate regime.

2 The Tuebingen experiment

A micro trap confines the atoms near the surface of the trap. Therefore it is necessary to place the micro trap inside a vacuum chamber. Our trap setup consisting of coils, wires and the microstructure is a complete *in vacuo* trap design at a base pressure below 2×10^{-11} mbar (Fig. 1). The micro trap is arranged between the transfer coils meanwhile for the operation of a MOT we use a second pair of coils (MOT coils). The separation guarantees free optical access for the MOT and allows high flexibility in mounting different micro trap geometries.

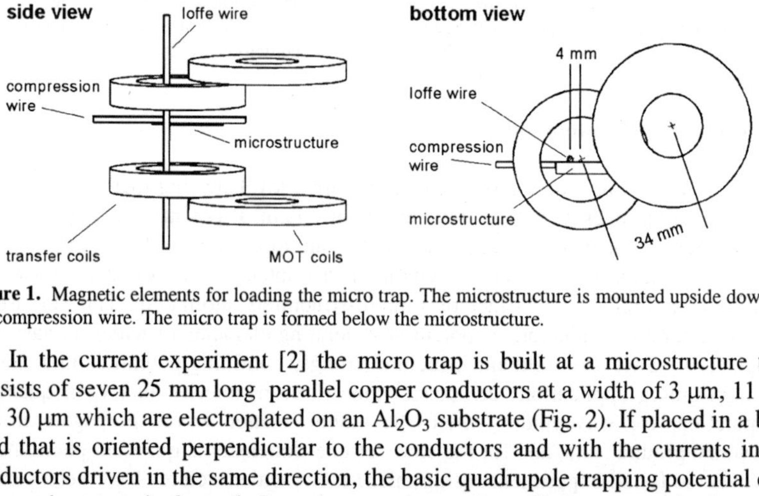

Figure 1. Magnetic elements for loading the micro trap. The microstructure is mounted upside down on the compression wire. The micro trap is formed below the microstructure.

In the current experiment [2] the micro trap is built at a microstructure that consists of seven 25 mm long parallel copper conductors at a width of 3 μm, 11 μm and 30 μm which are electroplated on an Al_2O_3 substrate (Fig. 2). If placed in a bias field that is oriented perpendicular to the conductors and with the currents in all conductors driven in the same direction, the basic quadrupole trapping potential of a linear micro trap is formed. By subsequently turning off the current in the outer conductors the field gradient is increased while the trap depth is kept constant. Further compression can be achieved by inverting the currents in the outer conductors. Then the bias field is formed only by the outer conductors and no external bias field is required. In the final configuration the magnetic field gradient exceeds 50 T/cm. To avoid Majorana spin flips, the linear quadrupole field is superimposed with a magnetic offset field oriented parallel to the direction of the wave guide. The offset field is generated by two extra wire loops that are mounted directly onto the microstructure substrate. With a longitudinal offset field of 1 G, the radial oscillation frequency of the trap reaches $\omega_r = 2\pi\times600{,}000$ Hz. By choosing a small axial oscillation frequency (e.g. $\omega_a < 5$ Hz), an aspect ratio of $\omega_r/\omega_a > 100{,}000$ can be achieved.

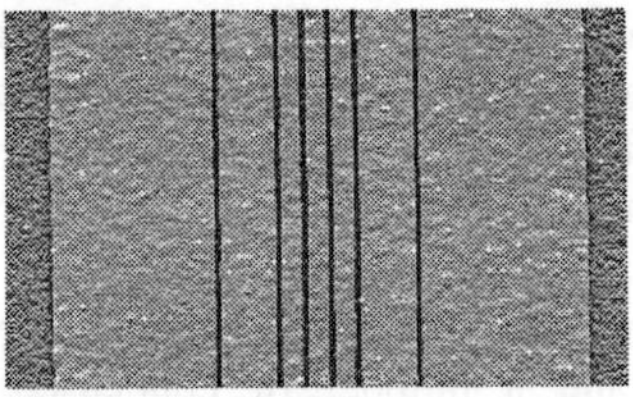

Figure 2. Scanning Electron Microscope (SEM) image of the microstructure which displays a part of the 25 mm long Al_2O_3 substrate with seven parallel conductors. The widths of the conductors are 30, 11, 3, 3, 3, 11, and 30 μm each. The height of the conductors is 2.5 μm. A variety of trap geometries and on chip interferometers can be realized by applying suitable bias field and current distribution.

The microstructure is mounted horizontally on a 2x2 mm^2 copper rod (compression wire) which is embedded in a heat sink at the bottom of the upper transfer coil. The conductors on the microstructure are oriented parallel to the compression wire. The setup is completed by a vertical copper wire with 2 mm diameter (Ioffe wire) that is oriented parallel to the symmetry axis of the transfer coils but displaced by 4 mm (Fig. 1).

The MOT is loaded from a pulsed thermal dispenser source located at a distance of 50 mm from the center of the MOT. The dispenser is heated by a 12 s long current pulse of 7 A. After the current pulse the MOT is operated for another 8 s. During this time the vacuum recovers and trap life times of 100 s have been routinely achieved. Up to 5×10^8 ^{87}Rb atoms are collected within a six-beam MOT configuration with a beam diameter of 20 mm and 20 mW laser power in each beam. After 5 ms polarisation gradient cooling we optically pump the atoms into the $|F=2, m_F=2\rangle$ hyperfine ground state and load a 70 μK cold cloud of 2×10^8 atoms into the spherical quadrupole trap formed by the MOT coils at field gradient of 45 G/cm. Because the MOT coils and the transfer coils overlap, an adiabatic transfer of the magnetically trapped atoms can be performed with the quadrupole potential minimum moving on a straight line from the center of the MOT coils to the center of the transfer coils. At a current of 3 A the transfer coils generate a spherical quadrupol field with a gradient of 58 G/cm along the symmetry axis. By increasing the current in the Ioffe wire, the center of the spherical quadrupol is shifted and transformed into a Ioffe-type trapping field. At a current of 13 A in the Ioffe wire the resulting harmonic trap potential is characterized by an axial oscillation frequency of $2\pi\times14$ Hz, a radial oscillation frequency of $2\pi\times110$ Hz and an offset field of 0.7 G. In the Ioffe-type trap the atoms are cooled for 20 s by radio frequency evaporation to a temperature of 5 μK. The precooled ensemble is now adiabatically compressed from the large volume Ioffe-type trap into the micro trap by changing the currents in the upper coil, lower coil and the compression wire that shifts the trap minimum to the microstructure (Fig. 3a). The compression is completed by inverting the current in the compression wire. Good conditions for condensation are achieved with a current of 2 A in the microstructure and -10 A in the compression wire which corresponds to trap frequencies of $\omega_r = 2\pi\times840$ Hz and $\omega_a = 2\pi\times14$ Hz. The trap minimum is then located at a distance of 270 μm from the surface. During transfer and compression the radio frequency is turned off. The compression heats the atomic cloud to 35 μK and boosts the elastic collision rate up to 600 s^{-1}. By now ramping the radio frequency from 10 MHz to 1 MHz within 7 s we reach condensation with 1×10^6 atoms at a critical temperature of 900 nK (Fig. 3b). For the pure condensate we obtain a chemical potential μ/k_B = 380nK, a density n_0 = 1 10^{15} cm^{-3} and a number of condensed atoms of N_0 = 400,000 (Fig. 3d). The life time of the condensate is limited by three body collisions to 100 ms. After relexation of the trapping potential the life time is increased to 1s. No heating occurred and the application of a radio frequency shield had no effect.

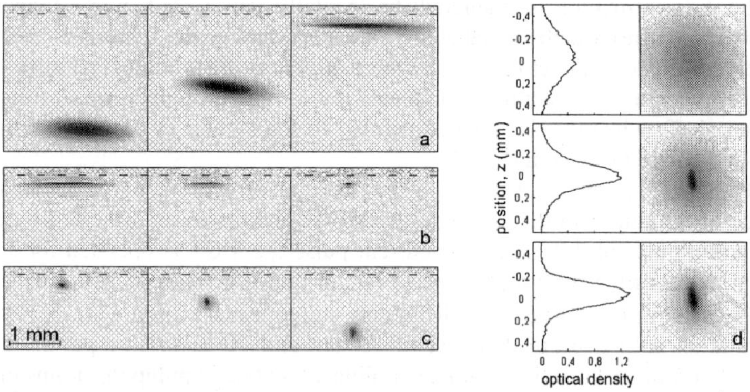

Figure 3. The movie on the left shows absorption images of the compression, the final cooling stage and the release of the condensate. The dashed line indicates the surface of the microstructure. (a) Transfer and compression of the Ioffe-type trap into the micro trap. (b) RF cooling in the micro trap. The last image of the series displays the condensate in the trap. (c) Release of the condensate. The images are taken after 5, 10, 15 ms time of flight, resp. (d) In the right view graph and absorption images the phase transition can be investigated. The absorption images are taken after 20 ms time of flight. Temperature, number of atoms, density and chemical potential are determined by fits to the data scans. For the almost pure condesate in the last image is T<500 nK, N=5×10^5, N_0=4×10^5, n_0=1×10^{15} cm^{-3}, μ/k_B=380 nK.

Our micro trap combines the advantage of a deep magnetic trapping potential with the possibility to vary the ratio of the oscillations frequencies over a wide range. This offers a promising testing ground for investigations of phase transition in highly anisotropic traps, collisional properties of the condensate in strongly anisotropic traps and quasi one-dimensional situations.

3 References

1. J. D. Weinstein and K. G. Libbrecht, Phys. Rev. A **52**, 4004 (1995)
2. H. Ott, J. Fortagh, G. Schlotterbeck, A. Grossmann, and C. Zimmermann, *accepted for publication in Phys. Rev. Lett., and references therein*

OBSERVATION OF IRROTATIONAL FLOW AND VORTICITY IN A BOSE-EINSTEIN CONDENSATE

G. HECHENBLAIKNER, E. HODBY, S.A. HOPKINS, O. MARAGÒ AND C.J. FOOT

Clarendon Laboratory, Department of Physics, University of Oxford, Parks Road, OX1 3PU, Oxford, United Kingdom
Tel +44-1865-272285, Fax +44-1865-272400

We have observed the expansion of vortex-free Bose-condensates after their sudden release from a slowly rotating anisotropic trap and found clear experimental evidence for irrotational flow. At higher rotation frequencies we nucleated vortices (singularities in the macroscopic wavefunction of the condensate). Our experiment used a purely magnetic TOP trap. This time-averaged orbiting potential was rotated by modifying the bias field which allowed us to study the conditions for vortex nucleation over a wide range of eccentricities and rotation rates.

1 Irrotational flow of a condensate

Recent experiments on the superfluidity of Bose-condensates include the observation of quantized vortices, the absence of dissipation below a critical velocity and the quenching of the moment of inertia in the scissors mode (for a review see [1,2]). In our experiment [3] the superfluidity of the condensate, together with the conservation of angular moment, leads to clearly observable consequences when a condensate is released from a slowly rotating trap, under conditions where no vortices are present.

A superfluid always has a moment of inertia less than that of a rigid body of the same mass distribution [4]. For the special case of cylindrical symmetry, the superfluid has zero moment of inertia about the symmetry axis. Thus a condensate with some angular momentum avoids having a circular shape when viewed along the rotation axis, since this would imply the unphysical situation of infinite angular velocity. We observed this behaviour in absorption images of the condensate taken along the direction of the rotational axis. The evolution of the condensate density distribution is calculated using hydrodynamic equations [5], and agrees well with our data.

In our experiment we have an anisotropic harmonic potential with three angular frequencies $\omega_x/2\pi = 60 < \omega_y/2\pi = 1.4 \times 60 < \omega_z/2\pi = 206$ Hz. The potential rotates about the z-axis with angular frequency Ω. In the hydrodynamic limit, a condensate rotating in this trap displays a quadrupolar flow pattern, as shown in Fig.1a. We produced a condensate of about 1.5×10^4 Rb atoms at a temperature of 0.5 T_c. We then made the trap eccentric in

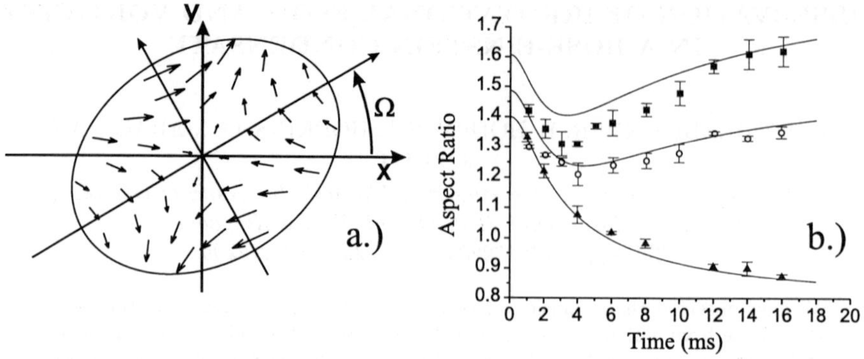

Figure 1. a.) Irrotational quadrupolar flow pattern of a rotating condensate. b.) The aspect ratio of the cloud (x-radius/y-radius) is plotted against time of flight. The three curves correspond to $\Omega/2\pi = 28$ Hz (upper), $\Omega/2\pi = 20$ Hz (middle) and release from a nonrotating trap (lower).

the x-y plane and rotated the condensate for 500 ms before it was released at a well defined point in the trap rotation. Fig.1b shows the theoretical prediction for the evolution of the condensate's aspect ratio in time of flight and experimentally measured values. The three curves correspond to $\Omega/2\pi =$ 28 Hz (upper), $\Omega/2\pi = 20$ Hz (middle) and release from a nonrotating trap (lower). The data clearly demonstrate how the aspect ratio for an initially rotating condensate decreases up to a critical point, which is reached after approximately 4 ms. Approaching this critical value in the aspect ratio (A_c), the condensate rotates more and more quickly, the rapid increase in angular velocity arising from the decrease of the moment of inertia and preservation of angular momentum [5]. However, energy conservation limits how fast the condensate can rotate and thus the aspect ratio cannot become smaller than a certain value. Beyond A_c, the condensate continues to expand, but more rapidly along its original long axis.

This behaviour is distinctly different to that of a thermal gas which can become spherical and still preserve angular momentum, since there are no constraints of irrotationality. Thus it is essentially the superfluid nature of the Bose-condensate which prevents it from becoming cylindrical. In contrast, a condensate released from a static trap has no velocity field which prevents it from becoming circular and the aspect ratio decreases steadily to a final value of less than one.

2 Vortex nucleation in a condensate

Another striking and fascinating signature of superfluidity is the occurrence of quantized vortices (for a review see [2]). In previous experiments vortices were nucleated using a laser beam to imprint a phase or to 'stir' the condensate. In our experiment [6] we nucleated vortices by rotating a purely magnetic potential as described in reference [7]. This gives very precise control over the deformation parameter $\epsilon = (\omega_x^2 - \omega_y^2)/(\omega_x^2 + \omega_y^2)$, which could be varied between $0 < \epsilon < 0.5$ with an accuracy of $\epsilon = 0 \pm 0.005$. We studied the nucleation threshold as a function of ϵ and the normalised trap rotation frequency $\bar{\Omega} = \Omega/\omega_\perp$, where $\omega_\perp^2 = (\omega_x^2 + \omega_y^2)/2$. This was done by keeping $\bar{\Omega}$ fixed and adiabatically ramping up ϵ to a certain value over 200 ms. The condensate was then held in the rotating anisotropic trap for a further 800 ms before being released. After 12 ms of free expansion the cloud was imaged along the axis of rotation using an absorption imaging system with 3 μm resolution. Figure 2a-d shows images of the expanded condensate at different stages during the nucleation process. Initially the cloud elongates, confirming that nucleation is being mediated by excitation of a quadrupole mode (Fig.2(a)) [8]. Then finger-like structures appear on the outside edge of the condensate which eventually close round and produce vortices, $\sim$ 800ms after rotation began (Fig. 2(b)). Approximately 200 ms later, these have moved to equilibrium positions within the bulk of the condensate. Figures 2(c) and (d) show typical, single-shot images of stable vortex arrangements. The maximum number of vortices we observed was seven, limited by the number of atoms in our condensate.

We determined the boundaries for vortex nucleation as a function of ϵ and $\bar{\Omega}$ and found that vortices could be nucleated within region 2 in Fig.2e. Vortices are nucleated with minimum eccentricity at $\bar{\Omega} = \bar{\Omega}_c$, when the trap is rotated at half the frequency of the $m = 2$ mode [8]. At $\overline{\Omega} > \overline{\Omega}_c$, the condensate follows the 'overcritical branch' of the quadrupole mode [9] with an elliptical density distribution which is orthogonal to the trap potential. Vortices are nucleated when the eccentricity is too large for this 'overcritical mode' to be a solution of the hydrodynamic equations. The boundary of the stable region 3 in the ϵ versus $\overline{\Omega}$ plot is given by $\epsilon = 2/\overline{\Omega}\left((2\overline{\Omega}^2 - 1)/3\right)^{3/2}$. We determined this relation from the solutions of the hydrodynamic equations for superfluids and it is plotted as a solid line in fig. 2e. This line agrees well with the experimental data for the critical conditions for nucleation for $\overline{\Omega} > \overline{\Omega}_c$ and a wide range of ϵ. Below $\overline{\Omega}_c$, the deformation needed to nucleate vortices appears to increase linearly with $\overline{\Omega}$. This boundary can be explained in terms of dynamic instabilities of the 'normal branch' of the quadrupolar modes [10].

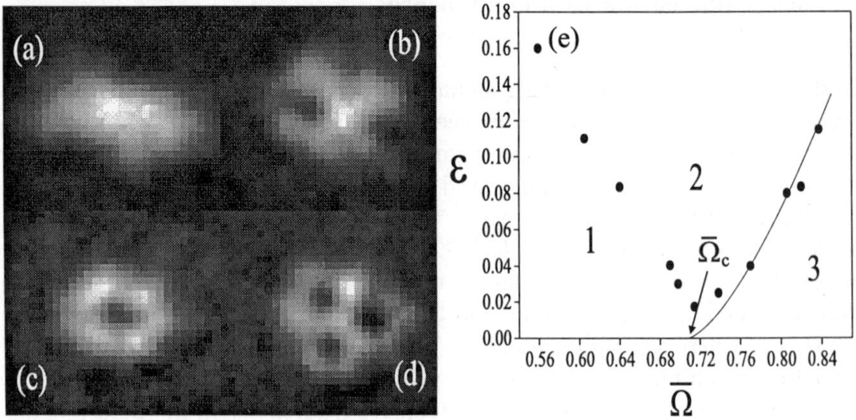

Figure 2. a,b,c,d) Images of the condensate at different stages during the vortex nucleation process. All images were taken with $\bar{\Omega} = 0.70$, $\epsilon = 0.05$ and after 12ms of free expansion. e) The critical conditions for vortex nucleation. The data points mark the minimum trap eccentricity for nucleation at a particular $\bar{\Omega}$. Vortices may be formed in region 2.

To make a quantitative prediction for the boundary between regions 1 and 2 shown in Fig. 2e will require a numerical calculation for our specific case. This lower limit for the nucleation frequency extends the results in [8] for very small eccentricities to higher ones. We observed that the thermal cloud destabilised the vortex arrays so that they lasted less than 1 second around 0.8 T_c but lived for several seconds at $T < 0.5\ T_c$.

We would like to thank K. Burnett, Y. Castin, J. Dalibard, M. Edwards, D. Feder and S. Sinha for fruitful discussions.

References

1. E. Timmermans, Contemp. Phys. **42**, 1 (2001);
2. A.L. Fetter et al., J. Phys: cond. matt. **13**, R135 (2001)
3. G. Hechenblaikner et al., cond-mat/0106182
4. F.Zambelli et al., Phys. Rev. A, **63**, 033602 (2001)
5. M. Edwards et al., cond-mat/0106080
6. E.Hodby et al., cond-mat/0106262
7. J. Arlt et al., J. Phys. B, **32**, 5861 (1999)
8. K.W. Madison et al. Phys. Rev. Lett. **86**, 4443 (2001);
9. A. Recati, Phys. Rev. Lett **86**, 377 (2001)
10. S. Sinha et al., cond-mat/0101292 (2001)

PHASE FLUCTUATIONS IN ELONGATED 3D-CONDENSATES

P. RYYTTY, D. HELLWEG, S. DETTMER, J. J. ARLT, W. ERTMER, AND K. SENGSTOCK

Institut für Quantenoptik, Universität Hannover, Welfengarten 1, 30167 Hannover, Germany

E-mail: ryytty@iqo.uni-hannover.de

We demonstrate the existence of phase fluctuations in elongated Bose-Einstein Condensates (BECs) and study the dependence of those fluctuations on the system parameters. A strong dependence on temperature, atom number, and especially on the trapping geometry is observed. In particular, we observe instances where the phase coherence length is significantly smaller than the condensate size. Our method of detecting phase fluctuations is based on their transformation into density modulations after ballistic expansion. We also discuss preliminary measurements on the growth of BEC in the regime of phase fluctuations.

Recently, there has been enormous interest in the coherence properties of Bose-Einstein condensates. In particular, the phase coherence has been under intensive theoretical and experimental study as it plays an essential role for the application of BEC in experiments on atom optics. For a trapped 3D condensate well below the BEC transition temperature T_c, experiments have confirmed the phase coherence, e.g., it was shown that the coherence length is equal to the condensate size[1,2,3]. However, it is expected that low-dimensional (1D and 2D) quantum gases differ qualitatively from the 3D case in this respect[4,5,6,7]. Recently, it was shown theoretically[8] that for very elongated condensates phase fluctuations can already be pronounced in the equilibrium state of the usual 3D ensemble, where the density fluctuations are suppressed. The phase coherence length in this case can be smaller than the axial size of the sample, which can have dramatic consequences for practical applications.

We have performed systematic studies on ^{87}Rb BECs in the regime where phase fluctuations are expected. This regime was achieved in highly anisotropic traps, leading to a strongly elongated shape of the condensate. The more elongated the condensate shape the more pronounced are phase fluctuations. The experiment was performed with Bose-Einstein condensates of up to $N_0 = 5\times 10^5$ atoms in the $|F=2, m_F=+2\rangle$ state of a cloverleaf-type magnetic trap. By combining the magnetic trap with the optical potential of a blue detuned Laguerre-Gauss mode (TEM^*_{01}) laser beam we were able to vary the radial trap frequencies ω_ρ between $2\pi \times 138\,\mathrm{Hz}$ and $2\pi \times 715\,\mathrm{Hz}$ corresponding to aspect ratios λ between 10 and 51. Further details of our

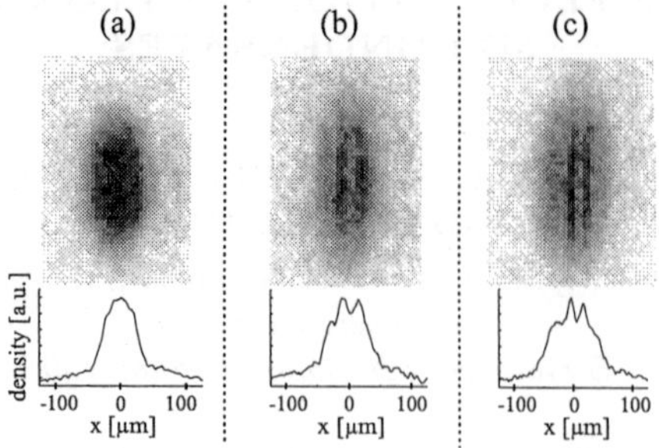

Figure 1. Absorption images and corresponding density profiles of BECs after 25 ms time-of-flight taken for various aspect ratios [$\lambda = 10$ (a), 26 (b), 51 (c)].

apparatus were described previously[9,10,11].

We observe the phase fluctuations by measuring the density distribution of the released cloud after ballistic expansion. Figure 1 shows examples of experimental data for various aspect ratios λ. The usual anisotropic expansion of the condensate related to the anisotropy of the confining potential is clearly visible in the absorption images. The line density profiles below reflect the parabolic shape of the BEC density distribution. Remarkably, we observe pronounced stripes in the density distribution in some cases shown in Fig. 1. On average these stripes are more pronounced at high aspect ratios of the trapping potential [Fig. 1(c)], high temperatures, and low atom numbers.

The appearence of stripes can be understood qualitatively as follows: Within the equilibrium state of a BEC in a magnetic trap the density distribution remains largely unaffected even if the phase fluctuates[8]. Transformation of local velocity fields provided by the phase fluctuations into modulations of the density is prevented by the mean-field interparticle interaction. However, after switching off the trap, the mean-field interaction rapidly decreases and the axial velocity fields are converted into a density distribution.

To determine the amount of phase fluctuations experimentally, we studied the structure of stripes in the atomic density distribution as a function of the trapping potential and temperature. For each realization of a BEC, the observed density distribution was integrated along the radial direction and then fitted by a bimodal function representing the condensate fraction and thermal cloud. For each image we obtained standard deviations σ_{BEC} of the experimental data from the fit in the central region of the condensate fraction (half width of full size).

Figure 2 shows a comparision of the observed average standard deviation

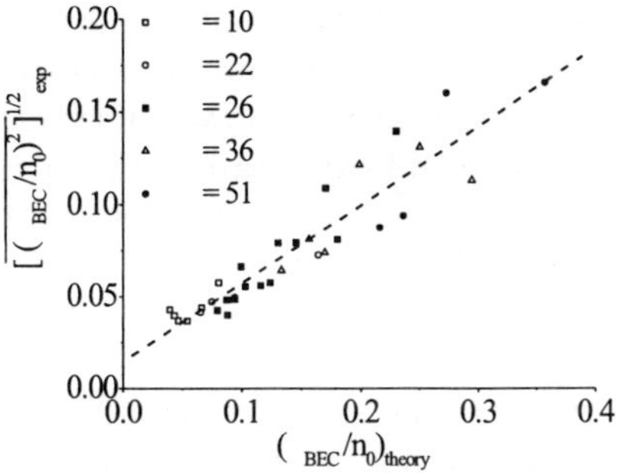

Figure 2. Average standard deviation of the measured line densities $[\overline{(\sigma_{\mathrm{BEC}}/n_0)^2}]^{1/2}_{\mathrm{exp}}$ compared to the theoretical value of $(\sigma_{\mathrm{BEC}}/n_0)_{\mathrm{theory}}$ obtained from Eq. (1). The dashed line is a fit to the experimental data.

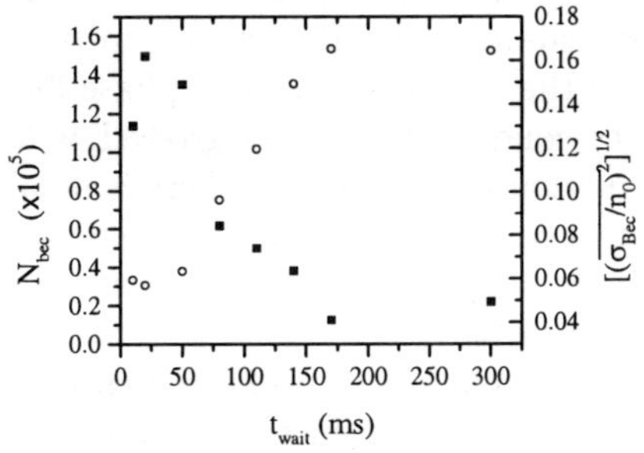

Figure 3. Growth of Bose condensate from thermal vapor. The condensate number is given by the open circles and the corresponding $[\overline{(\sigma_{\mathrm{BEC}}/n_0)^2}]^{1/2}$ value by the squares. The data correspond to $\lambda = 26$.

$[\overline{(\sigma_{\mathrm{BEC}}/n_0)^2}]^{1/2}_{\mathrm{exp}}$ to a theoretical value $(\sigma_{\mathrm{BEC}}/n_0)_{\mathrm{theory}}$ given by[12].

$$\left(\frac{\sigma_{\mathrm{BEC}}}{n_0}\right)^2 = \frac{T}{\lambda T_\phi}\sqrt{\frac{\ln\tau}{\pi}}\left(\sqrt{1+\sqrt{1+\left(\frac{\hbar\omega_\rho\tau}{\mu\ln\tau}\right)^2}}-\sqrt{2}\right), \tag{1}$$

where $\tau = \omega_\rho t$. In Eq. (1) we introduced a characteristic temperature $k_B T_\phi = 15(\hbar\omega_x)^2 N_0/32\mu$. For $T_\phi < T_c$ one expects the regime of quasicondensation for the initial cloud in the temperature interval $T_\phi < T < T_c$. This is the regime where the coherence length is smaller than the condensate size[8]. As can be seen from Fig. 2 the experimental results follow the expected dependence very well, and confirm the predicted general behavior of phase fluctuations in elongated BECs. Most of the observed experimental data, which exhibit fluctuations well above the noise limit, correspond to $T > T_\phi$. This implies that the measurements were performed in the regime of quasicondensation, i.e., the phase coherence length $l_\phi = LT_\phi/T$ of the initial condensate was smaller than the axial size L. For instance, for $\lambda = 51$, $T = 0.5\,T_c$, and $N_0 = 3\times10^4$, one obtains $l_\phi \approx L/3$.

In addition to studies of the equilibrium state of phase fluctuating BECs, we have performed preliminary measurements on the growth of a Bose condensate from thermal vapor[13,14]. An atomic sample was cooled to a temperature just above T_c followed by an abrupt reduction of the RF frequency to enter the BEC regime. Figure 3 shows typical data for the condensate number (open circles) and density modulations (squares) as a function of wait time after the abrupt change of RF frequency. At short wait times the condensate distribution deviated significantly from a Thomas-Fermi distribution, indicating the presense of strong phase fluctuations. As the wait time was increased the condensate number growed, which is accompined by decrease of the phase fluctuations. However, there is no abrupt jump in $[\overline{(\sigma_{\text{BEC}}/n_0)^2}]^{1/2}_{\text{exp}}$ but rather it decreases smoothly to the equilibrium value. We are currently performing more detaild studies on the temporal characteristics of the formation of phase fluctuations.

In conclusion, we have presented experimental and theoretical studies of a BEC state with fluctuating phase. The flexible method of ballistic expansion allows us to measure phase fluctuations under various experimental conditions, especially in various trap geometries. By measuring the phase fluctuations and comparing the temperature with T_ϕ we have demonstrated instances, where the phase coherence length was smaller than the axial size of the condensate, i.e., the initial cloud was in the quasicondensate state. Our results set severe limitations on applications of BECs in interferometric measurements, and for guided atom laser beams. Further studies of quasicondensates, e.g., with respect to superfluid properties, will be necessary to obtain a full understanding of the phase coherence properties of ultracold atomic gases.

References

1. J. Stenger et al., Phys. Rev. Lett. **82**, 4569 (1999).
2. E. W. Hagley et al., Phys. Rev. Lett. **83**, 3112 (1999).
3. I. Bloch, T. W. Hänsch, and T. Esslinger, Nature **403**, 166 (2000).
4. D. S. Petrov, G. V. Shlyapnikov, and J. T. M. Walraven, Phys. Rev. Lett. **85**, 3745 (2000).
5. D. S. Petrov, M. Holzmann, and G. V. Shlyapnikov, Phys. Rev. Lett. **84**, 2551 (2000).
6. Yu. Kagan et al., Phys. Rev. A **61**, 43608 (2000).
7. A. I. Safonov et al., Phys. Rev. Lett. **81**, 4545 (1998).
8. D. S. Petrov, G. V. Shlyapnikov, and J. T. M. Walraven, Phys. Rev. Lett. **87**, 050404 (2001).
9. K. Bongs et al., Phys. Rev. Lett. **83**, 3577 (1999).
10. S. Burger et al., Phys. Rev. Lett. **83**, 5198 (1999).
11. K. Bongs et al., Phys. Rev. A **63**, R31602 (2000).
12. S. Dettmer et al., cond-mat/0105525.
13. H.-J. Miesner et al., Science **279**, 1005 (1998).
14. M. Köhl, T. W. Hänsch, and T. Esslinger, cond-mat/0106642

FRANCIUM SPECTROSCOPY AND A POSSIBLE MEASUREMENT OF THE NUCLEAR ANAPOLE MOMENT

S. AUBIN, E. GOMEZ, J. M. GROSSMAN, L. A. OROZCO, M. R. PEARSON, G. D. SPROUSE

Dept. of Physics and Astronomy, State University of New York at Stony Brook, Stony Brook, NY 11794-3800, U.S.A.
E-mail: Luis.Orozco@sunysb.edu

D. P. DEMILLE

Dept. of Physics, Yale University, New Haven, CT 06520-8120, U.S.A.

The spectroscopy of francium performed in a magneto-optical trap permits quantitative comparison between *ab initio* calculations and measurements of energy levels, electronic lifetimes, and hyperfine splittings to an accuracy comparable to that achieved in other alkali elements. This understanding is fundamental for proposed anapole moment measurements in a chain of francium isotopes.

1 Introduction

Francium is the heaviest of the alkali atoms and has no stable isotopes. The longest lived isotope has a half-life of 22 minutes. Since 1995 we have carried out a systematic study of the atomic and nuclear properties of Fr at the Stony Brook Superconducting LINAC in our on-line magneto-optical trap (MOT)[1]. Because of the large number of constituent particles in Fr, electron correlations and relativistic effects are important, but its structure remains calculable with many-body perturbation theory (MBPT). Its more than two hundred nucleons and simple atomic structure make it an attractive candidate for studies of atomic parity non-conservation and nuclear anapole moment measurements.

2 Spectroscopy

We have been studying the atomic spectroscopy of Fr to test the *ab initio* atomic theory[2] that predicts a factor of 18 larger Parity Non-Conserving (PNC) effect in Fr than in Cs. Examples of our work include measurements of the lifetimes of the $7p$ levels[3] and of the hyperfine splitting of the $7P_{1/2}$ level[4] in five different isotopes $^{208-212}$Fr. The precision in the former (better than 0.5%) tested MBPT calculations of the dipole matrix elements, while the precision of the latter (200 ppm) was enough to resolve the hyperfine anomaly.

3 New Apparatus

We have constructed a new apparatus for on-line magneto-optical trapping of radioactive Fr atoms at the Stony Brook LINAC for the next generation of Fr experiments. The new apparatus separates the production, trapping, and measurement regions of the experiment.

An ion beam with an energy of ~100 MeV from the accelerator strikes a target of gold or isotopically enriched platinum mounted on the end of a tungsten rod[5]. A heavy-ion fusion reaction [^{16,18}O(^{197}Au,xn) or ^{19}F(^{198}Pt,xn)] selectively produces the isotopes $^{206-213}$Fr in the target room of the accelerator. The beam power and an auxiliary heater coil heat the target to just below its melting point, allowing the Fr to diffuse rapidly to its surface. The target ionizes the Fr as it exits. We then transport the Fr as ions to a different room for subsequent neutralization and capture in a MOT. Because the transport system is electrostatic, it is mass independent, allowing us to transport any isotope of Fr and other alkali elements. We have tested the entire apparatus from the target to the MOT with ^{85}Rb. We have produced an ion beam containing in excess of 1.4×10^7 Fr/sec from the target. A small constriction separates the target and transport region from the trap, permitting differential pumping of the trap. The captured atoms can then be transferred to a second chamber for further study.

4 Anapole moment

The anapole moment of a nucleus is a PNC, time reversal conserving moment that arises from weak interactions between the nucleons (see the recent review by Haxton and Wieman[6]). It can only be detected in a PNC electron-nucleus interaction and reveals itself in the nucleon spin dependent part of the PNC interaction. The measurement of the Cs anapole moment by Wood *et al.*[7], by looking at the changes in atomic PNC as a function of the hyperfine energy levels, showed that atomic PNC is a unique probe for the neutral weak interactions inside the nucleus.

The anapole moments in Fr arise from the weak interaction between the valence nucleons and the core. Flambaum, Khriplovich and Sushkov[8] developed a simple model to estimate anapole moments

$$\mathbf{a} = \frac{1}{e}\frac{Gg}{\sqrt{2}}\frac{K\mathbf{I}}{I(I+1)}\kappa_a \,, \qquad \kappa_a = \frac{9}{10}\frac{\alpha\mu}{mr_0}A^{2/3} \tag{1}$$

where $G = 10^{-5}$ m^{-2} is the Fermi constant of the weak interaction, and m, I, and μ are the mass, spin and magnetic moment of the valence nucleon. g is a dimensionless constant of order one describing the strength of the P-odd coupling between nucleons[9]. K is given by $K = (I + \frac{1}{2})(-1)^{I+\frac{1}{2}-\ell}$ and ℓ is the orbital angular momentum of the valence nucleon.

We have used the simple model above with the shell model parameters from our work on the hyperfine anomaly[4] to estimate the anapole moments of the light Fr isotopes. We find $\kappa_a(^{211}\mathrm{Fr}) = 0.45$, and that the anapole moment for the even-neutron Fr isotopes, ^{209}Fr and ^{211}Fr ($I = 9/2$), will be essentially the same since both have a single unpaired nucleon, an $h_{9/2}$ proton.

We focus on ^{211}Fr to calculate the direct excitation between the $(7s, F = 4) \rightarrow (7s, F = 5)$ hyperfine $E1$ transition allowed by the anapole moment. We use our estimated anapole moment, the measured electric dipole values[3], and first order perturbation theory (truncated at the $7p_{1/2}$ states) to calculate the electric dipole matrix element between the $7s$ hyperfine states. For example, with linear polarization transverse to the quantization axis, we calculate an expected value for the magnitude of the matrix element for $m_F = I - 1/2 \rightarrow m_F = I + 1/2$ of $4.4 \times 10^{-12} ea_0$ (see Ref. 9), a factor of ten larger than the similar matrix element in ^{133}Cs. The ratio between the anapole induced PNC $E1$ transition to the allowed $M1$ hyperfine transition is 1.4×10^{-9}.

In this proposed measurement, the suggestion of Fortson[11] to use an ion placed at the antinode of a standing optical wave is modified to place many trapped atoms at the antinode of a standing microwave, so the requirements of stability of the sample are relaxed. The atoms will have to be localized well within the antinode of the standing wave ($\nu \approx 50$ GHz, $\lambda \approx 6$mm). This approach has been suggested in the literature in the past[12,13,14], and it consists of measuring the probability of transition by the direct excitation through the E1 transition (allowed mainly through the nucleon spin dependent term) between the ground hyperfine states.

The preparation of the sample of Fr will take place in the MOT. From this first trap, we will transfer the Fr to a second chamber, where the atoms will be held in a purely optical trap (dipole) at the electric antinode of a standing wave microwave cavity, so as to maximally drive the $E1$ transition. To enhance the small probability of excitation, we will use interference with an allowed transition, such as the $M1$, in the presence of a coordinate system with reversible handedness. Given all the constraints for preparation and transfer of the Fr atoms, we will use a Fabry-Perot configuration for the cavity, where combinations of RF and optical pulses will prepare the appropriate superposition of states. Optical interrogation, through a cycling transition, after the excitation pulses will determine how many atoms have performed the $E1$ transition.

Recent work related to time-reversal invariance tests with traps[15,16] point to the many potential problems of combining traps with tests of fundamental symmetries, indicating further work is needed.

Acknowledgments

This work has been supported by the NSF. E. G. acknowledges support from CONACYT and the Fulbright-García Robles foundation.

References

1. J. E. Simsarian, A. Ghosh, G. Gwinner, L. A. Orozco G. D. Sprouse, and P. A. Voytas, *Phys. Rev. Lett.* **76**, 3522 (1996).
2. V. A. Dzuba, V.V. Flambaum, and O. P. Sushkov, *Phys. Rev. A* **51**, 3454 (1995).
3. J. E. Simsarian, L. A. Orozco, G. D. Sprouse, and W. Z. Zhao, *Phys. Rev. A* **57**, 2448 (1998).
4. J. S. Grossman, L. A. Orozco, M. R. Pearson, J. E. Simsarian, G. D. Sprouse, and W. Z. Zhao, *Phys. Rev. Lett.* **83**, 935 (1999).
5. A. R. Lipski, L. A. Orozco, M. R. Pearson, J. E. Simsarian, G. D. Sprouse, and W. Z. Zhao, *Nucl. Instr. Meth. A* **438**, 217 (1999).
6. W. C. Haxton, C. E. Wieman, arXiv:nucl-th/0104026.
7. C. S. Wood, S. C. Bennett, D. Cho, B. P. Masterson, J. L. Roberts, C. E. Tanner, and C. E. Wieman, *Science* **275**, 1759 (1997).
8. V. V. Flambaum, I. B. Khriplovich and O. P. Sushkov, *Phys. Lett.* **146**B, 367 (1984).
9. I. B. Khriplovich, *Parity Nonconservation in Atomic Phenomena* (Gordon and Breach Science Publishers, Philadelphia, 1991).
10. After the ICOLS01 conference we received the preprint: S. G. Porsev and M. G. Koslov, arXiv:physics/0107016 with a similar calculation.
11. E. N. Fortson, *Phys. Rev. Lett.* **70**, 2383 (1993).
12. C. E. Loving, and P. G. H. Sandars, *J. Phys. B.* **10**, 2755 (1977).
13. V. G. Gorshkov, V. F. Ezhov, M. G. Kozlov, and A. I. Mikhailov, *Sov. J. Nucl. Phys.* **48**, 867 (1988).
14. D. Budker in *Physics Beyond the Standard Model*, ed. P. Herczeg, C. M. Hoffman, and H. V. Klapdor-Klinfrothaus (World Scientific, Singapore, 1998).
15. M. V. Romalis and E. N. Fortson, *Phys. Rev. A.* **59**, 4547 (1999).
16. C. Chin, V. Leiber, V. Vuletic, A. J. Kerman, and S. Chu, *Phys. Rev. A* **63**, 033401 (2001).

MERGING TWO INDEPENDENT FEMTOSECOND LASERS INTO ONE

LONG-SHENG MA*, ROBERT K. SHELTON, HENRY K. KAPTEYN, MARGARET M. MURNANE, JOHN L. HALL, AND JUN YE

JILA, National Institute of Standards and Technology and University of Colorado, Department of Physics, University of Colorado, Boulder, Colorado 80309-0440

Ye@jila.colorado.edu

Two independent mode-locked femtosecond lasers are synchronized and phase-locked to an unprecedented precision. The timing jitter between the two fs lasers is less than 5 fs rms, observed within a 160-Hz bandwidth over minutes. The beat frequency between the two synchronized fs lasers has a standard deviation of 0.15 Hz at 1-s averaging time under phase-locked conditions. Coherence between the two lasers is demonstrated via spectral interferometry and second order field cross-correlation when the two fs lasers are tightly synchronized and phase-locked. The auto-correlation measurement of the combined pulse reveals a narrower and larger amplitude "synthesized' pulse.

1 Introduction

Precision control of CW laser light in the frequency domain has been a prominent area for atomic, molecular, solid-state physics and fundamental measurements in physics. It has improved spectral resolution to the sub-Hertz level and the detection sensitivity of weak absorption to the $< 10^{-12}$ level. In recent years, the precision control of the repetition rate and phase of femtosecond lasers has improved so much that the fs laser comb has became a superior technique [1-4] for optical frequency metrology, carrier-envelope phase control and for implementing an optical clock. Many exciting areas of current research stimulate us to make greater efforts to synchronize and phase lock two independent femtosecond lasers. For example, one exciting field is "coherent control". Light pulses that have been precisely shaped in amplitude and phase can selectively "drive" a chemical reaction [5,6], molecular vibration [7], or other process such as the nonlinear-optical conversion of light into the extreme ultraviolet region of the spectrum [8]. Often the quantum transitions for a process of interest in coherent control are concentrated in a few disparate regions of the spectrum. In this case, it would be desirable to be able take two separate laser systems, generating light with distinct optical properties, and precisely synchronize the output of both lasers, essentially generating a single, composite coherent light field from two separate sources. Clearly, the ability to precisely synchronize separate pulsed laser sources is an important step in the road toward the ultimate, 'arbitrary light wave-form generator'. It is also important for a number of other technologies, such as mid-infrared light generation through difference frequency mixing, and for experiments requiring synchronized laser light and x-rays or electron beams.

To date, previous work in synchronizing separate mode-locked Ti:Sapphire lasers has demonstrated a timing jitter of at best a few hundred femtoseconds (fs) [9]. In addition, to our knowledge, there has been no report of phase locking between two separate fs lasers. In this paper, we demonstrate robust synchronization of pulse trains from two separate fs lasers, with a timing jitter of < 5 fs, at a bandwidth of 160 Hz, observed over an interval of one minute. Due to our precision synchronization of two independent fs lasers, we are now able to tightly phase lock the two lasers to each other. Measuring the beat frequency between two fs lasers gives us a standard deviation of 0.15 Hz at 1-s averaging time. Coherence between the two lasers is demonstrated via spectral interferometry and second order field cross-correlation when the two fs lasers are tightly synchronized and phase-locked. The auto-correlation measurement of the combined pulse reveals the "synthesized" pulse to be narrower and of larger amplitude.

2 Experimental setup and results

The experimental configuration is shown in Figure 1. Two independent mode-locked Ti:Sapphire lasers operate at 760 nm and 810 nm respectively [10], with ~ 100 MHz repetition rates. Two separate pump lasers are used. Our synchronization and phase-locking setup employs two high-speed photo-diodes to detect the two pulse trains from their respective lasers. There are four phase locked loops (PLL) in the system that establish synchronization and phase locking between the two fs lasers. We synchronize the repetition rate of laser #1 to a stable RF source using the first PLL, while laser #2 is synchronized to laser #1 using a second PLL. This second PLL compares and locks the fundamental frequencies of the two lasers at 100 MHz. The phase shift between the two 100 MHz signals can be used to control the timing offset between the two pulse trains. The third PLL at 8 GHz compares the phases of the 80th harmonics of the two repetition frequencies. When the two pulse trains are nearly overlapped, the third PLL at 8 GHz is gradually activated while the second PLL at 100 MHz is gradually deactivated. This represents an electronic realization of a "differential micrometer" -- the 100 MHz loop provides the full dynamic range of timing offset between two pulse trains, while the 8 GHz loop produces enhanced phase stability of the repetition frequency. The two PLL loops at 8 GHz actuate fast piezo-transducers (PZT) mounted to the laser end-mirrors. We use a combination of a fast, small PZT and a slow, long PZT to achieve a high servo bandwidth (> 50 KHz) and a large tuning range. To further characterize the timing jitter of our system, we focus the two pulse trains so that they can cross in a 500 micrometer thick, room temperature, BBO crystal for nonlinear frequency generation (Type-I). When the two pulses are overlapped in space and time, sum frequency generation (SFG) is enabled. We can thus use the

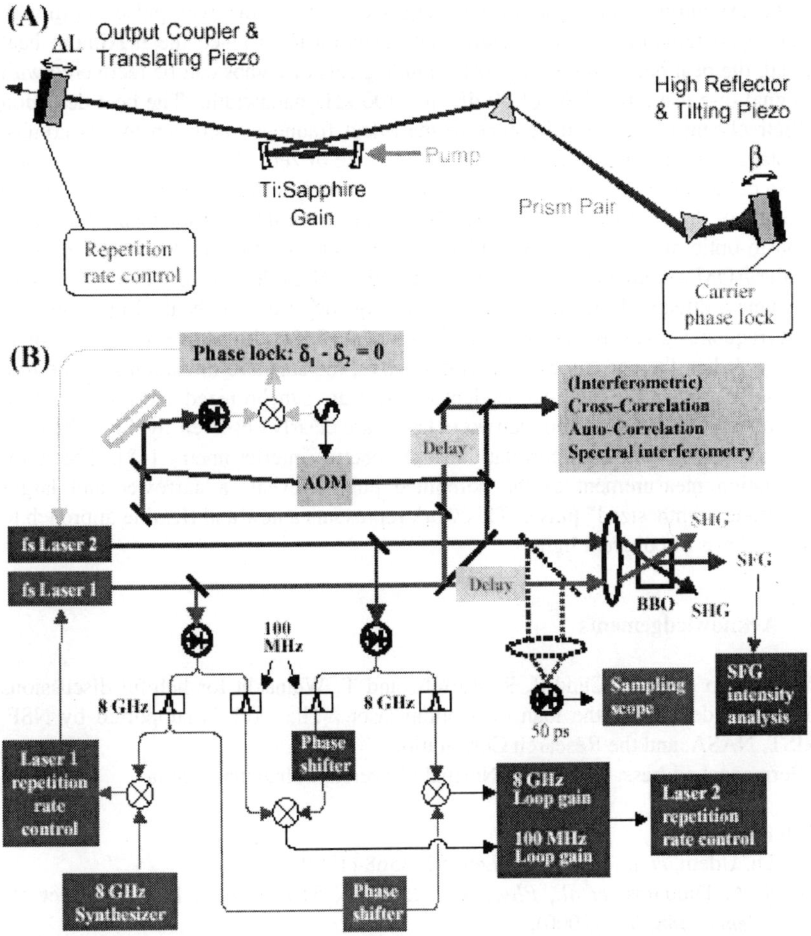

Figure 1. Experimental setup for timing synchronization and phase locking of two fs lasers. (A) Femtosecond laser and its control elements. (B) The phase locked loops for synchronization, along with the signal analysis scheme. Another phase locked loop is used to track the two carrier frequencies.

SFG intensity as a diagnostic tool to study our system performance. The SFG intensity fluctuation is proportional to the timing jitter, particularly when the two pulses are offset in time by ~ ½ the pulse width. Using this method, the timing jitter between the two independent fs lasers is estimated to be 4.3 fs rms observed within a 160 Hz bandwidth [11]. When the two lasers are well synchronized and phase shift in the synchronization lock loop is adjusted to have the two pulses optimally overlapped temporally at the heterodyne detection to produce the maximum beat signal, the beat between the two corresponding sets of combs can be recovered with a signal-to-noise ratio (*S/N*) of 60 dB in a 100 kHz bandwidth. The beat detection effectively measures the difference in the offset frequency between two fs combs. By stabilizing the beat frequency to a mean value of zero Hertz, the carrier-envelop phase evolution dynamics of one laser will be closely matched by the second laser. Locking of this beat frequency to zero Hz can be conveniently implemented using an acousto-optic modulator (AOM). One laser beam passes through the AOM, picking up the AOM's frequency offset. The beat is then phase locked to the AOM's drive frequency, effectively removing the AOM frequency from the beat. Measuring the beat frequency gives us a standard deviation of 0.15 Hz with an averaging time of 1-s [12]. When the phase lock loop is not active, the standard deviation goes up to a few MHz. When the two independent fs lasers are synchronized and phase locked, the coherent effects between the two fs lasers are clearly observed by the techniques of cross-correlation, auto-correlation, and spectral interferometry [12]. The auto-correlation measurement of the combined pulses reveals a narrower and larger amplitude "synthesized" pulse. This work represents a new and flexible approach to the synthesis of coherent light.

3 Acknowledgements

We wish to thank S. Cundiff, R. Bartels, and T. Weinacht for helpful discussions and D. Anderson for the loan of a pump laser. The work is supported by NSF, NIST, NASA, and the Research Corporation.

* Permanent address: East China Normal University, Shanghai, China

References

1. Th. Udem, *et al., Phys. Rev. Lett.* **82**, 3568 (1999).
2. S. A. Diddams, *et al., Phys. Rev. Lett.* **84**, 5102 (2000); D.J. Jones, et al., *Science,* **288**, 635 (2000).
3. J. Ye, *et al., Op. Lett*, **25**, 1675 (2000).
4. S.A. Diddams, *et al., Science* **293**, 875, (2001).
5. R. Judson and H. Rabitzet, *Phys. Rev. Lett.* **68**, 1500 (1992).
6. A. Assion, *et al., Science* **282**, 919 (1998).
7. A.M. Weiner *et al., Science* **247**, 1317 (1990).
8. R. Bartels, *et al., Nature* **406**, 164 (2000).
9. D.E. Spence, *et al., Opt. Lett.*, **19**, 481, (1994).
10. M.T. Asaki *et al., Opt. Lett.* **18**, 977 (1993).
11. L.-S. Ma *et al.*, Phys. Rev. A **64**, 021802 (R), (2001).
12. R.K. Shelton *et al., Science*, **293**, 1286, (2001).

FERROMAGNETIC WAVEGUIDES FOR ATOM INTERFEROMETRY

WILBERT ROOIJAKKERS, MUKUND VENGALATTORE AND MARA PRENTISS
Center for Ultracold Atoms, Harvard University, Physics Department, Lyman Laboratory, Cambridge MA 02138, USA
E-mail: wilbert@atom.harvard.edu

Ferromagnetic materials are used to create atomic waveguides with tighter confinement than what is typically achieved in current carrying wire guides. This may lead to the integration of sophisticated atom optics experiments on a small surface. As an example we describe an experiment in which the atom waveguide can be split and recombined by changing the magnitization electronically. This corresponds to adiabatically altering a single well into a double well. We apply this in new ideas to construct large area (1 cm^2) reciprocal interferometers based on ferromagnetic guides.

1 Introduction

Due to the much weaker holding forces, tight confinement is much harder to realize for neutral atoms than for ions. Recently we have demonstrated that magnetizable materials can provide very tight confinement for neutral atoms [1]. They are also suitable for engineering complicated structures, potentially providing a leading technology for atom optics on a substrate. The purpose of this paper is to demonstrate the feasibility of atom optics with magnetic confinement, exploiting the wave like nature of atomic matter. As an example we describe an experiment in which we demonstrate electronically "crafting" the shape of a confining potential adiabatically going from a single well to a double well. Ferromagnetic waveguides may also be used to build ring interferometers, quantum point contacts, Josephson junctions and other atom-optical devices.

2 Experimental setup

As details of the experimental setup can be found elsewhere [1], we will only give a short overview of the basic features. Our most recent waveguide consists of four long strips (length 15 cm, depth 1 cm) of 0.5 mm thick μ-metal, wound with 10 loops of kapton isolated copper wire (Fig. 1). The application of ferromagnetic material amplifies the magnetic field at the end of the coil, comparable to the effect of filling a long solenoid with a ferromagnetic core. An 150 μm thin gold coated mirror is placed on top of the coils. We create a surface magneto-optical trap (surface-MOT) [2] above the mirror, resulting in a cloud that is long and stretched along the guide. We obtain 5 cm long clouds (limited by the diameter of the laser beam) of 20 μK cold cesium atoms, applying a current of 50 mA to each coil with alternating sign for neighboring coils. All necessary fields are generated in the substrate with no external field present. The height of the cloud above the surface (typically 0.5 mm) can be controlled by changing the ratio of the current in the inner (I_{inner}) and outer (I_{outer}) pair of coils. By switching off the MOT beams and increasing the magnetic field gradient atoms have been transferred efficiently into a magnetic guide ("in situ" loading [1]).

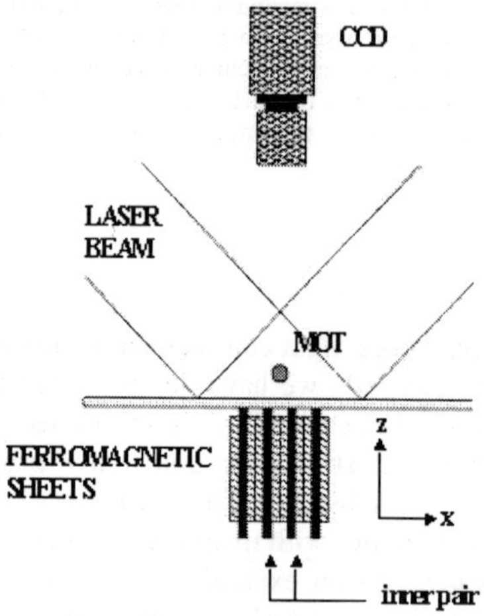

Figure 1. Experimental setup: a surface magneto-optical trap of cold cesium atoms is created in the field of four ferromagnetic coils, located underneath a mirror surface. Another pair of laser beams (not shown) propagates along the 15 cm long guide in the y-direction. An external bias field may be created in the x-direction using a pair of Helmholtz coils.

3 Double Well Magneto-Optical Trap / Magnetic Trap

By applying an external bias field (B_{bias}) a new system of two parallel guides is formed above the surface. By applying the appropriate circular polarization of the trapping laser beams, these parallel guides can be loaded with atoms simultaneously. Fig. 2a shows a fluorescence image of two long clouds (separation 1.9 mm) trapped magneto-optically at the same time with only one set of laser beams (I_{inner}=200 mA, I_{outer}= 300 mA, B_{bias}= 6 G). The distance between the clouds can be controlled adiabatically, changing only I_{inner}. The two clouds merge in the limit where I_{inner}=0

(Fig 2b). The continuous deforming of a double well potential into a single well potential and vice versa is reproduced by a numerical calculation.

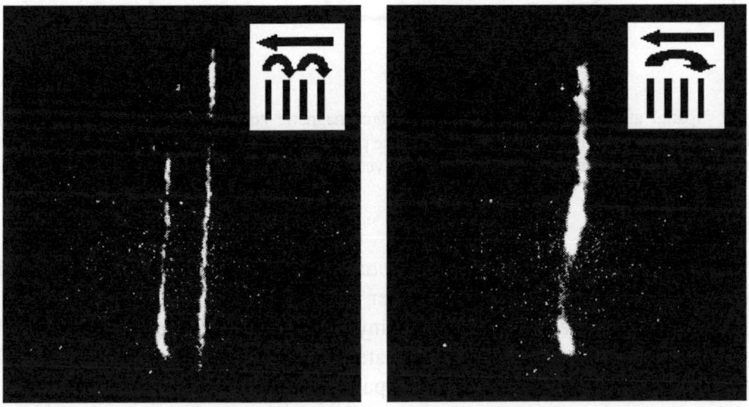

Figure 2. Fluorescence images (2 cm x 2 cm) of a double well MOT (left, I_{inner}=200 mA) and a single well MOT (right, I_{inner}=0 mA). The magnetic field gradient in the single well MOT is 32 G/cm. CCD camera mounted as in Fig. 1. Only extreme situations are shown; the separation between the clouds (1.9 mm for the left image) can be tuned with I_{inner}. Inset: field minima occur where external bias field (straight arrow) cancels the field due to the ferromagnetic coils (curved arrows).

4 Atom Interferometry

Our ferromagnetic guides may be used to realize an interferometer in *time* [3]. Consider an initial situation with all atoms in the ground state of the single well. By increasing the barrier the even and odd states of the single well become nearly degenerate. Applying a phase shift in one of the minima of the double well, but not the other, will change the population in even and odd states after the minima are recombined. In principle it is possible to completely invert the population. Population inversion in a degenerate gas to this degree of purity has not been realized before and may have important physical consequences. Although contrast may eventually vanish, one can still observe fringes even when more than one quantum state is populated initially.

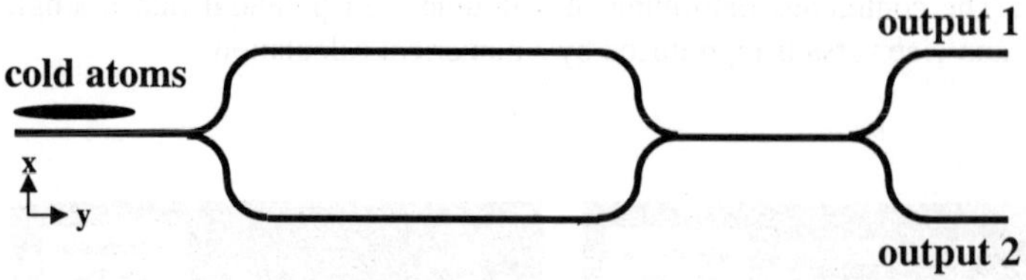

Figure 3. Spatial atom interferometer, based on ferromagnetic guiding structure (see text). Atoms produced at the input can follow two indistinguishable paths. Applying a controlled phase shift to one of the arms or controlled rotation changes the number diverted to each output port.

Interferometry may also be done by *spatially* varying the magnetic field. In the experiments described earlier the parameter controlling the height of the barrier in the double well potential is I_{inner}. Using tiny Helmholtz coils that compensate the effect of I_{inner} locally, it is possible to create X and Y-beamsplitters. This can be used to form a Sagnac-ring [4,5] with input and output ports (Fig. 3). From our experimental parameters we infer that a surface area of 5 cm x 2 mm = 1 cm^2 can be achieved. Recently Zimmermann et al. and Reichel et al. [6] have successfully applied evaporative cooling to obtain Bose Einstein Condensation (BEC) of atoms in a microstructure, hence populating a single quantum state. Propagating these atoms through the ring structure outlined above may lead to rotation gyros with unprecedented accuracy. We note again that preparation in a single quantum state initially is helpful but not not necessary and that fringes with reduced contrast may be observed in multimode-propagation [7].

References

1. M. Vengalattore, W. Rooijakkers and M. Prentiss, preprint server LANL physics/0106028 (2001).
2. J. Reichel, W. Hänsel and T.W. Hänsch, Phys. Rev. Lett. **83**, 3398 (2000).
3. E. A. Hinds, C.J. Vale and M.G. Boshier, Phys. Rev. Lett. **86**, 1462 (2001).
4. T. Gustavson, P. Bouyer and M. Kasevich, Phys. Rev. Lett. **78**, 2046 (1997).
5. A. Lenef, T. Hammond, E. Smith, M. Chapman, R. Rubenstein and D. Pritchard, Phys. Rev. Lett. **78**, 760 (1997).
6. Post-deadline posters at the ICOLS XV conference, Snowbird, Utah (2001).
7. E. Andersson et al., preprint server LANL quant-ph/0107124 (2001)

COHERENT MANIPULATION OF COLD ATOMS IN OPTICAL LATTICES FOR A SCALABLE QUANTUM COMPUTATION SYSTEM

CHENG CHIN, VLADAN VULETIĆ, ANDREW J. KERMAN AND STEVEN CHU

Physics department, Stanford University, CA 94305, USA
E-mail: cchin@stanford.edu

We obtain a coherence time of up to $0.4s$ on the cesium clock transition in a 1D Nd:YAG optical lattice. The coherence time is limited by both collision shifts and light shifts. We demonstrate the coupling between atomic internal and external degrees of freedom using microwave transitions. We propose a quantum computation scheme in 3D optical lattices with strongly suppressed decoherence. The entanglement between two arbitrary atoms can be realized by microwave-induced Feshbach resonances. We estimate that 10^6 quantum operations can be achieved with a coherence time of $10s$ and an operation time of $10\mu s$.

1 Quantum coherence and manipulations in optical lattices

The coherence time of a system determines the energy resolution of measurements and the performance of quantum manipulations. In particular, atomic fountains with a coherence time close to $0.5s$ are widely applied in various precision measurements.[1] In far-off-resonance dipole traps, a much longer coherence time is possible,[2] which might provide a new environment for precision measurements and quantum manipulations.[3]

In this article, we demonstrate Rabi and Ramsey spectroscopy on cesium atoms in a 1D YAG lattice.[4] The coherence time is found to be limited by collision shifts and tensor light shifts. Using microwave transitions, we perform quantum manipulations which couple the internal and external degrees of freedom. These manipulations are needed in our quantum entanglement and computation proposal described below.

We typically load $1 \sim 3 \times 10^8$ atoms in the YAG 1D optical lattice. 98% of the atoms are optically pumped into $|F = 3, m_F = 3\rangle$. The peak density ranges from $1 \times 10^{12} cm^{-3}$ to $2 \times 10^{13} cm^{-3}$, and the temperature from $0.4\mu K$ to $5\mu K$ depending on the trap parameters. To measure the coherence time, we apply microwaves to induce Rabi flopping on the clock transition $|F = 3, m_F = 0\rangle$ to $|F = 4, m_F = 0\rangle$ in a bias field of $50mG$. From the Rabi flopping decay, we deduce a coherence time of up to $t_{1/e} = 0.4s$ at a density of $2 \times 10^{11} cm^{-3}$, temperature of $0.8\mu K$ and trapping intensity of $1kW cm^{-2}$. Ramsey fringes and their phase shifts are also observed and shown in Fig.1.

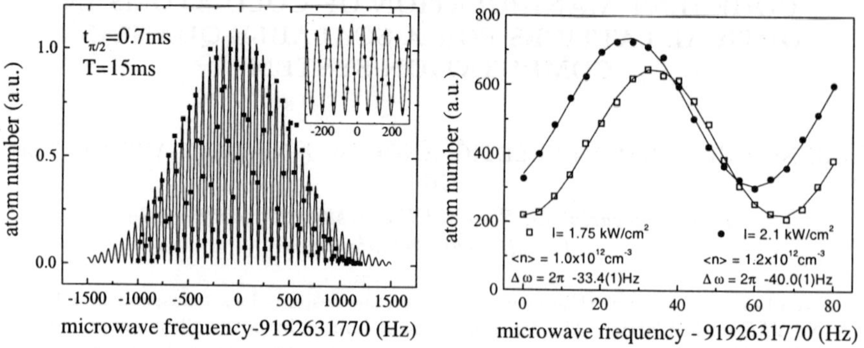

Figure 1. Ramsey spectroscopy of cesium atoms in a linearly polarized 1D YAG lattice at a temperature of $1.0\mu K$. Ramsey fringes shown in the left figure are obtained by two short $\pi/2$ pulses separated by $T = 15ms$ on the $|F = 3, m_F = 0\rangle \rightarrow |F = 4, m_F = 0\rangle$ transition. In the right figure, we zoom in on the central fringe and observe a phase shift for different trap parameters and atom density.

The frequency shifts are due to both tensor light shifts and collision shifts, as verified by adjusting the beam intensity and the atomic density, respectively.

Microwave transitions also provide the tool to manipulate the external degrees of freedom of trapped atoms. By slightly changing the polarization of the Nd:YAG dipole trap, we spatially shift the trapping potential wells for different sublevels. Therefore, microwave transitions from $|F = 3, m_F; \nu\rangle$ to $|F = 4, m'_F; \nu \pm 1\rangle$ are allowed, see Fig.2.

2 Proposal of quantum computation in optical lattices

Quantum computation poses stringent experimental requirements. These include a long coherence time and short operation time, the scalability to $N \gg 10$ qubits and the ability to entangle any two nearby or distant qubits.

A system based on 3D optical lattices can satisfy the above conditions.[5] Cold atoms strongly localized in singly occupied lattice sites do not collide and a long coherence time is expected, while for two atoms in one lattice site, the high local density and the available molecular structure allow us to manipulate and entangle two atoms with high efficiency. Furthermore, due to the compactness of the optical lattices, 1000 qubits can be stored in a small volume of $(10\lambda/2)^3$, where λ is the optical wavelength.

We propose using two sets of optical lattices with different colors L_α and L_β to localize two atomic species individually: α as qubits and β as

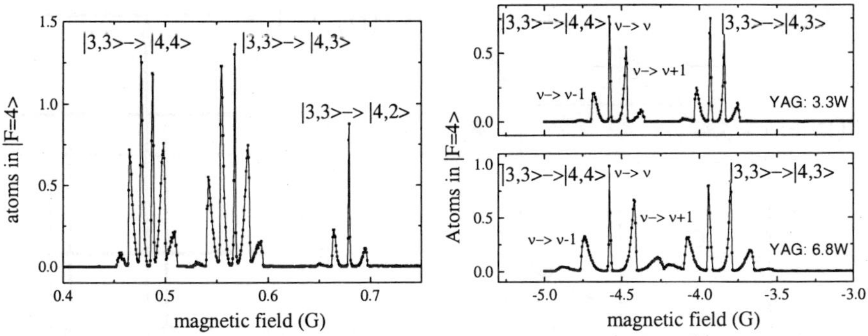

Figure 2. Microwave spectroscopy of cesium atoms in a 1D YAG lattice. Atoms are initially polarized to $|F = 3, m_F = 3; \nu\rangle$, where ν is vibrational quantum number. Excitation to $|F = 4, m'_F; \nu'\rangle$ are induced by microwave transition. In the right figure, the vibrational sidebands are shown to depend on the trapping laser intensity since the vibration frequency is higher in a deeper trap.

operators. The α atoms, having unity occupancy number in lattice L_α, carry the quantum information, while the β atoms, having low occupancy number $\epsilon \ll 1$, induce quantum operations when the β atom is overlapped with a specific α atom, see Fig.3 (a). The overlap of two atoms is made possible by controlling the phases of the two standing waves, which moves the lattice sites relative to each other.[a] Specifying two internal ground states of the atoms as $|0\rangle$ and $|1\rangle$, we can consider the two-atom system in the basis of $|\chi_\alpha, \chi_\beta\rangle$, where $\chi = 0, 1$. An additional bound molecular state $|m\rangle$, available only when two atoms are close, is used to mediate the quantum manipulations.

Shown in Fig.3 is the proposed quantum manipulation scheme. To induce an operation on an α atom, we bring a β atom to overlap the α atom. Swapping their quantum information, necessary for the read/write processes, is achieved by two π pulses on the $|0,1\rangle \to |m\rangle$ and $|m\rangle \to |1,0\rangle$ transitions. Information can then be brought into or taken out of the qubits. A controlled-NOT gate consists of two π pulses on the $|0,1\rangle \to |m\rangle$ and $|m\rangle \to |1,1\rangle$ transitions, which flip the state of the α atom only when the β atom is in $|1\rangle$. Single qubit rotations can be performed similarly.

Since the hyperfine splitting and the molecular binding energy are in the range of GHz, the microwave transitions can be driven in a time shorter than $0.1\mu s$. The time for a single operation is limited by the requirement of

[a]To precisely ovelap the target atoms in the system, the wavelengths of the lattice beams can be chosen to be commensurate by a phase-locking or frequency-doubling scheme.

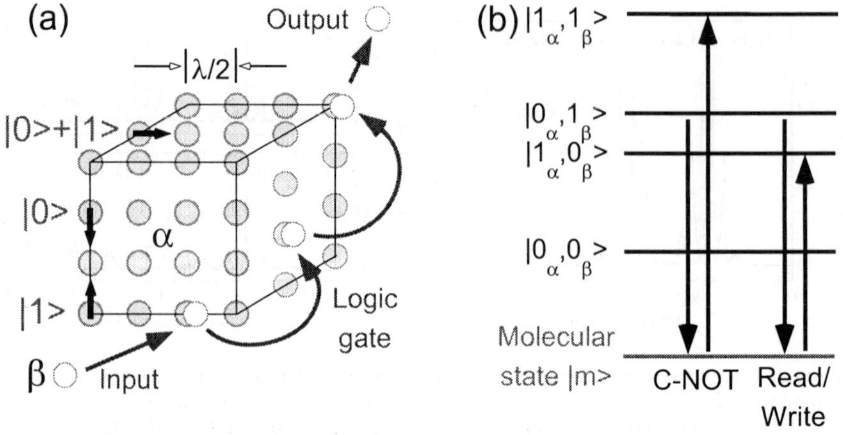

Figure 3. In (a), white circles symbolize the operators and gray circles the qubits. The arrows on the qubits specify their quantum states. Manipulations on a specific qubit are carried out when the operator is brought overlapped with that qubit. In (b), energy structure of the two atoms is shown. Entanglement and read/write processes are implemented by microwaves via a molecular bound state $|m\rangle$.

adiabatic motion of the operator atom from one site to another site. This time is estimated as $T_t > \eta/\omega$, where η is the Lamb-Dicke parameter, and ω is the trap vibration frequency. Typical optical lattices have $\eta = 0.2$, $\omega = 2\pi\, 100kHz$ and $T_t > 0.3\mu s$. Given a projected coherence time of $10s$, $> 10^6$ quantum operations can be implemented.

References

1. B. Young, M. Kasevich, and S. Chu, in *Atom interferometry*, edited by P. R. Berman (Academic Press, 1997), Chap. 9, pp. 363-406.
2. N. Davidson, H. J. Lee, C. S. Adams, M. Kasevich and S. Chu, *Phys. Rev. Lett.* **74**, 1311 (1995).
3. C. Chin, V. Leiber, V. Vuletić, A. J. Kerman, and S. Chu, *Phys. Rev. A* **63**, 033401 (2001).
4. V. Vuletić and C. Chin and A. J. Kerman and S. Chu, *Phys. Rev. Lett.* **81**, 5768 (1998).
5. D. Jaksch, H. -J. Criegel, J. I. Cirac, C. W. Fardiner, and P. Zoller, *Phys. Rev. Lett.* **82**, 1975 (1999).

PUMP-PROBE SPECTROSCOPY AND VELOCIMETRY OF A SLOW BEAM OF COLD ATOMS

G.DI DOMENICO, G.MILETI AND P.THOMANN

Observatoire cantonal, 58 rue de l'Observatoire, CH-2000 Neuchâtel, Switzerland
E-mail: gianni.didomenico@ne.ch

We have observed Raman, Rayleigh and recoil induced resonances (RIR) in a continuous beam of slow and cold cesium atoms extracted from a 2D MOT with the moving molasses technique. We use the RIR to measure the velocity distribution, therefore the average speed (0.6–4 m/s) and temperature (50–500 μK) of the atomic beam. Compared to time of flight (TOF), this technique has the advantage of being local, more sensitive in the low-velocity regime ($v < 1$ m/s) and it gives access to transverse velocities and temperatures. It may be extended to measure atomic velocities in the 2D MOT source of the atomic beam. It is an additional tool to study and optimize the cooling, the extraction and the (transverse) post-cooling of slow atomic beams.

Pump-probe spectroscopy is a powerful tool for high resolution measurements and to get informations on laser cooling mechanisms [1,2,3,4,5]. In particular one can measure atomic velocity distributions by using the recoil induced resonance [6]. This technique has been demonstrated either in a cloud of cold atoms in the dark [7] or in a 1D molasses [8]. In both situations the average velocity of the atomic sample was zero ($\overline{v} = 0$).

We have used the same technique to measure the velocity distribution in a beam of slow and cold cesium atoms where $\overline{v} \neq 0$. As shown in the upper part of figure 1, the continuous beam is produced with a 2D magneto-optical trap operating as a moving molasses in the vertical direction [9]. The atoms are launched downward and the velocity distribution measurement takes place in the atomic beam under the source using pump-probe spectroscopy. The pump (ω,$\vec{k}$) and probe (ω_p,$\vec{k}_p$) beams cross in the atomic beam at a small angle θ.

The probe transmission spectrum is shown in figure 2. It displays amplification and absorption sideband resonances associated to stimulated Raman transitions between adjacent vibrational levels in the optical potential wells along with a central resonant structure whose shape depends on probe polarisation. When probe polarisation is parallel to that of the co-propagating pump beam, the dispersion-like central structure is a recoil induced resonance (RIR). This resonance is associated with a Raman transition between two different momentum states and one can show that it is proportional to the derivative of the atomic momentum distribution along the z direction [6]. Thus its center is a measure of the z component of the average velocity $\overline{v}_z$ and its

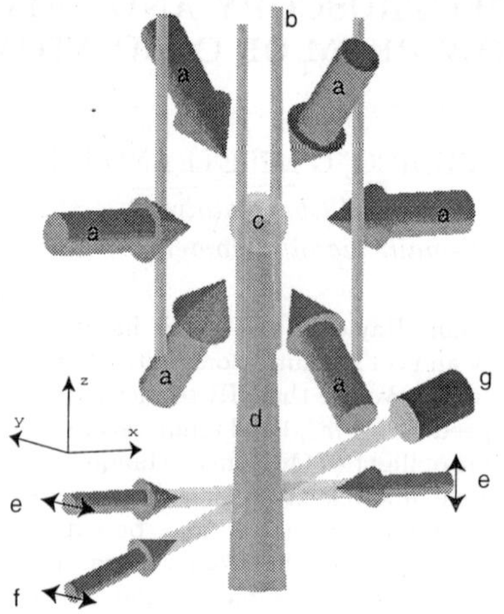

Figure 1. Experimental setup. (a) Cooling beams. (b) 2D magnetic gradient wires. (c) Cold atomic source. (d) Continuous beam of cold atoms. (e) Pump beams. (f) Probe beam. (g) Photodetector. The pump and probe beams have linear polarisations.

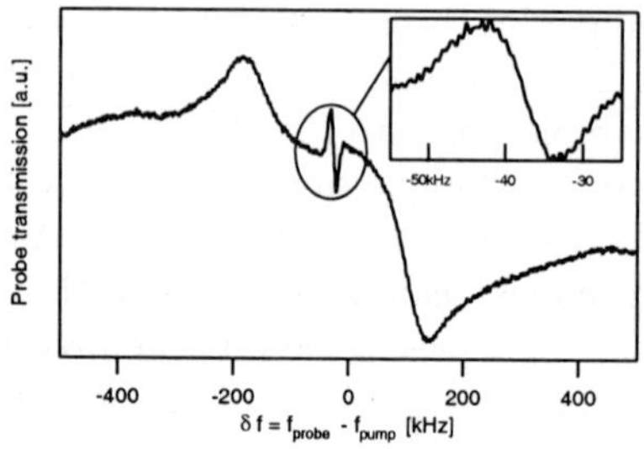

Figure 2. Probe transmission spectrum in an atomic beam launched at 1.205 m/s (1MHz detuning). There is a recoil induced resonance near the spectrum center (see the inset).

width is a measure of the longitudinal temperature T of the atomic beam.

In order to validate this velocimetry technique, we have measured the RIR for moving molasses detunings between 0.5 and 3.25 MHz i.e. for launching velocities between 0.6 and 3.9 m/s. Then we have computed $\overline{v}_z$ and T from the center and distance between peaks of the RIR and the results are presented in figure 3. The average velocity $\overline{v}_z$ is in good agreement with the result obtained by the time of flight technique (TOF) [9]. But the longitudinal

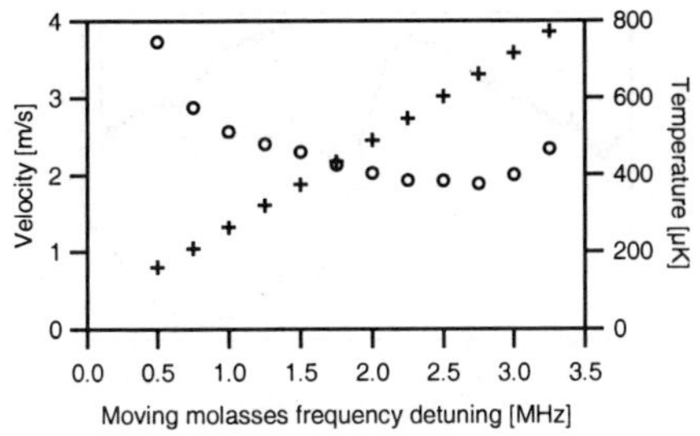

Figure 3. Mean velocity $\overline{v}_z$ (+) and longitudinal temperature T (∘) of the atomic beam as a function of moving molasses detuning $\Delta/2\pi$. These values of $\overline{v}_z$ and T have been computed from the center and distance between peaks of the RIR.

temperature T is higher than the temperature measured by TOF. The difference is explained by the transverse heating caused by the pump. We have measured this heating, and when it is substracted from the RIR temperature, one finds the temperature measured by TOF [9].

In principle, the observation of a RIR requires a free atomic motion in the recoil direction, which precludes velocity measurement in a 3D molasses. However, preliminary theoretical considerations suggest that the RIR should still be visible in the presence of a transverse friction force provided that $k\theta\Delta v \gg \alpha/M$ where α is the friction coefficient and Δv is the velocity spread [10]. Therefore this method may be extended to measure atomic velocities in the 2D MOT source of the continuous beam. In this case we use one of the cooling beams (for example the Ox beam) as a pump beam. Thus it is sufficient to add a probe beam and to record its transmission spectrum. The spectrum obtained is displayed in figure 4. It is more complicated because of the 3D geometry of the cooling beams. Moreover $\sigma^+ - \sigma^-$ polarization produces wider Raman resonances. Nevertheless a small RIR is still visible in the spectrum center (as shown in the inset) and a displacement of this resonance is observed when we operate the source as a moving molasses.

We have used the RIR center and width to evaluate the mean velocity and longitudinal temperature of the atoms in the source. The results cannot be compared with TOF because the domains of validity of the two techniques do not overlap[a]. Nevertheless, we can compare the measured velocity with the theoretical launching velocity and we observe that they are in agreement

[a]Indeed for such small moving molasses frequency detunings (<0.5MHz) the TOF technique is not operational anymore and for bigger detunings (>0.5MHz) the RIR disappears, embedded by other resonances (Raman and Rayleigh)

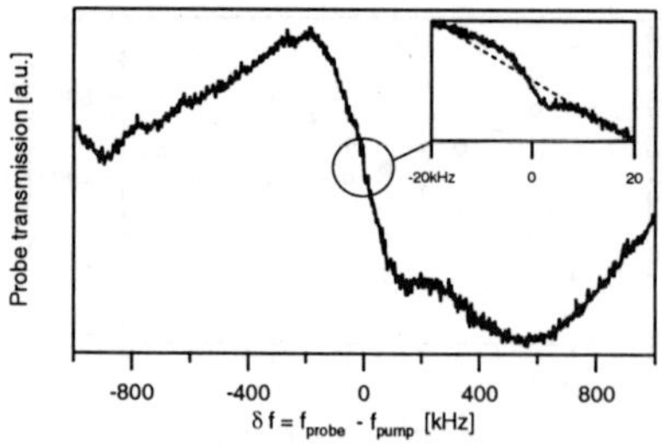

Figure 4. Probe transmission spectrum in the 2D MOT source for zero moving molasses frequency detuning. It is still possible to distinguish a small recoil induced resonance in the spectrum center as shown in the inset.

(error<5%). The measured temperature is lower in the source than in the atomic beam because the Ox cooling beam acts as the pump and consequently there is no more transverse heating. The temperature reaches a minimum of 20μK for zero moving molasses frequency detuning.

Compared to TOF, this velocimetry technique has the advantage of being local, more sensitive in the low-velocity regime ($v < 1$ m/s) and it gives access to transverse velocities and temperatures. We have shown that it is usable in the 2D MOT source of the atomic beam. It is a useful tool to understand and optimize the processes involved in the cooling, the extraction and the (transverse) post-cooling of slow atomic beams.

This work was supported by the Swiss National Science Foundation.

References

1. J.W.R. Tabosa *et al*, *Phys. Rev. Lett.* **66**, 3245 (1991).
2. D. Grison *et al*, *Europhys. Lett.* 15,149,(1991)
3. J.-Y. Courtois, G. Grynberg, *Phys. Rev.* A **46**, 7060 (1992), *Phys. Rev.* A **48**, 1378 (1993).
4. P. Verkerk *et al*, *Phys. Rev. Lett.* **68**, 3861 (1992).
5. B. Lounis *et al*, *Phys. Rev. Lett.* **69**, 3029 (1992).
6. J. Guo, P.R. Berman, B. Dubetsky, *Phys. Rev.* A **46**, 1426 (1992).
7. D. R. Meacher *et al*, *Phys. Rev.* A **50**, R1992 (1994).
8. J.-Y. Courtois *et al*, *Phys. Rev. Lett.* **72**, 3017 (1994).
9. P. Berthoud, E. Fretel, P. Thomann, *Phys. Rev.* A **60**R42411999.
10. Y. Castin, private communication.

GROUND STATE LASER COOLING OF TRAPPED ATOMS USING ELECTROMAGNETICALLY INDUCED TRANSPARENCY

J. ESCHNER,[1] G. MORIGI,[2] C. KEITEL,[3] C. ROOS,[1] D. LEIBFRIED,[1] A. MUNDT,[1] F. SCHMIDT-KALER,[1] R. BLATT[1]

[1] *Institut für Experimentalphysik, Universität Innsbruck, A-6020 Innsbruck, Austria*
[2] *Max-Planck-Institut für Quantenoptik, D-85748 Garching, Germany*
[3] *Fakultät für Physik, University of Freiburg, D-79104 Freiburg, Germany*
E-mail: Juergen.Eschner@uibk.ac.at

A laser cooling method for trapped atoms is presented which achieves ground state cooling by exploiting quantum interference in a Λ-shaped arrangement of atomic levels driven by two lasers.[1] The scheme is technically simpler than existing methods of sideband cooling, yet it can be significantly more efficient, in particular when several motional modes are involved. We have applied the method to a single Calcium ion in a Paul trap,[2] coupling a single laser to the Zeeman structure of its $S_{1/2} \to P_{1/2}$ dipole transition at 397 nm. We have achieved more than 90% ground-state occupation probability. By suitably tuning the laser parameters, we obtain simultaneous ground-state cooling of two oscillator modes. This is of great practical importance for the implementation of quantum logic schemes with trapped ions.

The cooling method. We consider three atomic levels, $|g\rangle$ (ground), $|e\rangle$ (excited), and $|r\rangle$ (metastable), arranged in a Λ-scheme and laser-excited as in Fig. 1a. The transition $|r\rangle \to |e\rangle$ is strongly driven by a blue-detuned laser (coupling laser) with Rabi frequency Ω_r and detuning Δ_r. The absorption spectrum of a probe (cooling) laser driving the transition $|g\rangle \to |e\rangle$ at detuning Δ_g shows a zero at $\Delta_g = \Delta_r$ (dark resonance, EIT-condition), a narrow resonance at $\Delta_g = \Delta_r + \delta$, and a broad resonance at $\Delta_g = -\delta$, where $\delta = (\sqrt{\Delta_r^2 + \Omega_r^2} - |\Delta_r|)/2$ is the light shift due to the coupling laser (inset in Fig. 1a).

We assume that the atomic center-of-mass motion is harmonic at frequency ν and that the Lamb-Dicke regime holds, i.e. the size of the motional state is much smaller than the laser wavelength. Let $|n\rangle$, $n = 0, 1, 2, \dots$, be the energy eigenstates of the motion. The probe absorption spectrum can then be decomposed into transitions between different motional states, of which the significant contributions are the $|n\rangle \to |n\rangle$ "carrier" transition at Δ_g and the $|n\rangle \to |n \pm 1\rangle$ "sideband" transitions at $\Delta_g \mp \nu$. By tuning the probe laser to $\Delta_g = \Delta_r$, the absorption probability on these transitions is such that: (i) Carrier absorption is suppressed, since it corresponds to the dark resonance

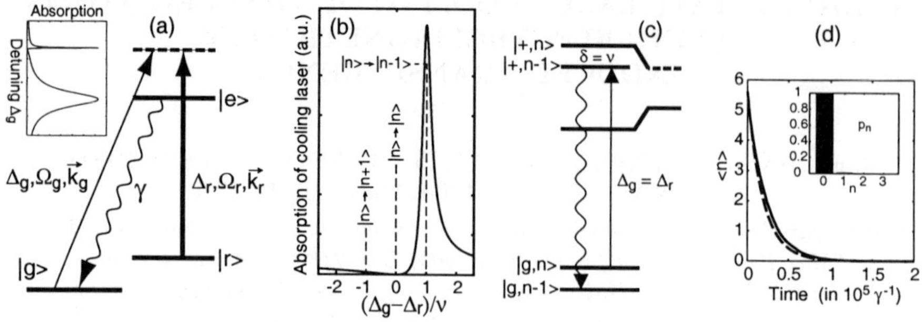

Figure 1. (a) Levels and transitions of the EIT-cooling scheme. The inset shows schematically the absorption rate on $|g\rangle \rightarrow |e\rangle$ when the atom is strongly excited above resonance on $|r\rangle \rightarrow |e\rangle$. (b) Absorption of cooling laser around $\Delta_g = \Delta_r$ (solid line); dashed lines mark the probabilities of carrier ($|n\rangle \rightarrow |n\rangle$) and sideband ($|n\rangle \rightarrow |n \pm 1\rangle$) transitions when $\Delta_g = \Delta_r$. (c) Dressed state picture: the cooling laser excites resonantly transitions from $|g, n\rangle$ to the narrow dressed state denoted by $|+, n-1\rangle$ which preferentially decays into $|g, n-1\rangle$. (d) Quantum Monte-Carlo simulation of the cooling dynamics (solid line) and rate equation approximation (dashed line), calculated with $\Omega_r = \gamma$, $\Delta_r = 2.5\gamma$ and $\nu = \gamma/10$. The inset shows the population of the lowest vibrational states after cooling (from Ref. [1]).

(EIT); (ii) blue-sideband absorption ($|n\rangle \rightarrow |n+1\rangle$) has very small probability, falling on the left of the dark resonance, while (iii) red-sideband absorption ($|n\rangle \rightarrow |n-1\rangle$) is strongly enhanced, and it is maximum when the condition $\delta \simeq \nu$ holds (Fig. 1b). This is the principle of EIT cooling. Note that, in absence of the coupling laser, the dipole transition of width $\gamma \gg \nu$ would only permit Doppler cooling.

The procedure can also be understood as optical pumping between $|g\rangle$ and the dressed states of the driven $|r\rangle \rightarrow |e\rangle$ transition, see Fig. 1c. While this picture looks similar to ordinary sideband cooling, EIT cooling can in fact provide *higher* efficiency, in particular for cooling in 3 dimensions.[1] This is a result of the total suppression of the carrier transition $|g, n\rangle \rightarrow |+, n\rangle$ due to the EIT condition. In Fig. 1d the cooling efficiency is evaluated numerically, showing that near-unity occupation of the ground state is achieved.

Experimental realization. The scheme has been applied to a single $^{40}Ca^+$ ion trapped in a Paul trap, whose $S_{1/2} \leftrightarrow P_{1/2}$ dipole transition at 397 nm forms a four-level system (Fig. 2a). Three of the levels, $|S, \pm\rangle$ and $|P, +\rangle$, together with the σ^+ and π-polarized laser beams, form the Λ scheme (Fig. 2b). The ion is stored in a trap with $(\nu_x, \nu_y, \nu_z) = 2\pi \times (1.69, 1.62, 3.32)$ MHz and first Doppler cooled on $S_{1/2} \rightarrow P_{1/2}$ to mean excitation numbers $\bar{n}_z = 6.5(1.0)$ and $\bar{n}_y = 16(2)$. Then, EIT-cooling is applied with a pulse of σ^+- and π-light,

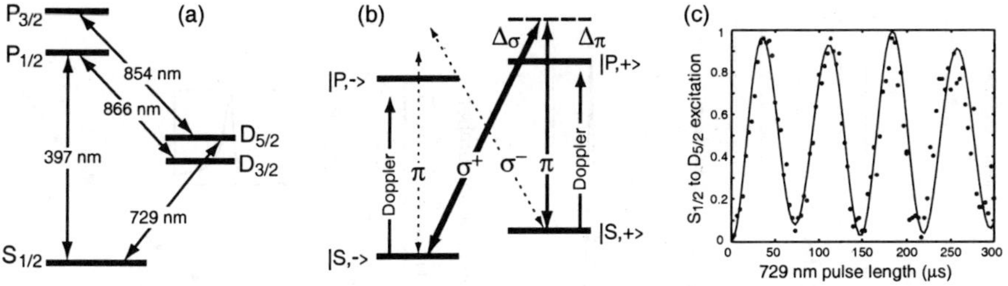

Figure 2. (a) Levels and transitions in $^{40}Ca^+$ used in the experiment. The $S_{1/2}$ to $P_{1/2}$ transition is used for Doppler cooling and for EIT-cooling, and the scattered photons are observed to detect the ion's quantum state. The narrow $S_{1/2} \rightarrow D_{5/2}$ transition serves to investigate the vibrational state. (b) Zeeman sublevels $|S, \pm\rangle$, $|P, \pm\rangle$ of $S_{1/2}$ and $P_{1/2}$, respectively, and laser frequencies (solid lines) which are relevant for EIT-cooling. Dashed lines: transitions which involve the $|P, -\rangle$ level. Due to a non-ideal polarization of the π-light the σ^--transition slightly counteracts the cooling. (c) Rabi oscillations excited on the blue z-mode sideband of the $S_{1/2} \leftrightarrow D_{5/2}$ transition (points). A mean phonon number of $\bar{n}_z = 0.1$ was determined from the fit (line).

blue-detuned to $\Delta_\sigma = \Delta_\pi \approx 75$ MHz (3.5γ) and with $\Omega_\sigma = 2\pi \times 21.4$ MHz and $\Omega_\pi = 2\pi \times 3$ MHz.[2] The motional state after cooling is analyzed by spectroscopy on the $S_{1/2} \rightarrow D_{5/2}$ quadrupole transition at 729 nm, using an electron shelving technique[3] (Fig. 2c). For $\delta = 2\pi \times 3.3$ MHz, a mean vibrational number of $\bar{n}_z = 0.1$ was obtained, corresponding to 90% occupation of the ground state of the axial motion.

The particular shape of the absorption spectrum (Fig. 1b) allows to simultaneously cool all modes whose frequencies lie around the value of the chosen δ. We have simultaneously cooled the z- and y-modes by setting Ω_σ such that $\delta \approx 2.6$ MHz, roughly halfway between the mode frequencies. Repeating the procedure outlined above, the modes were found to be cooled to $p_0^y = 58\%$ and $p_0^z = 73\%$ ground state probability (Fig. 4).

These results could be extended to cooling the motion of ion strings in linear ion traps. Here, the collective motion is described by the normal modes of the crystals. EIT-cooling seems to be particularly suited for cooling these modes, since it allows simultaneous cooling of several modes at different frequencies. We have estimated the cooling performance for a 10-ion string,[4] finding that all vibrational modes may be cooled to a mean phonon number below one. This is promising for the application of the procedure to cold ion strings for quantum information processing.

Finally, EIT-cooling is not restricted to harmonic motion: particles in anharmonic potentials can also be efficiently cooled, provided that the motional sideband frequencies are not too different from the light shift δ. For

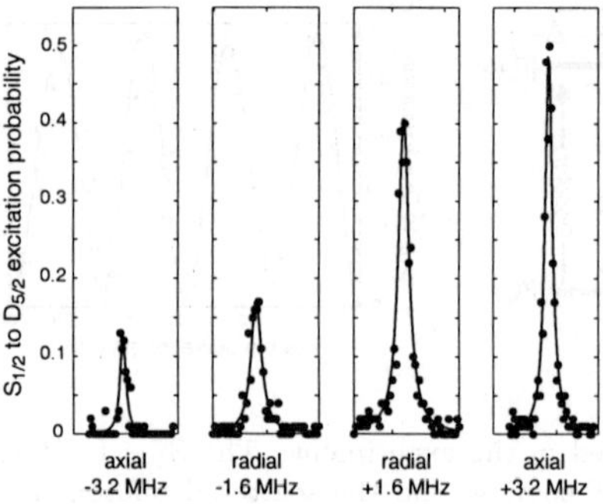

Figure 3. EIT-cooling of two modes at 1.6 MHz and 3.2 MHz simultaneously. From the sideband excitation rate after cooling we deduce a ground state occupation number of 73% for the axial mode (3.2 MHz) and 58% for the radial mode (1.6 MHz).

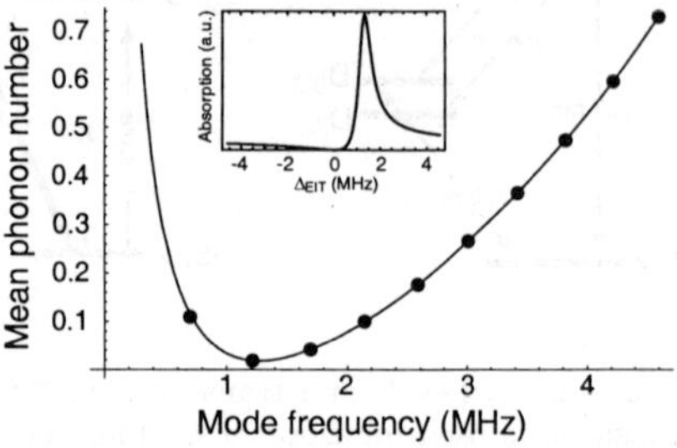

Figure 4. Mean phonon number (black dots) of the axial modes of a 10-ion string *vs* the mode frequencies, after EIT-cooling has been applied to the level scheme of Figs. 2a,2b. Here, the c.o.m. axial trap frequency is assumed to be 0.7 MHz, while $\Omega_\sigma = 20$ MHz, $\Omega_\pi = 0.5$ MHz, and $\Delta_\sigma = \Delta_\pi = 75$ MHz, giving a light shift $\delta \sim 1.3$ MHz (see inset).

this reason, EIT-cooling seems to be also suitable for optical lattices.

Acknowledgments

We gratefully acknowledge support by the European Commission (TMR network QSTRUCT, ERB-FMRX-CT96-0077), by the Austrian Science Fund (FWF, SFB15), and by the Institut für Quanteninformation GmbH.

References

1. G. Morigi, J. Eschner, C. H. Keitel, PRL **85**, 4458 (2000)
2. C. F. Roos, D. Leibfried, A. Mundt, F. Schmidt-Kaler, J. Eschner, R. Blatt, PRL **85**, 5547 (2000)
3. D. M. Meekhof, C. Monroe, B. E. King, W. M. Itano, D. J. Wineland, PRL **76**, 1796 (1996)
4. F. Schmidt-Kaler, J. Eschner, G. Morigi, C. F. Roos, D. Leibfried, A. Mundt, R. Blatt, submitted to Appl. Phys. B, quant-ph/0107087.

DISSOCIATION DYNAMICS OF A H_2^+ IONIC BEAM IN INTENSE LASER FIELDS - HIGH RESOLUTION OF THE FRAGMENTS' KINETIC ENERGY

H. FIGGER, D. PAVIČIĆ, K. SÄNDIG AND T. W. HÄNSCH

Max Planck Institut für Quantenoptik, Hans-Kopfermannstrasse 1,
85748 Garching, Germany
E-mail: hartmut.figger@mpq.mpg.de

H_2^+ molecules in an ionic beam were exposed to fs-laser pulses of intensities up to $3 \cdot 10^{14}$ W/cm^2. H^+ as well as H fragment 2-dimensional momentum distributions were measured. In the Coulomb explosion channel clear structure was observed for the first time and was related to charge-resonance enhanced ionisation (CREI).

1 Introduction

New exciting effects like above threshold ionisation (ATI) were discovered when atoms and small molecules were exposed to pulsed laser fields of fs-lengths and intensities of 10^{12} to 10^{16} W/cm^2. These effects depend nonlinearly on the light intensity and are mostly initiated by multiphoton absorption. Much of the experimental work was done on the H_2 molecule, while theoretical work was concentrated on the H_2^+ ion because its potential curve system can be approximated by a relatively simple two state potential system. [1]

In the interpretation of the experiments on H_2 it was mostly assumed that this molecule is ionised at the leading edge of the laser pulse and thereby in the ion H_2^+ a vibrational population distribution corresponding to the Franck-Condon factors was achieved and that this molecular ion is then fragmentised. More recent work shows that this assumption is inadequate. [2]

2 Experiment

In order to have better defined starting conditions for H_2^+ than in the usual experiments where a gas probe is formed by H_2 leaking into high vacuum, in our experiment a beam of molecular H_2^+ was generated in a dc-discharge, accelerated to 11.2 kV, mass selected and collimated to the width of about 50 μm. The diameter of the laser beam in the focus was approximately 100 μm, which ensured a small intensity variation over the ion beam width. The commercial Ti:Sapphire laser had a repetition rate of 1000 pulses per second, a pulse energy of up to 2 mJ and a shortest pulse duration of 80 fs. Since

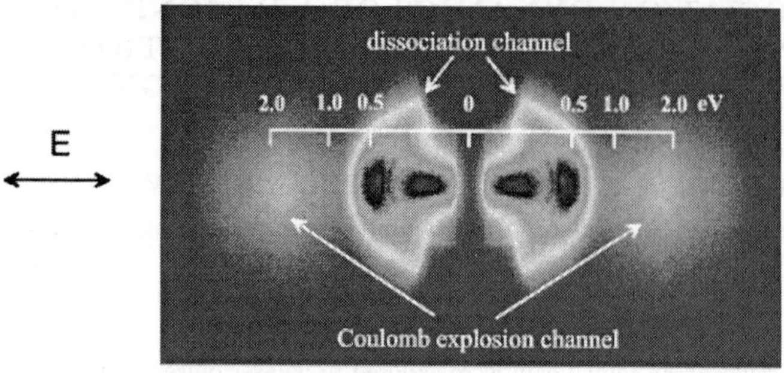

Figure 1. Two-dimensional momentum distribution of the photofragments at intensity of $2.7 \cdot 10^{14}$ W/cm^2 and pulse length of 80 fs. The arrow on the left side shows the direction of laser polarisation and the axis along this direction denotes the fragment energies in electronvolts. The left half of the image is obtained by mirroring the experimentally recorded right half. The characteristically different angular distributions of the two fragmentation channels are striking.

its polarisation direction was perpendicular to the ion beam, the fragments have additional velocity component which can be projected on an on-axis 2-dimensional detector. The detection system consisting of a multichannel plate (MCP), a fluorescent screen and a CCD camera allows a relative momentum resolution better than 1 percent. The vibrational population distribution of H_2^+ in the ion beam was determined in an additional experiment. [3]

3 Results

In earlier experiments, two fragmentation channels were already discerned and identified as [1]:

$$
\begin{aligned}
H_2^+ + nh\nu &\rightarrow H + H^+ && (a) \\
H_2^+ + nh\nu &\rightarrow H^+ + H^+ + e^- && (b)
\end{aligned}
\qquad (1)
$$

Here n is the number of absorbed photons, hν their energy and e^- the ionised electron. In the "dissociation" channel (a), up to 3 photons were already found to be absorbed and the two fragments H and H^+ were measured to escape each with a kinetic energy up to 0.5 eV. In channel (b), which is Coulomb explosion after ionisation of the electron by multiphoton absorption, the fragments escape with a kinetic energy up to 4.5 eV [2]. Our experimental setup clearly shows these two channels (a) and (b), with their characteristically different angular distribution (Fig. 1). Consequently in the dissociation channel (a) we were able to distinguish for the first time between fragments from different vibrational levels of H_2^+ [3].

Dissociation: Predictions of the light induced potential theory (LIP) for the dissociation rate in (a) of single vibrational levels of H_2^+ as a function of the laser light intensity could be tested for the first time. So bond-softening and energetic downshifting [1] of the vibrational levels situated below the crossing of the LIP potential curves could be investigated in detail. In particular the "trapping" of the vibrational levels $v = 11, 12$ in the new light induced potential minimum was observed at wavelength of 780 nm. [3]

The angular distribution of the H fragment was measured on the multichannel plate detector and strong narrowing for the levels below the light induced crossing observed while those in and above the crossing seem to approximate a $\cos^2$ distribution. This was already discussed in more detail. [3]

Coulomb explosion: More recently, also the channel (b) was investigated. It opens up at about $5 \cdot 10^{13}$ W/cm^2 and becomes the dominant one when the intensity is increased to 10^{15} W/cm^2. Experiments were performed at 790 nm with 80-fs pulses of peak intensities up to $3 \cdot 10^{14}$ W/cm^2.

The H^+ momentum distribution of channel (b) was completely measured here for the first time (Fig. 1). The original 3-dimensional distribution can be obtained from its projection by an inverse Abel transformation. It differs characteristically from that of channel (a): while the angular distribution of fragments in channel (a) for v=6-8 follows $\cos^2$ dependence, the one of channel (b) can be characterised by much narrower $\cos^{18}$ distribution. This narrowing of the protons' distribution in direction of laser polarisation is due to the strong nonlinearity of the ionisation rate in the intensity. However, also an alignment of the molecules contributes to this effect.

Reducing the laser intensity to about $9 \cdot 10^{13}$ W/cm^2, at a pulse length of 80 fs, several clear maxima in the fragment distribution along the polarisation direction appear, the strongest ones at about 1.3, 1.7 and 2.2 eV.

Theoretical simulations on H_2^+ in intense femtosecond fields based on the time-dependent Schrödinger equation predict enhanced ionisation at certain, so-called critical distances. This effect is known as charge-resonant enhanced ionisation (CREI) [4].

If we assume that H_2^+ is stretched by the dissociation mechanism, the energy of a proton fragment is $E = E_{DISS} + e^2/2R$, where for E_{DISS} is taken the average energy of channel (a), $E_{DISS} = 0.26$ eV, and R is the internuclear separation at the moment of ionisation. Our data transformed in this way (Fig. 2a) indicate that the ionisation probability is enhanced at certain

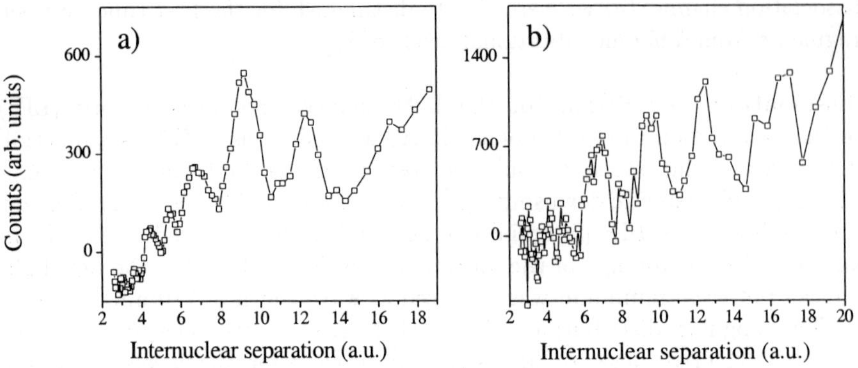

Figure 2. Ionisation signal as a function of the internuclear separation. a) H_2^+, $I = 9 \cdot 10^{13}$ W/cm^2; b) D_2^+, $I = 7 \cdot 10^{13}$ W/cm^2. The plots were derived from the corresponding momentum distributions using Coulomb law and taking into account the fragments' average dissociation energy of 0.26 eV for H_2^+ and 0.28 eV for D_2^+.

internuclear separations. The three dominant peaks are at 6.8, 9.2 and 12.4 a.u.

A crucial experiment concerning the interpretation of these peaks is the analogous experiment on D_2^+. According to CREI, the critical distances in D_2^+ should be similar to those in H_2^+. That is because the electron sees the potential of the two deuterons which is approximately that of the two protons in H_2^+. Indeed, the measurement on D_2^+ (Fig. 2b) at intensity of $7 \cdot 10^{13}$ W/cm^2 resulted in nearly the same critical distances.

References

1. A. Giusti-Suzor *et al*, *J. Phys.* B **28**, 309 (1995).
2. T. D. G. Walsh *et al*, *Phys. Rev.* A **58**, 3922 (1998).
3. K. Sändig, H. Figger and T. W. Hänsch, *Phys. Rev. Lett.* **85**, 4876 (2000).
4. T. Zuo and A.D. Bandrauk, *Phys. Rev.* A **52**, R2511 (1995).

QUANTUM COMPUTATION IN A ONE-DIMENSIONAL CRYSTAL LATTICE WITH NUCLEAR MAGNETIC RESONANCE FORCE MICROSCOPY

J. R. GOLDMAN,[1] T. D. LADD,[1] F. YAMAGUCHI,[1] AND Y. YAMAMOTO[1,2]

[1]*Quantum Entanglement Project, ICORP, JST*
Edward L. Ginzton Laboratory, Stanford University, Stanford, California 94305-4085, USA
[2]*NTT Basic Research Laboratories, 3-1 Morinosato-Wakamiya Atsugi, Kanagawa, 243-0198, Japan*

E. ABE[3] AND K. M. ITOH[3,4]

[3]*Department of Applied Physics and Physico-Informatics, Keio University, Yokohama, 223-8522, Japan,* [4]*PRESTO-JST*

A proposal for a quantum computer using a nuclear spin network in a crystal lattice is presented. Magnetic resonance force microscopy is used for readout of an ensemble of nuclear spins. Two systems are proposed: One uses fluorine nuclear spins in a natural solid crystal of fluorapatite [$Ca_5F(PO_4)_3$], which approximates a one-dimensional structure, while the other uses isotope engineering to construct a more idealized one-dimensional system of ^{29}Si (spin-1/2) nuclear spins in a ^{28}Si (spin-0) matrix.

The design and construction of a scalable quantum computer has become an important goal of modern physics and engineering. A quantum information processor could potentially simulate quantum systems[1], and implement fast quantum algorithms for factoring[2] and searching databases[3]. There has been considerable interest in developing a quantum computer in the solid state[4]. Kane proposed the use of phosphorous impurities in silicon[5]. The main difficulties with this proposal are the need to detect a single nuclear spin state and the limitations of current fabrication capabilities. Our proposal, based in part on Yamaguchi and Yamamoto[6], makes use of ensemble measurement of many nuclear spins in a solid crystal lattice.

A generalized schematic for a solid-crystal quantum computer is shown in Fig. 1. Each chain of n nuclei with spin 1/2 is a quantum computer with each nucleus possessing a distinct frequency of precession due to the presence of a large static magnetic field gradient. This field gradient is engineered to have a large variation in one-dimension with no inhomogeneity in the other two dimensions[7]. This allows N parallel copies of quantum computers to exhibit the same behavior. Thus, single nuclear spin detection is not required; as in the very successful case of solution NMR quantum computing[8], a weak

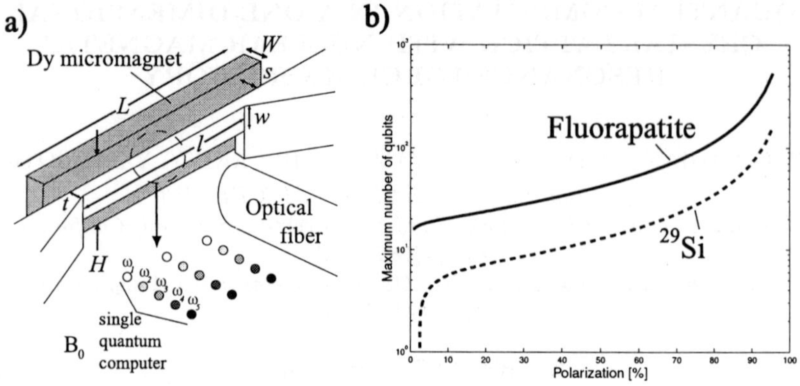

Figure 1. (a) The crystal, of length 10 μm and width roughly 0.1 μm, is mounted or grown epitaxially in the middle of a bridge structure. A dysprosium micromagnet with dimensions $H = 10$ μm, $W = 4$ μm, $L = 400$ μm is deposited nearby and produces field contours which are aligned parallel to the crystal planes. A field gradient of 1.4 T/μm is produced at the crystal, a distance of 2.1 μm above the magnet. An optical fiber is used to observe oscillations in the bridge structure whose dimensions are $l = 300$ μm, $w = 4$ μm, and $t =$ 0.25μm. The different shades indicate different resonant frequencies. (b) The maximum number of qubits that can be used versus the percent polarization is plotted for both fluorapatite and ^{29}Si. Ref. [13] discusses the formula used for this plot.

ensemble measurement is made.

The network of dipole interactions among distinct qubits (differing Larmor frequencies) and 'equivalent qubits' (identical Larmor frequencies) can be used for carrying out required logical operations. The Hamiltonian for the interaction among the qubits can be nulled by the introduction of both narrowband and broadband decoupling RF pulses[9,10]. RF pulses can be modified to selectively restore coupling between two particular qubits and thus allow for two qubit logic operations. The adverse effect of this is to introduce undesirable but unavoidable coupling between distinct qubits from different one-dimensional computers. This has motivated the search for one-dimension nuclear spin systems[11].

Two candidate systems which provide one-dimensional nuclear structures are the spin 1/2 fluorine nuclei in a natural crystal of fluorapatite[12] [$Ca_5F(PO_4)_3$] and spin 1/2 ^{29}Si nuclear spins[13] which are embedded via molecular beam epitaxy into chains in a matrix of pure^{28}Si. Fluorine nuclei are separated by 0.34 nm in the dimension parallel to the field gradient and by 0.94 nm in the transverse directions. Larger spacing between neighboring computers can be made by depositing chains of ^{29}Si atoms on a slightly miscut[14] ^{28}Si wafer oriented in the (111) direction as outlined in a recent proposal for a quantum computer[15].

The field gradient is made to have a uniform field contour in the directions transverse to the field gradient. However, the homogeneity can only be maintained for small areas less than the dimensions of the magnet, and the field gradient must be of order 10 G/nm, which demands a micromagnet[7]. The small number of equivalent qubit spin with a finite polarization requires the introduction of a sensitive measuring device to detect the nuclear spin states. Detection of a single nuclear spin using magnetic resonance force microscopy[16] requires a force sensitivity in a cantilever of 10^{-21} N/$\sqrt{\text{Hz}}$ in a field gradient of 10 G/nm. Single crystal silicon cantilevers have been fabricated with a force sensitivity[17] of 10^{-18} N/$\sqrt{\text{Hz}}$. For an all-silicon quantum computer, 10^5 equivalent qubits can be implemented while for a fluorapatite quantum computer 10^8 nuclear spins can be incorporated. The reduced number of equivalent qubits with the all-silicon system requires a more sensitive probe or a means of achieving higher nuclear spin polarizations. Figure 1(b) shows a plot of the scalability[13] of the a quantum computers using fluorapatite and ^{29}Si assuming these numbers for equivalent qubits and a minimum detectable force of 10^{-17} N.

In semiconductors, circularly polarized light can be used to optically create spin-polarized conduction electrons[19], which can interact via hyperfine couplings with nuclei and transfer the electron spin polarization to the nuclei[18]. Thus, the all-silicon quantum computer has this capability readily available while the fluorapatite quantum computer requires polarization transfer from remote electrons. Additional cooling can be provided with an algorithm which improves the net polarization of a subset of qubits[20]. This 'algorithmic cooling' technique has been shown to be efficient in the time required to cool a subset of the available qubits. Finally, pseudo-pure state techniques can be used when perfect polarization is not reached[21].

Acknowledgments

This work was also partially supported by NTT Basic Research Laboratories. T. D. L. was supported by the Fannie and John Hertz Foundation.

References

1. R. P. Feynman, Int. J. Theor. Phys. **21**, 467 (1982); S. Lloyd, *Science* **273**, 1073 (1996).

2. P. W. Shor, in *Proceedings of the 35th Symposium on the Foundations of Computer Science* (IEEE Computer Society Press, Los Alamitos, 1994), p. 124.
3. L. K. Grover, *Phys. Rev. Lett.* **79**, 325 (1997).
4. M. A. Nielsen, I. L. Chuang, *Quantum Computation and Quantum Information*, Cambridge Univ. Press, p. 346 (2000).
5. B. Kane, *Nature*, **393**, 133, (1998).
6. F. Yamaguchi and Y. Yamamoto, *Appl. Phys. A* **68**, 1 (1999)
7. J. R. Goldman, T. D. Ladd, F. Yamaguchi, and Y. Yamamoto, *Appl. Phys. A* **71**, 11 (2000).
8. N. A. Gershenfeld and I. Chuang, *Science* **275**, 350 (1997); D. G. Cory, A. F. Fahmy, and T. F. Havel, *Proc. Natl. Acad. Sci.* USA **94**, 1634 (1997).
9. D. W. Leung, I. L. Chuang, F. Yamaguchi, and Y. Yamamoto, *Phys. Rev. A* **61**, 042310 (1999).
10. U. Haeberlin, *High Resolution NMR in Solids: Selective Averaging* (Academic Press, New York, 1976); M. Mehring, *High Resolution NMR in Solids* (Springer-Verlag, Berlin, 1983).
11. T. D. Ladd, J. R. Goldman, F. Yamaguchi, and Y. Yamamoto, *Appl. Phys. A* **71**, 27 (2000).
12. T. D. Ladd, J. R. Goldman, F. Yamaguchi, and Y. Yamamoto, *e-print* quant-phy/0009122.
13. T. D. Ladd *et al*, *e-print* quant-phy/0109039.
14. J. Viernow, *et al.*, Appl. Phys. Lett. **72**, 948 (1998); J.-L. Lin, *et al.*, J. Appl. Phys. **84**, 255 (1998).
15. I. Schlimak, V.I. Safarov, and I.D. Vagner, *J. Phys.: Condens. Matter* **13**, 6059 (2001).
16. J. A. Sidles, *Appl. Phys. Lett.* **58**, 2854 (1991); D. Rugar, C. S. Yannoni, and J. A. Sidles, *Nature* **360**, 563 (1992).
17. T. D. Stowe *et al.*, Appl. Phys. Lett. **71**, 288 (1997).
18. R. Tycko, *Solid State Nuclear Magnetic Resonance* **11** 1 (1998).
19. F. Meier and B. P. Zakharchenya, *Optical Orientation*, Elsevier Sci. Publ. (1998).
20. L. J. Schulman and U. V. Vazirani, Proc. 31st ACM Symp. on Theory of Computing, 322 (1999); D. E. Chang, L. M. K. Vandersypen, and M. Steffan, Chem. Phys. Lett. **338**, 337 (2001).
21. S. L. Braunstein *et al.*, *Phys. Rev. Lett.* **83**, 1054 (1999).

SIDEBAND COOLING AND SPECTROSCOPY OF STRONTIUM ATOMS IN THE LAMB-DICKE CONFINEMENT

TETSUYA IDO*, MAKOTO KUWATA-GONOKAMI*,
AND HIDETOSHI KATORI*†

**Cooperative Excitation Project, ERATO, JST*
KSP D-842, 3-2-1 Sakado Takatsu-ku, Kawasaki, 213-0012, Japan
† Engineering Research Institute, the University of Tokyo
Bunkyo-ku, Tokyo 113-8656, Japan
E-mail: ido@ap4.t.u-tokyo.ac.jp

We have demonstrated sideband cooling and recoil-free spectroscopy on the $^1S_0 - {}^3P_1$ intercombination transition of ^{88}Sr atoms. One-dimensional optical lattice was created for the lower and upper electronic states to realize the Lamb-Dicke confinement for the fast axis. More than 90% of the trapped atoms are sideband cooled to the vibrational ground state in this direction. The elastically scattered photons off the confined atoms provided a narrow Doppler free spectrum of $\sim$ 20kHz. The light-shift cancellation technique applied in the optical lattice, combined with the Lamb-Dicke confinement, may offer an alternative approach for optical precision spectroscopy.

As successfully demonstrated in ion-trap experiments, a quantum absorber in the Lamb-Dicke confinement, enabling long interaction times as well as recoil-free absorption, has so far provided the finest spectrum in the visible region[1]. In these approaches, the number of measurable absorbers is limited, owing to the ion's strong coulomb repulsion forces and the influence of micromotion in Paul traps. It would, therefore, lead to a dramatic improvement in frequency measurement, if $N(\gg 1)$ quantum absorbers could be simultaneously prepared in the Lamb-Dicke regime. In the case of neutral atoms, while they allow spectroscopic measurements on a large ensemble of particles because of their weaker interactions, precise measurements can be only accomplished during their free flights or in atomic fountains at the expense of extended interaction time. This is because their trap potentials employ position dependent modulation of their electronic states via the Zeeman or Stark effect, which causes strong perturbations and uncertainties in the observed spectrum. These uncertainties, however, can be avoided by designing the trapping potential so as to produce an equal amount of frequency shifts in the probed upper and lower electronic states.

We report on a novel spectroscopic technique for neutral strontium atoms confined in the Lamb-Dicke regime. In a previous experiment [2,3], we have developed a Far-Off Resonant optical dipole Trap (FORT), in which the $5s^2\ {}^1S_0$

ground state as well as the $5s5p\ ^3P_1$ state experiences equal a.c. Stark shifts, thus leaving the atomic resonance frequency unchanged. This optical trapping scheme is realized by coupling these two states to the upper singlet and triplet states, respectively, by a laser tuned to $\lambda_T \sim 800$ nm. Figure 1(a) shows the relevant energy levels for strontium atoms. Based on this FORT scheme, we have demonstrated an optical sideband cooling and spectroscopy on the $^1S_0 - ^3P_1$ "clock transition".

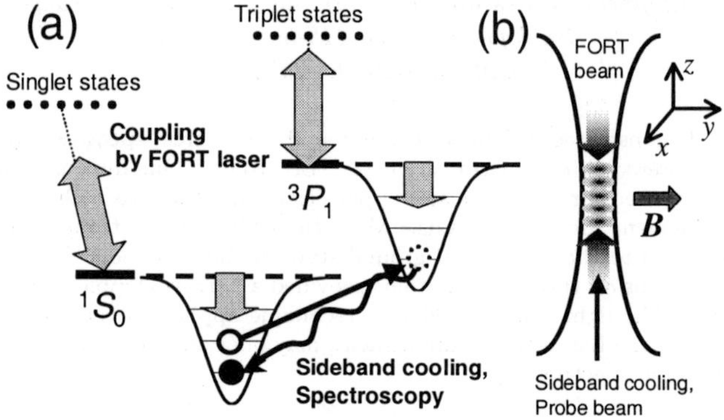

Figure 1. (a) Relevant energy levels for sideband cooling and spectroscopy on the $^1S_0 - ^3P_1$ transition. A FORT laser tuned to 800 nm produces the same amount of a.c. Stark shift in the $^1S_0$ and $^3P_1$ states by coupling these states to the upper singlet and triplet states, respectively. (b) Configuration of a one-dimensional optical lattice. We applied a magnetic field of $B = 0.5$ G perpendicular to the lattice beam. Sideband cooling and spectroscopy were performed on the $^1S_0 - ^3P_1(m_J = 0)$ transition to avoid line broadening due to stray magnetic fields.

The Lamb-Dicke confinement of neutral atoms is achieved by loading atoms into a one-dimensional optical lattice as shown in Fig. 1(b). We introduced a FORT laser vertically with its beam waist located in the region of the magneto-optically trapped atom cloud. The trapping beam was retro-reflected to form a standing wave or a 1D optical lattice in the z-direction. In our experimental conditions, the vibrational frequencies of each micro trap in the lattice was $\nu_z = 50$ kHz and $\nu_r = 100$ Hz for the axial and radial directions, respectively. Since ν_z is sufficiently larger than the natural linewidth $\gamma/2\pi = 7.1$ kHz of the $^3P_1$ state as well as the photon recoil energy of $E_R/h = 5$ kHz, the condition for resolved sideband cooling in the Lamb-Dicke regime is fulfilled for the vertical direction. After loading magneto-optically cooled and trapped atoms into this lattice potential [4], we sideband-cooled the atoms for 30 ms along the z-direction, by irradiating a single laser beam that excited the $^1S_0 - {}^3P_1$ transition. The atom temperature was measured by a time-of-flight technique as a function of the laser detuning, which is shown in Fig. 2(a). At a laser detuning of $\delta = -50$ kHz, corresponding to the excitation of the first lower sideband of $-\nu_z$, the axial temperature was reduced to 1.3 μK, which is limited by the zero-point energy of the vibrational ground state. This temperature infers that more than 90% of the atoms were in the vibrational ground state in z-direction.

In order to demonstrate the feasibility of using these tightly confined atoms for high resolution spectroscopy, we measured laser induced fluorescence of a weak probe laser propagating along z-direction. To avoid residual Zeeman shifts, we applied a magnetic field of 0.5 G perpendicular to z-axis and probed the $^1S_0 - {}^3P_1(m_J = 0)$ transition. Figure 2(b) shows the change of fluorescence as a function of laser frequency; two peaks, an elastic peak at $\delta = 0$ and an inelastic peak at $\delta = +\nu_z$, were clearly resolved. The first lower sideband at $\delta = -\nu_z$ was not visible, since most of the atoms occupied the ground state. The observed linewidth of the elastic peak was 20 kHz, equivalent to a few times the natural linewidth $\gamma/2\pi$. This broadening may be attributed partly to the frequency fluctuation of the probe beam and partly to an imperfect matching of the a.c. Stark shifts in the probed electronic states.

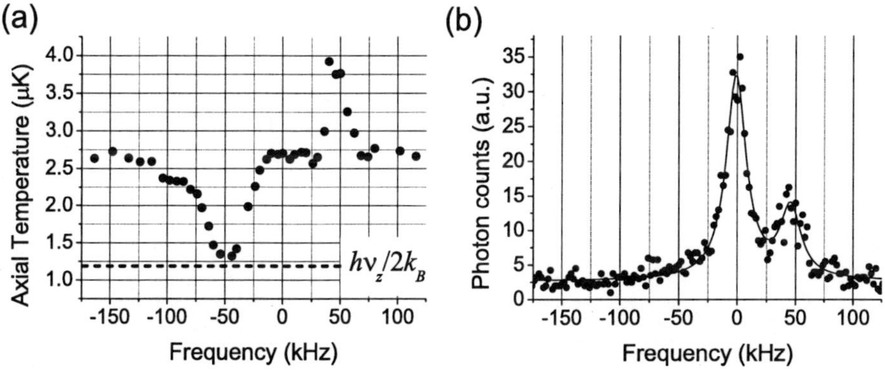

Figure 2. (a) The temperature of atoms measured by a time-of-flight technique. Sideband cooling to the vibrational ground state was achieved at a laser detuning of $\delta = -\nu_z$. (b) Recoil-free spectroscopy of atoms in the Lamb-Dicke regime. A Lorentian fit to the central elastic scattering peak gives a FWHM of 21 kHz. A slightly broader inelastic peak, corresponding to the excitation of $n = 1$ vibrational level in the $^3P_1(m_J = 0)$ state. was also visible at the laser detuning of $\delta = +\nu_z$.

Ultimately, frequency accuracy below 1 Hz level could be expected in such a FORT potential, if the trap laser wavelength is adjusted within 10^{-3} nm, since the a.c. Stark shift difference $\nu = \nu(^3P_1) - \nu(^1S_0)$ is rather insensitive to the trap laser wavelength λ ($\delta\nu/\delta\lambda \sim 300$ Hz/nm at $\lambda \sim 800$ nm). Furthermore, the use of three-dimensional optical lattices could effectively diminish the collisional frequency shifts by isolating each atom in the lattice potential.

References

1. R. J. Rafac *et al.*, Phys. Rev. Lett. **85**, 2462 (2000).
2. H. Katori, T. Ido, and M. K-Gonokami, J. Phys. Soc. Jpn. **68** 2479 (1999).
3. T. Ido, Y. Isoya, and H. Katori, Phys. Rev. A **61** 061403(R) (2000).
4. H. Katori, T. Ido, Y. Isoya, and M. K-Gonokami, Phys. Rev. Lett. **82**, 1116 (1999).

SYMPATHETIC COOLING OF LITHIUM BY LASER-COOLED CESIUM

S. KRAFT, M. MUDRICH, K. SINGER, R. GRIMM*, A. MOSK AND M. WEIDEMÜLLER

Max-Planck-Institut für Kernphysik, 69117 Heidelberg, Germany
E-mail: m.weidemueller@mpi-hd.mpg.de

We present first indications of sympathetic cooling between two neutral trapped atomic species. Lithium and cesium atoms are simultaneously stored in an optical dipole trap formed by the focus of a CO_2 laser, and allowed to interact for a given period of time. The temperature of the lithium gas is found to decrease when in thermal contact with cold cesium. The timescale of thermalization yields an estimate for the Li-Cs cross-section.

Mixtures of atomic gases enrich the field of physics with ultracold gases by making many interesting phenomena available for experimental study. A prime example is the exchange of thermal energy between different components in the mixture. This cross-thermalization can be used for determining inter-species collision cross-sections, and, more importantly, for sympathetic cooling. The process of sympathetic cooling, where the gas of interest is cooled through thermalization with a coolant gas, is especially advantageous if the cooled gas (Li in our case) is hard to cool by other methods, while the coolant gas (Cs) allows for efficient laser cooling. The Li can only be optically cooled to approximately the Doppler temperature of 0.14 mK, and evaporative cooling of this atom in its stable $F = 1$ ground state is extremely difficult due to its anomalously low scattering length [1]. Cesium, on the other hand, allows sub-Doppler temperatures ($< 3\,\mu$K in free space) to be reached by simple polarization-gradient cooling. This opens the perspective of reaching a stable Bose-Einstein condensate (BEC) of Li in the $F = 1$ state (as in the experiment by Schreck *et al.* [1]) without evaporative cooling.

Our apparatus is described in detail in a previous paper [2]. The thermalization experiments take place in an optical dipole trap, which is based on a CO_2 laser emitting light at 10.6 μm. As the laser frequency is detuned far below any optical resonances of the atoms, this trap is referred to as a quasi-electrostatic trap (QUEST) [3]. Our QUEST employs a 140 W laser beam which is focused to a beam waist of 90 μm. This yields a depth of 0.4 mK for Li and 1.0 mK for Cs. The axial and radial oscillation frequencies for Cs

*PERMANENT ADDRESS: UNIVERSITY OF INNSBRUCK, A-6020 INNSBRUCK, AUSTRIA

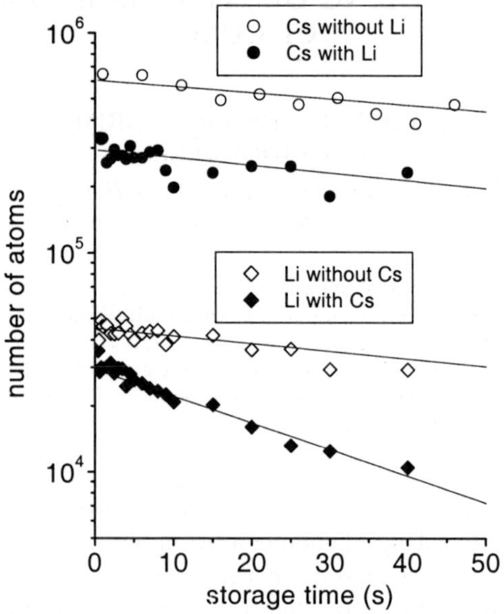

Figure 1. Evolution of the number of trapped Li (F=1) and Cs (F=3) atoms in simultaneous trapping experiments (closed symbols). For comparison, the open symbols show the evolution of separately trapped gases. Thin lines: Exponential fits. Time constants are 150 s for Cs with or without Li, 125 s for Li without Cs, and 35 s for Li trapped simultaneously with Cs.

atoms are 18 and 800 Hz, respectively, the corresponding frequencies for Li are higher by a factor 2.8 due to the difference in mass and polarizability.

The atoms are loaded into the QUEST from a combined magneto-optical trap (MOT) [4]. We load up to 10^6 Cs atoms, at a temperature of $\sim 30\,\mu$K and up to 10^5 Li atoms, which have a thermal energy of the order of the trap depth.

In a typical experiment, we trap a mixed sample and let the atoms interact for a variable time, and then turn off the QUEST. We subsequently measure the temperature and density of the Cs gas by absorption imaging (projection of the shadow of the gas cloud on a camera). The Li particle number is measured by fluorescence measurements after recapture in the MOT. The optical density of the Li gas is insufficient to use absorption imaging, therefore the temperature of the Li cannot be determined directly. The evolution of the number of atoms in the trap is shown in Fig. 1. When stored separately, both Li and Cs gases decay due to rest gas collisions in 125 and 150 s, re-

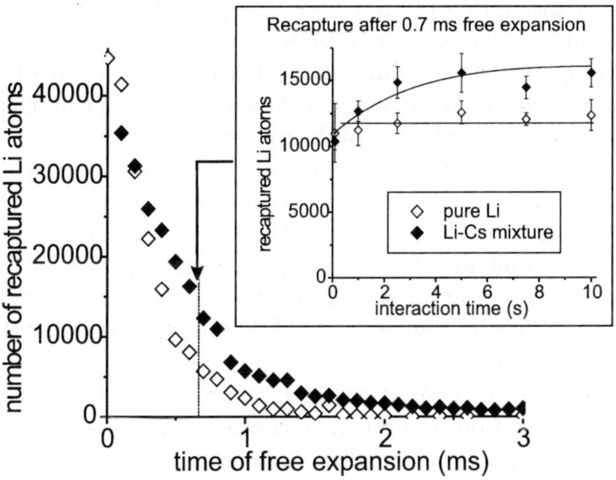

Figure 2. *Main graph:* Example of release-recapture measurements. The number of recaptured lithium atoms is plotted against the free expansion time. Open symbols: a Li sample just after loading from the MOT. Closed symbols: a sympathetically cooled Li sample (interaction time 15 s). *Inset:* Evolution of the number of recaptured atoms (free expansion time 0.7 ms). Closed symbols: number of recaptured Li atoms after sympathetic cooling. Open symbols: reference measurement with no Cs present. The lines serve as a guide to the eye.

spectively. There is no indication of evaporation in either case: Cs does not evaporate as its temperature is far below the trap depth. Li evaporation is energetically possible but the cross section for Li-Li collisions is so small that the characteristic timescale is > 1000 s.

When the gases are stored simultaneously, the Li atom number shows a non-exponential initial behavior followed by an approximately exponential decay with a characteristic time of 35 s, which is much faster than the 125 s rest gas induced decay. We interpret this as evaporation due to elastic collisions with the Cs. From the evolution of the particle number we can determine inelastic and elastic heteronuclear collision rates [2]. However, the clearest signature of sympathetic cooling, reduction of the temperature of the cooled gas, cannot be obtained this way.

To determine the lithium temperature change due to sympathetic cooling, we have performed release-and-recapture measurements, in a manner analogous to experiments by O'Hara *et al.* [5]. In this type of measurement, the QUEST is turned off and the atoms are released in free space for a short time.

After a free expansion time of up to 2 ms the QUEST is turned on again. The fraction of Li atoms that were recaptured in the QUEST is a measure of the Li temperature, since slow atoms take a longer time to leave the trapping region. A more quantitative analysis of these release-recapture measurements is being performed and will be published elsewhere.

In figure 2 release-recapture thermometry is demonstrated on two different Li samples: one which has been loaded from the MOT and stored in the QUEST for a short time without Cs, and one which has been stored together with a cold Cs sample for 15 s. The figure clearly shows that in the latter case more atoms are recaptured at long times, i.e., a larger number of slow atoms is present. This indicates that the lithium is cooled sympathetically by the laser cooled cesium atoms.

The timescale of thermalization can be estimated from the inset in Fig. 2, where a single point from the release-recapture curve was measured for different interaction times. A thermalization timescale of roughly 3 seconds can be estimated. From a rate equation model [2] one estimates elastic scattering cross-section of $\sigma_{\mathrm{CsLi}} \sim 3 \times 10^{-12}\,\mathrm{cm}^2$ which is in satisfactory agreement with the earlier estimate we obtained from evaporation measurements [2]. Note that the *absolute* number of slow atoms increases during thermalization, indicating an increasing phase space density.

Further efforts will concentrate on lowering the temperature of the coolant gas and increasing the number of trapped Li atoms, thereby increasing the degeneracy parameter. By repeated pulsed optical cooling one can extract the entropy that the Li transfers to the Cs, thereby providing a truly lossless cooling mechanism towards BEC of Li. For a sample of 10^6 Li atoms, the BEC transition temperature is $\sim 3\,\mu$K in our trap.

This work has been supported in part by the Deutsche Forschungsgemeinschaft. A.M. is supported by a Marie-Curie fellowship from the European Community programme IHP under contract number CT-1999-00316. We are indebted to D. Schwalm for encouragement and support.

References

1. F. Schreck *et al.*, Phys. Rev. Lett. **87**, 080403 (2001).
2. A. Mosk *et al.*, submitted to Appl. Phys. B; arXiv:physics/0107075
3. T. Takekoshi and R. J. Knize, Opt. Lett. **21**, 77 (1996).
4. U. Schlöder *et al.*, Eur. Phys. J. D **7**, 331 (1999).
5. K. M. O'Hara *et al.*, Phys. Rev. Lett. **85**, 2092 (2000).

DETERMINISTIC DELIVERY OF A SINGLE ATOM

STEFAN KUHR, WOLFGANG ALT, DOMINIK SCHRADER,
MARTIN MÜLLER, VICTOR GOMER, DIETER MESCHEDE

Institut für Angewandte Physik, Universität Bonn,
Wegelerstr. 8, D-53115 Bonn, Germany
E-mail: kuhr@iap.uni-bonn.de

Using optical dipole forces we have realized a conveyor belt for single atoms. This device can transport a single or any desired small number of neutral atoms over a distance of a centimeter with sub-micrometer precision. A standing wave dipole trap is loaded with a prescribed number of cesium atoms from a magneto-optical trap. Mutual detuning of the counter-propagating laser beams moves the interference pattern, allowing us to accelerate and stop the atoms at preselected points along the standing wave with a transportation efficiency close to 100 %. The trapping field can also be accelerated to eject a single atom into free flight with well-defined velocities.

1 Introduction

The manipulation of individual atomic particles is considered a key factor in the quantum engineering of microscopic systems. In an experimental realization it is necessary to achieve full control of all degrees of freedom of the physical system under study. Recently, techniques have been developed to load an optical dipole trap with single atoms only[1,2]. In our case, we have combined controlled manipulation of the trapping potential with deterministic loading of a dipole trap with a prescribed small number of atoms[3,4].

2 An optical dipole trap for single atoms

Our dipole trap consists of two counter-propagating laser beams (Fig. 1) with equal intensities and optical frequencies ν_1 and ν_2 producing a position-dependent dipole potential $U(z,t) = U_0 \cos^2[\pi(\Delta\nu\ t - 2z/\lambda)]$, where $\Delta\nu = \nu_1 - \nu_2 \ll \nu_1, \nu_2$. Both dipole trap laser beams are derived from a single Nd:YAG laser ($\lambda = 1064$ nm) which is far red detuned from the D-lines of cesium ($\lambda_{D1} = 894$ nm, $\lambda_{D2} = 852$ nm). An atom initially trapped in the stationary standing wave ($\Delta\nu = 0$) is moved along the optical axis by changing the frequency difference $\Delta\nu$ which causes the potential wells to move. The value of $\Delta\nu$ is controlled by means of two acousto-optic modulators (AOMs) driven by a digital dual-frequency synthesizer with two phase-synchronized RF-outputs. A standard six-beam MOT serves as a source of single cold

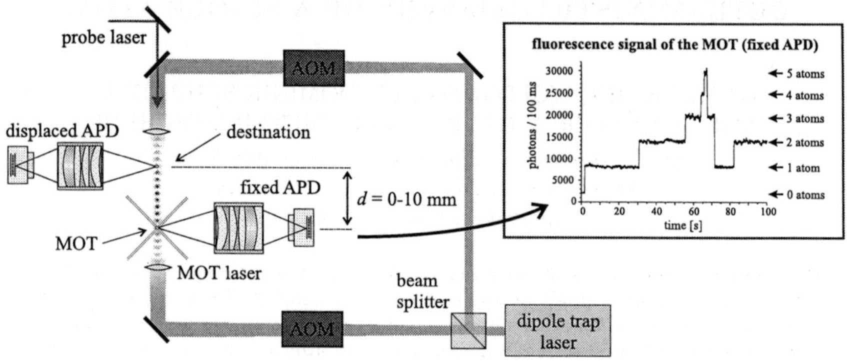

Figure 1. Single-atom conveyor belt. The two counter-propagating beams of the dipole trap form an interference pattern. AOMs vary the laser frequencies to move the pattern along the beam axis. Atoms initially trapped in the MOT thus can be displaced over a distance $d = 0..10$ mm.

atoms[1]. A high magnetic field gradient of 400 Gauss/cm localizes the trapped atoms to a region of diameter 30 μm. The fluorescence light of the atoms is collected by a diffraction limited objective[5] and projected onto an avalanche photodiode, which yields a photon count rate of $6 \cdot 10^4 s^{-1}$ per atom. This allows us to determine the exact number of trapped atoms in real time (Fig. 1). To transfer cold atoms from the MOT into the dipole trap, both traps are simultaneously operated for several milliseconds before we switch off the MOT (Fig. 2a).

3 The single-atom conveyor belt

Our device allows us to accelerate an atom in the dipole trap and bring it to a stop at preselected points along the standing wave. We detect a displaced atom within the dipole trap by illumination with a weak resonant probe laser. The resulting fluorescence signal (about 40 photons in 50 ms) is collected by imaging optics which are displaced by the exact transportation distance. This detection scheme serves to prove the deterministic delivery of a single atom to a desired spot. The measured probability to observe the transported atom as a function of the displacement (Fig. 2b, circles) shows that for small distances, the fraction of detected atoms is above 90 %.

However, the decrease of the trap depth for larger displacements from the laser focus limits the detection efficiency. During resonant excitation the atom is heated by scattering photons. The fluorescence signal lasts until the atom is ejected out of the trap. We detect the presence of the atom if a peak of fluorescence photons substantially exceeds (more than 5 photons) the stray light background (2 photons on average) of the Nd:YAG and the probe laser beams. In our case, at $d > 3$ mm, the probe laser can evaporate the atom out of the dipole trap before enough fluorescence photons have been detected.

The actual transportation efficiency, however, is much higher than that shown by the resonant illumination detection. To demonstrate this, we use the MOT to detect the atom with 100 % efficiency. Without resonant illumination, the displaced atom is transported back to $z = 0$ before we switch on the MOT lasers to reveal the presence or absence of the atom. The results of this measurement are shown in Fig. 2b as filled circles. Even for distances as large as 10 mm, the two-way transportation efficiency remains above 80 %. At a distance of 15 mm, however, the transportation efficiency drastically decreases to 16 %. The atoms are lost at this distance because gravity reduces the effective potential depth[4].

The investigation of various accelerations at a constant displacement of 1 mm yields a constant transportation efficiency of more than 95 % until the acceleration exceeds a value of 10^5 m/s^2. At larger accelerations, the efficiency rapidly decreases due to two different effects. Firstly, there is a heating effect due to abrupt changes of the acceleration[4]. Secondly, the finite bandwidth

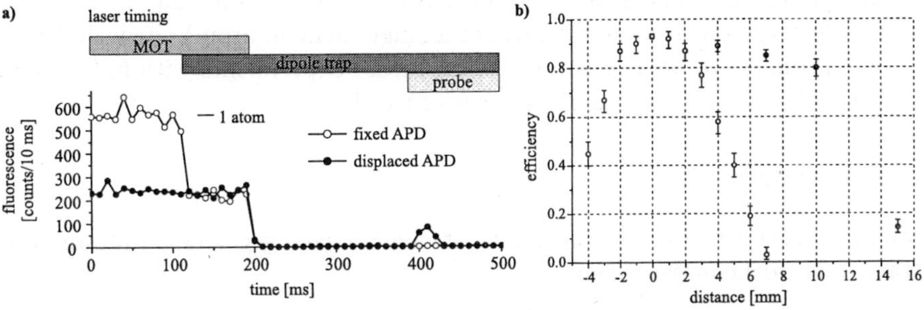

Figure 2. **a)** Deterministic delivery of a single atom. The fixed APD initially confirms the presence of the atom in the MOT. During transfer into the dipole trap the MOT fluorescence is decreased due to the light shift. Initially, the displaced APD (filled circles) does not see the trapped atom but only detects stray light of the MOT laser beams. The peak at $t = 400$ ms originates from the same atom displaced by 1 mm, which is illuminated in the dipole trap with a resonant probe laser. **b)** Transportation efficiency of the optical conveyor-belt for a constant acceleration of 500 m/s^2. Each data point results from ~100 shots performed with one atom each. Empty circles: The atom is detected by resonant illumination at its new position. Filled circles: More efficient detection by moving the atom back and recapturing it into the MOT.

of the AOMs reduces the dipole trap laser power during a frequency sweep. However, we achieve accelerations of 10^5 m/s^2 which allow us to change the atomic velocity from zero to 10 m/s in 100 μs.

We also used the conveyor belt to realize a catapult for single atoms. For this purpose, the atom is initially displaced by 3 mm, before we move the standing wave in the opposite direction, accelerating the atom back over 0.5 mm towards the position of the MOT. The atom is now released into free flight 2.5 mm away from the MOT with a velocity of 2 m/s by rapidly switching off the standing wave. If the MOT lasers are turned on exactly after the time of flight, the atom is always recaptured by the MOT, demonstrating the working principle of the catapult.

4 Outlook

The single-atom conveyor belt will soon serve to place a desired number of atoms into a resonator of high finesse. Quantum logic gates can be implemented by entangling neutral atoms through their simultaneous coupling to the optical field of this resonator. The axial oscillation frequencies of about 400 kHz open a route to the application of Raman sideband cooling techniques. Additional cooling of the atoms will improve the performance of the conveyor belt with respect to both transportation efficiency and atomic localization. Long spin relaxation times of many seconds in combination with the state-selective detection at the level of a single neutral atom[1] show that this system promises to be a versatile tool for future experiments with full control of internal and external atomic degrees of freedom.

Acknowledgments

We have received support from the Deutsche Forschungsgemeinschaft and the state of Nordrhein-Westfalen.

References

1. D. Frese *et al.*, *Phys. Rev. Lett.* **85**, 3777 (2000)
2. N. Schlosser *et al.*, *Nature* **411**, 1024 (2001)
3. S. Kuhr *et al.*, *Science* **293**, 278 (2001), Published online June 14 2001; 10.1126/science.1062725
4. D. Schrader *et al.*, submitted to *Appl. Phys.* B, *quant-ph/0107029*
5. W. Alt, submitted to *Rev. Sci. Instr.*, *physics/0108058*

CAVITY-QED WITH A SINGLE TRAPPED ^{40}CA^{+}-ION

GERHARD R. GUTHÖHRLEIN, MATTHIAS KELLER, WOLFGANG LANGE, AND HERBERT WALTHER

Max-Planck-Institut für Quantenoptik, Hans-Kopfermann-Straße 1, 85748 Garching, Germany
E-mail: Gerhard.Guthoehrlein@mpq.mpg.de

KAZUHIRO HAYASAKA

Communications Research Laboratory, 588-2 Iwaoka, Nishi-ku, Kobe 651-24, Japan
E-mail: hayasaka@crl.go.jp

Radio-frequency ion traps provide ideal conditions for storing and manipulating single particles and localize them with a precision far below their resonance wavelength. By combining an optical cavity with the excellent position control provided by the trap, we have implemented a completely deterministic coupling of ions and the electromagnetic field. In this system we can investigate single-particle cavity QED dynamics with a predefined interaction strength and interaction time, which is not possible in atom-based systems.

The interaction of a single atom with a single mode of a high-finesse cavity has been the subject of a number of experiments in the field of cavity QED in the microwave[1] and optical region[2,3]. While providing new insights into the dynamics of atoms and fields under strong coupling conditions, a drawback of the systems investigated so far is the lack of control over the position of the atom, which results in non-deterministic fluctuations of the coupling between atoms and field. We have overcome this problem by employing a single ^{40}Ca^{+}-ion instead of an atom and storing it in a radio frequency trap. In order to avoid the coating and charging of the mirrors in the ionization process, we use a trap of linear geometry and load it in a region separated from the cavity by 25 mm (see Fig. 1). Subsequently, the ion is shuttled along the trap axis to the cavity region. We carefully compensate offset voltages in the system in order to position the ion at the node of the trapping rf-field. In this way we minimize the motion of the ion driven by the rf-field and provide optimum localization conditions.

In order to test the degree of control we have over the ion's position, we utilize the ^{40}Ca^{+}-ion as a near-field probe of the optical cavity field[4]. The cavity is tuned close to the $S_{1/2}$-$P_{1/2}$ resonance transition in ^{40}Ca^{+} at a wavelength of $\lambda = 397$ nm. We scan the relative position of ion and field and at the same time monitor the fluorescence emitted by the ion. The intensity of the fluorescent light is a measure of the local intensity of the optical field

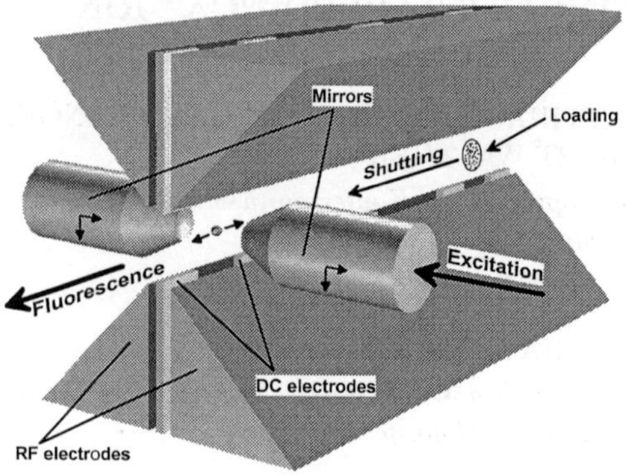

Figure 1. Linear ion trap with the cavity mirrors mounted near the front end. The ion is confined to the trap axis in an rf quadrupole-field generated by a set of four electrodes. After loading the trap at the far end, the ion is moved into the cavity region by using auxiliary electrodes placed along the trap axis. Arrows indicate the directions in which the ion and the mirrors are translated to scan the cavity field.

in the cavity, and thus provides information on its mode structure.

In the direction of the trap axis, the ion is confined in a DC potential well, which is approximately harmonic with an oscillation frequency of $\omega_z/2\pi \approx$ 300 kHz. By applying slightly asymmetric voltages, the minimum of the potential well, and thus the equilibrium position of the ion, is moved along the trap axis. In this way, we were able to map one-dimensional cross sections of the transverse mode structure of the cavity.

Figure 2 shows scans of the first four TEM_{0n} modes of the cavity. The fluorescence distribution is not entirely symmetric, because the principal axis of the cavity eigenmodes is slightly rotated with respect to the trap axis. In each plot, an inset shows the theoretical field distribution and indicates the path along which the ion is scanned. The solid curves in Fig. 2 are obtained from a fit using Hermite-Gauss functions and take into account saturation of the ion's transition. The influence of saturation is apparent in Fig. 2c,

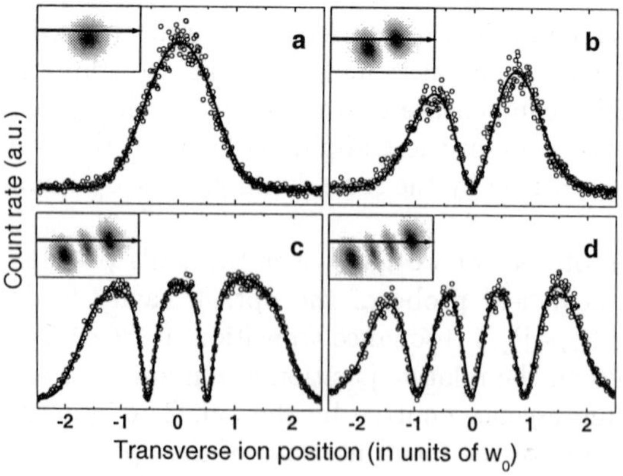

Figure 2. Transverse profiles of the Hermite-Gauss modes of the cavity, obtained by monitoring the ion's fluorescence while scanning over a range of 5 times the beam waist w_0 (120 μm). The solid line is a fit including saturation of the transition. The inset shows the calculated intensity distribution of the mode with the scan path indicated. The scanned modes are a) TEM_{00} b) TEM_{01} c) TEM_{02} d) TEM_{03}.

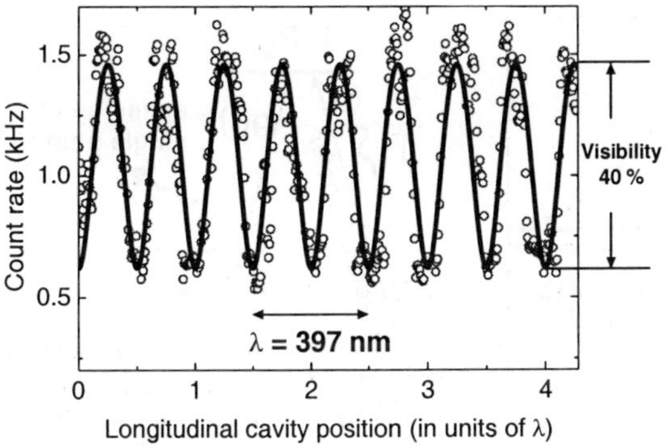

Figure 3. Single-ion mapping of the longitudinal structure of the cavity field. The visibility is determined by the residual thermal motion of the Doppler-cooled ion. It corresponds to a resolution of 60 nm. The center of the ion's wave packet in this measurement is localized to a precision of 16 nm, determined by the signal-to-noise ratio and the averaging time (200 ms per point).

where a slightly higher intensity was injected into the cavity. In each case, the correspondence with the measured fluorescence is excellent.

Perpendicular to the trap axis, the ion is confined by the rf-field, resulting in an effective harmonic potential with an oscillation frequency $\omega_r/2\pi \approx$ 1.1 MHz. This is larger than the axial frequency ω_z, and therefore field structures in the radial direction of the trap are better resolved. Since the ion must be confined to the trap axis, to avoid exposure to the rf-field, the other dimensions of the cavity field were mapped by keeping the ion stationary and scanning the position of the cavity. To this end, the mirrors were mounted on a three-axis piezo-electric stage. To determine the spatial resolution of our method, we have probed the standing-wave structure of the cavity field by moving the cavity parallel to its axis. Figure 3 shows a measurement of the ion's fluorescence obtained in this way. A pronounced standing-wave pattern is observed with a period of $\lambda/2$ and a visibility of 40%. This value corresponds to a resolution of 60 nm or 15% of the transition wavelength λ.

The sub-wavelength control over the position of the ion in the cavity field provides us with a means of pre-determining the coupling g between ions and field; we will employ this to deterministically generate single photon pulses[5]. For this purpose, the optical cavity is coupled to the $D_{3/2} - P_{1/2}$ transition in $^{40}Ca^+$ with a wavelength $\lambda = 866$ nm. In this region, mirror losses as low as 10 ppm are feasible, corresponding to a cavity decay rate smaller than $\kappa/2\pi = 0.5$ MHz. Using the adiabatic passage technique with a pump pulse on the $S_{1/2} - P_{1/2}$ transition, a photon is transferred to the cavity mode and finally emitted from the cavity as a single photon pulse with an efficiency of 97% (Fig. 4). A second application is the realization of lasing with a single ion as the gain medium. To this end we pump the $P_{1/2}$ level continuously and empty the $D_{3/2}$ level via the $P_{3/2}$ state. For our experimental parameters, an intracavity field of 10 photons may be generated. The cw-radiation

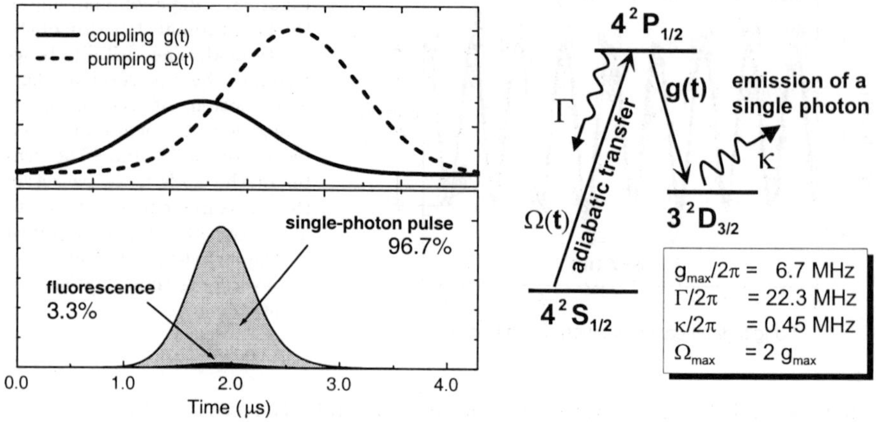

Figure 4. Calculation of the single photon emission on the $P_{1/2}$ to $D_{3/2}$ transition, using the adiabatic passage technique. A single photon is emitted with a probability of 97%. The graph on the right shows the relevant levels of the $^{40}Ca^{+}$-ion.

emitted from the cavity is expected to show non-classical properties such as sub-poissonian photon statistics and antibunching[6]. Beyond its application as a novel light source, our system holds great promise for quantum information processing: when two ions are placed in the cavity, the optical field provides the necessary controlled coupling between them.

References

1. *Cavity quantum electrodynamics*, P. R. Berman, Ed., (Advances in atomic, molecular and optical physics, supplement 2, Boston, 1994).
2. C. J. Hood, T. W. Lynn, A. C. Doherty, A. S. Parkins, and H. J. Kimble, *The atom-cavity microscope: Single atoms bound in orbit by single photons*, Science **287**, 1447-1453 (2000).
3. P. W. H. Pinkse, T. Fischer, P. Maunz, and G. Rempe, *Trapping an atom with single photons*, Nature **404**, 365-368 (2000).
4. G. R. Guthöhrlein, M. Keller, K. Hayasaka, W. Lange, and H. Walther, *A single ion as a nanoscopic probe of an optical field*, Nature, in print.
5. C. K. Law and H. J. Kimble, *Deterministic generation of a bit-stream of single-photon pulses*, J. Mod. Opt. **44**, 2067 (1997).
6. G. M. Meyer, H. J. Briegel, and H. Walther, *Ion-trap laser*, Europhys. Lett. **37**, 317 (1997).

TRIGGERED SINGLE PHOTONS FROM A QUANTUM DOT

CHARLES SANTORI, MATTHEW PELTON, GLENN S. SOLOMON*, YSEULTE DALE, AND YOSHIHISA YAMAMOTO†

Quantum Entanglement Project, ICORP, JST, E.L. Ginzton Laboratory, Stanford University, Stanford, California 94305, USA
E-mail: chars@stanford.edu

We demonstrate a single-photon source by exciting a single InAs quantum dot with resonant laser pulses, and spectrally selecting the last photon emitted in the subsequent decay process. Correlation measurements show a reduction of the two-photon probability to 0.12 times the value for a Poisson-distributed photon stream of the same intensity.

Practical sources of single photons are needed for quantum information applications, such as quantum cryptography[1] and quantum computation.[2] Several systems have been demonstrated as useful single-photon sources, including atoms and ions,[3], micropost p-i-n structures,[4] molecules,[5] and color centers in diamond.[6] Quantum dots have several advantages as single-photon sources. They are relatively easy to hold in place and incorporate into larger structures, do not require milli-Kelvin temperatures, do not suffer from photobleaching effects, and have narrow emission linewidths. Here, we present spectral and temporal properties of a single semiconductor quantum dot, and then demonstrate that this dot may be used to generate single photons.[7,8]

Optically active quantum dots confine electrons and holes to regions so small that their energy levels are quantized.[9] If several electrons or holes are placed in a dot at the same time, they will, to a first approximation, occupy single-particle states as allowed by the Pauli exclusion principle. Electrostatic interactions play an important role in multi-particle states, however, and lead to significant energy shifts that can be seen in optical emission spectra.[10] These shifts provide two mechanisms by which a quantum dot, excited by a laser pulse, can generate a single photon. If the laser pulse is resonant with creation of a single electron-hole pair, then absorption of a second photon may be suppressed. However, even if a "multi-excitonic" state containing more than one electron-hole pair is produced, the subsequent decay process produces a sequence of photons having different energies, so that the last emitted photon can be spectrally isolated.[11] If the last two photons are collected, a source of single polarization-entangled photon pairs may even be possible.[12]

A sample was fabricated containing self-assembled InAs quantum dots (11 μm^{-2}) in a GaAs matrix, and etched into 0.2 μm posts. All optical

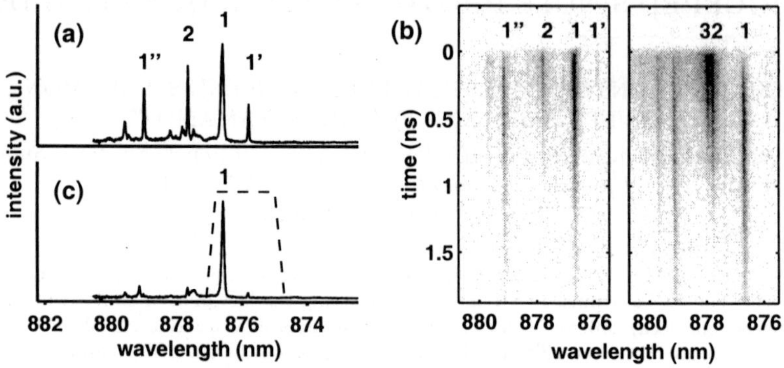

Figure 1. (a) Emission spectrum of a single dot under CW, above-band (655 nm) excitation. The peaks labeled 1 and 2 correspond to one-exciton and biexciton emission, respectively. (b) Time-resolved spectra from the same dot under pulsed, above-band (708 nm) excitation, with weak (27 μW, left) and moderate (108 μW, right) excitation power. Under moderate excitation power, the one-exciton emission is seen to occur only after the multi-excitonic emission has diminished. (c) Emission spectrum of the same dot under CW, resonant (857.5 nm) excitation. The dotted line indicates approximately the filter function used for the single-photon source.

measurements were performed at 5K. The spectra shown in Fig. 1 are from a mesa containing a single dot. Under continuous-wave excitation above the GaAs bandgap (Fig. 1a), several lines are seen. Peaks 1 and 2 correspond to emission following the addition of one and two electron-hole pairs to the dot, respectively. This can be determined from the quadratic pump-power dependence of line 2, or from time-resolved spectra, as shown in Fig. 1b. These time-resolved spectra show emission following an excitation pulse at time zero, averaged over about 2×10^{10} cycles. Under weak excitation (left), usually only a single electron-hole pair is added to the dot, and the one-exciton emission (line 1) appears quickly and decays exponentially following the laser excitation pulse. With moderate excitation, however, usually more than one electron-hole pair is added to the dot, and the one-exciton emission is delayed until the multi-excitonic emission (lines 2, 3) has occurred. When multiple electron-hole pairs are present in the dot, they appear to decay independently.[13] The other lines, 1' and 1", come from the same dot, and most likely correspond to charged states[14]. With resonant excitation (Fig. 1c), the charge of the dot is unlikely to change, and these lines almost disappear.

To demonstrate a single-photon source, we excited this mesa every 13 ns with 2.9 ps pulses from a mode-locked Ti-sapphire laser, tuned to an absorp-

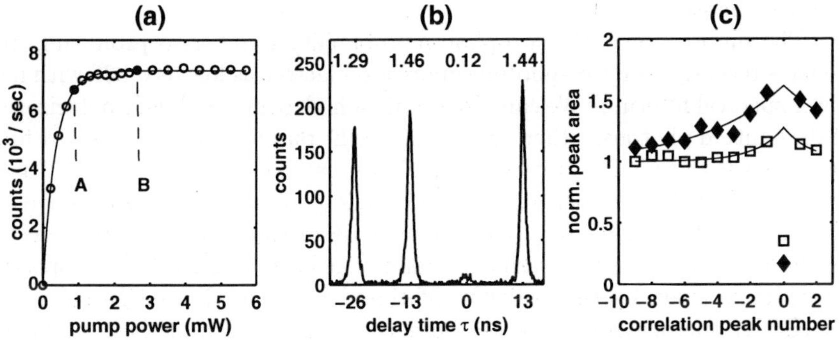

Figure 2. Results from the single-photon source: (a) Photon count rate versus pump power, under pulsed, resonant excitation. Since the multi-excitonic emission is rejected through spectral filtering, a clear saturation behavior is seen. (b) Raw correlation data under 0.88 mW excitation power (point A on saturation curve). The reduced size of the central peak at $\tau = 0$ demonstrates two-photon probability suppression. The numbers above the peaks are peak areas, normalized by those of an ideal Poisson-distributed source of the same intensity. (c) Normalized peak areas versus peak number, for 0.88 mW excition power (solid diamonds, point A on saturation curve) and 2.63 mW (squares, point B on saturation curve), showing again antibunching, as well as evidence for blinking on a 100 ns time scale.

tion resonance at 857.5 nm. The emission was analyzed by a Hanbury Brown and Twiss-type setup, containing a beamsplitter, monochrometer-type spectral filters, two photon counters, and a time interval counter. The 2 nm-wide measurement bandwidth was defined to accept one-exciton emission and reject multi-exciton emission, as illustrated in Fig. 1c. Under this filtering condition, a clear saturation behavior is seen in the photon count rate (Fig. 2a). A correlation histogram collected with the dot excited close to saturation is shown in Fig. 2b. A series of peaks is seen, separated by the laser repetition period. The peak areas, normalized by those of an ideal Poisson-distributed source of the same intensity, are shown above the peaks. The peak at $\tau = 0$ corresponds to multi-photon emission events, and from its area we calculate that the multi-photon probability is only 0.12 times what it would be for a Poisson source of the same intensity. The peaks to the side have areas larger than one, and this behavior is better seen in Fig. 2c. Here, we plot normalized peak area versus peak number for excitation close to saturation (solid diamonds) and well into saturation (squares). The exponential decay of the peak areas towards one suggests a blinking behavior, with a time scale of roughly 100 ns, and with an "on" to "off" ratio that improves as the pump power is increased.

The suppression of the two-photon probability seen here is promising, and we have recently seen two-photon suppression factors as low as 0.036 with narrower spectral filtering. We can also obtain a high degree of linear polarization of the emitted photons, either by optimizing the pump polarization, or by finding dots with naturally preferred emission polarizations. But two challenges remain before this can become an ideal single-photon source: to improve the photon collection efficiency (only about 0.5% of the photons were collected by the first lens in this experiment), and to achieve a source whose spectral linewidth is given by the radiative lifetime limit. Distributed Bragg reflector (DBR) microcavities have the potential to solve both of these problems, through modification of the spontaneous emission pattern and lifetime.[15]

References

[*] Also at Solid-State Photonics Laboratory, Stanford University, CA.
[†] Also at NTT Basic Research Laboratories, Atsugishi, Kanagawa, Japan.

1. G. Brassard *et al.*, *Phys. Rev. Lett.* **85**, 1330 (2000).
2. E. Knill, R. Laflamme, and G. J. Milburn, *Nature* **409**, 46 (2001).
3. H. J. Kimble, M. Dagenais, and L. Mandel, *Phys. Rev. Lett.* **39**, 691 (1977); F. Diedrich and H. Walther, *Phys. Rev. Lett.* **58**, 203 (1987).
4. J. Kim, O. Benson, H. Kan, and Y. Yamamoto, *Nature* **397**, 500 (1999).
5. B. Lounis and W.E. Moerner, *Nature* **407**, 491 (2000);
6. C. Kurtsiefer *et al.*, *Phys. Rev. Lett.* **85**, 290 (2000); R. Brouri *et al.*, *Opt. Lett.* **25**, 1294 (2000).
7. C. Santori, M. Pelton, G. S. Solomon, Y. Dale, and Y. Yamamoto, *Phys. Rev. Lett.* **86**, 1502 (2001).
8. P. Michler *et al.*, *Science* **290**, 2282 (2000).
9. D. Bimberg *et al.*, *Quantum Dot Heterostructures* (John Wiley & Sons, Chichester, 1999).
10. A. Kuther *et al.*, *Phys. Rev. B* **58**, R7508 (1998); H. Kamada, H. Ando, J. Temmyo, and T. Tamamura, *Phys. Rev. B* **58**, 16243 (1998).
11. J.-M. Gérard and B. Gayral, *J. Lightwave Technol.* **17**, 2089 (1999).
12. O. Benson, C. Santori, M. Pelton, and Y. Yamamoto, *Phys. Rev. Lett.* **84**, 2513 (2000).
13. C. Santori, G. S. Solomon, M. Pelton, and Y. Yamamoto, Preprint cond-mat/0108466 at xxx.lanl.gov (2001).
14. R. J. Warburton *et al.*, *Nature* **405**, 926 (2000).
15. G. S. Solomon, M. Pelton, and Y. Yamamoto, *Phys. Rev. Lett.* **86**, 3903 (2001).

"SUPERLUMINAL" AND SUBLUMINAL PROPAGATION OF AN OPTICAL PULSE IN A HIGH-Q OPTICAL MICRO-CAVITY WITH A FEW COLD ATOMS

YUKIKO SHIMIZU, NORITUGU SHIOKAWA, NORIAKI YAMAMOTO, MIKIO KOZUMA, AND TAKAHIRO KUGA

Institute of physics, University of Tokyo, 3-8-1 Komaba, Meguro-ku, Tokyo 153-8902, Japan
E-mail: shimizu-yukiko@aist.go.jp

LU DENG AND ED W. HAGLEY

National Institute of Standard and technology, Gaithersburg, MD 20899, USA

Propagation of a light pulse through a high-Q micro-cavity which contains only a few cold atoms ($\overline{N}$ = 6.4) is investigated experimentally. Rubidium atoms (T=100 μK) are launched into the cavity and subjected to interaction with a Gaussian probe light pulse during its transit time. Depending on the detuning of the probe frequency from the atomic resonance, the propagation of the light pulse is "advanced" or retarded in time compared to the case when no atoms are present in the cavity mode. The strong coupling condition allows the optical pulse to interact with the same atom as many times as the cavity finesse and 165ns advance and 436ns delay in the pulse propagation are obtained.

When a light pulse propagates in a dispersive medium, its group velocity can be described by $v_g=c/(n+\omega dn/d\omega)$, where c is the speed of light in vacuum, n is the refractive index of the medium, and ω is the angular frequency of the wave. This simple relation provides an insight on how an optical pulse may propagate with an abnormal group velocity in the medium whose refractive index changes rapidly as a function of frequency [1]. Recent experiments on the Gaussian light pulse propagation have shown significant slow down in the Bose-Einstein condensates [2] as well as "superluminal" propagation in the gain-assisted medium [3].

The group velocity critically depends on the change of dispersion in a narrow frequency range. This change dictates the magnitude of the group velocity, whereas its sign determines the propagation characteristics. Compared with the speed of light in vacuum c, the group velocity of an optical pulse in medium is usually said to be subluminal when $dn/d\omega > 0$, and "superluminal" when at least $dn/d\omega < 0$.

It is so challenging and also physically important to realize the same effects with only a single atom. For example let us consider the three-level atomic system that has two grand states $|g1\rangle$, $|g2\rangle$, and an excited state $|e\rangle$. If the atom is first prepared in a superposed state of $|g1\rangle$ and $|g2\rangle$, and interacts with a laser pulse in resonance with the $|g1\rangle \leftrightarrow |e\rangle$ transition, the light pulse will be split into two propagating parts, normal and super/sub-luminal ones. In this case the atomic and light states are entangled, forming a Schroedinger cat state. Such a system could work as a novel type of quantum gate [4, 5]. Unfortunately, the effect by a single atom is difficult to be detected because of extremely small change in dispersion.

It is, however, possible to enhance such a single-atom effect by using a high-finesse optical cavity that allows the light pulse to interact with the atom as many times as its finesse. To achieve this one must work in the strong-coupling regime so that appreciable interaction takes place before photons are scattered from the cavity mode. Recent progresses in high-reflectivity mirror technology and atomic manipulation techniques have provided the strong coupling condition [6-11]. We report the successful observation of large propagation time shifts of the order of several hundreds nanoseconds with a small number of atoms ($\overline{N} = 6.4$).

The arrival time of a pulse after traversing the cavity is shifted by $\Delta T = FL_c / v_g - L_c / c$ compared with that of a pulse traversed an empty cavity. In this expression F and L_c are the finesse and length of the cavity, respectively. Note that an empty cavity can also shift the pulse arrival time by $\Delta T'$. However, the strong coupling condition requires very narrow mirror separation and $\Delta T'$ is negligible.

The experimental setup is depicted in Fig. 1. ^{85}Rb atoms are cooled and trapped with conventional magneto-optical trapping (MOT) with 6 circularly polarized laser beams in a cubic arrangement. The D2-line $5^2S_{1/2}$ F=3 $\leftrightarrow$ $5^2P_{3/2}$ F=4 transition is used for cooling cycle. Typically, about 10^8 atoms are captured in a few seconds from a background vapor at a pressure of 10^{-9} Torr. By removing the downward MOT beam, we launch the atoms into an optical micro-cavity located 33 mm above. The velocity of launched atoms at the cavity region is about 3 m/s.

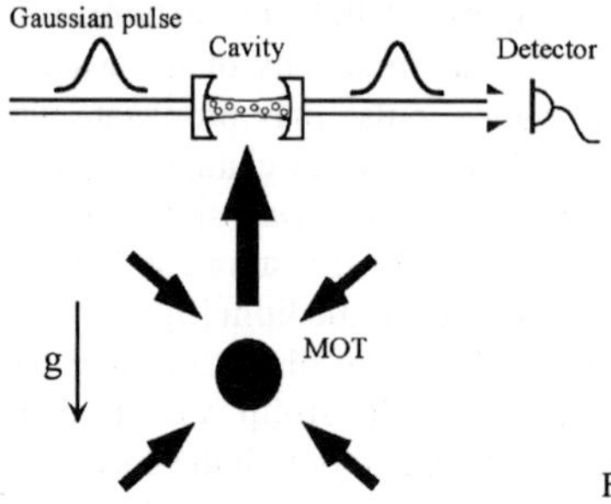

Figure 1. A schematic diagram of the experiment.

The optical micro-cavity consists of two concave mirrors with the reflectivity of 99.9980%. The cavity length, finesse, and decay rate are L_c=70 μm, F=2.1×10^5, and $\kappa/2\pi$ = 5 MHz, respectively. The beam waist for TEM_{00} mode of the cavity is w_0= 31 μm. The atom-field coupling constant, i.e. the vacuum Rabi splitting, is $g_0/2\pi$ = 19.6 MHz. This was calculated for the F=3 $\leftrightarrow$ F'=4 transition with the dipole moment μ. The decay rate of atomic polarization is $\gamma / 2\pi$ = 3 MHz. The strong coupling condition is well satisfied.

When atoms reach the cavity region the Gaussian probe pulse whose frequency is properly detuned from the atomic resonance is sent into the cavity. Intra-cavity photon number at the pulse peak is n=0.4 . We set the situation that cavity mode is continuously filled with successively coming launched atoms during a few milliseconds. In Fig. 2, we show the heterodyne detected signals with and without atoms in the cavity. When the probe beam detuned by –53MHz with respect to the atomic resonance (where $dn/d\omega > 0$), 436 ns delay of the light pulse is clearly observed. Similarly at the detuning of -45 MHz (where $dn/d\omega < 0$), pulse advancement of 165 ns is observed. Here both signals are averaged over 200 trials.

The effective intra-cavity atomic number $\overline{N}$ for a 3-second MOT loading is evaluated by measuring the probe transmittance as a function of detuning [7]. When atoms are present in the cavity, the light transmission through the cavity at the frequency ω_p is given by [7]

$$T = T_0 \left| \frac{\kappa[\gamma + i(\omega_a - \omega_p)]}{(\omega_p - \lambda_+)(\omega_p - \lambda_-)} \right|^2 \tag{1}$$

where T_0 is the peak transmission, ω_a is the atomic transition frequency and $\lambda_\pm = (\omega_a \pm g) - i(\gamma + \kappa)/2$. When $\overline{N}$ atoms cooperatively interact with the same light field, the Rabi splitting is expressed by $g \equiv \sqrt{g_0^2 N - (\gamma - \kappa)^2/4}$ and is well approximated by $g = g_0\sqrt{N}$ in the present case. Fitting the data in Fig. 3 with Eq.(1), we obtain the effective atomic number of $\overline{N}$ =6.4.

Two types of numerical simulations are carried out. A signal shape of the delayed/advanced Gaussian pulse as a function of propagation time and effective atomic number $\overline{N}$ is given by

$$A(t) = \sum_N T(\overline{N}) \exp\left[-\frac{(t - T_D)^2}{2\sigma^2} \right] \tag{2}$$

where σ is the temporal width of the Gaussian light pulse. The effective atomic number is used as a fitting parameter in the simulation. In both cases of Fig. 2a and 2b, the simulation yields $\overline{N}$ =6.4 . These results are in good agreement with $\overline{N}$ =6.4 obtained from the Rabi splitting in Fig. 3.

However, the data are accumulated over 200 trials and $\overline{N}$ may fluctuate among individual trials. Therefore, we apply the method of Monte Carlo simulation. The atoms are spatially distributed throughout the cavity mode function $\psi(\mathbf{r}_j)$ with mode volume $V = \pi w_0^2 L_c/4$,

$$\psi(\mathbf{r}_j) = \sin kz_j \exp[-(x_j^2 + y_j^2)/w_0^2] \tag{3}$$

For a random distribution of atoms at different positions r_j, the effective intra-cavity atomic number is $\overline{N} \equiv \Sigma_j|\Psi(r_j)|^2$. A thousand atoms are placed in the region of $\pi(10w_0)^2 \times L_c$ around the cavity. (The number of atoms and the volume of region are estimated from the total number of atoms in MOT cloud and the solid angle of the launching target.) The number of atoms in the cavity mode can be calculated. Figure 4 shows the "advance" or delay as a function of atomic number calculated by using Monte Carlo simulation. In this figure we find that the number of effective atoms giving the observed advance and delay is about 9.4.This result is in good agreement with $\overline{N}$=6.4 obtained from Fig. 3 and the simulation by using Eq. (2).

We have confirmed that only a few cold atoms (between six and ten) are enough to cause an observable advance/delay in light pulse propagation.

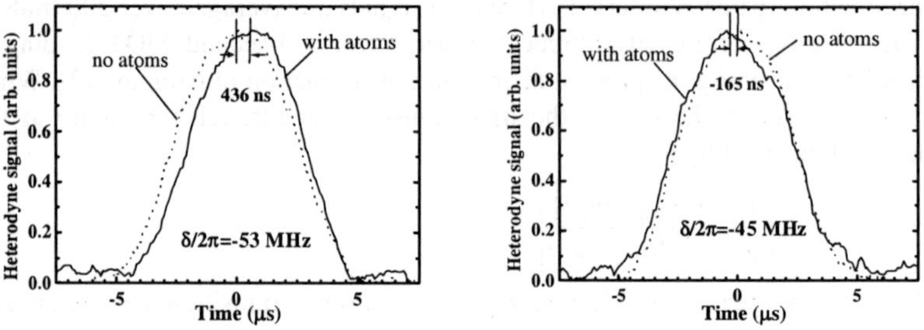

Figure 2. The observed (a) subluinal and (b) superluminal transmitted Gausssian light pulse.

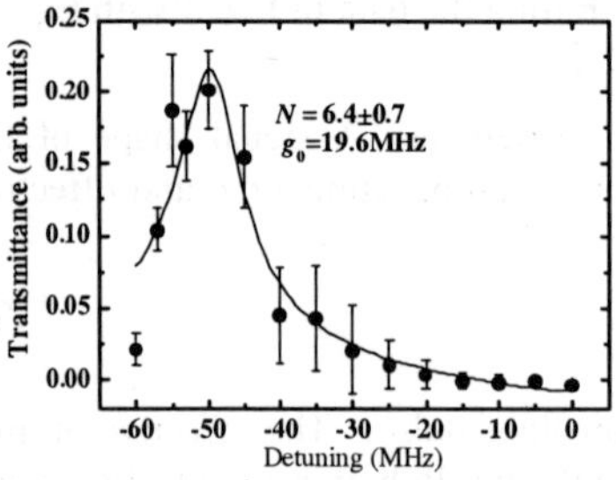

Figure 3. Observed and calculated Rabi-splitting.

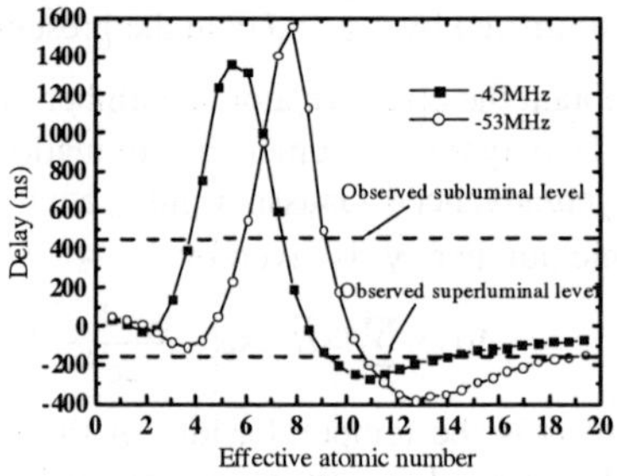

Figure 4. The delay/advance time as a function of effective atomic number calculated by Monte Carlo simulation.

References

1. L. Brillouin *"Wave Propagation and Groupe Velocity"* (Academic Press, New York, 1960).
2. Lene Vestergaard Hau, S.E.Harris, Z.Dutton, and C.H.Behroozi, Nature, **397**, 594 (1999).
3. L.J.Wang, A.Kuzmich, and A.Dogarlu, Nature, **406**,277 (2000).
4. L.Davidovich, A.Maali, M.Brune, J.M.Raimond, and S.Haroche, Phys. Rev. Lett. **71**, 2360 (1993).
5. K.M.Gheri, and H.Riysch, Phys.Rev.A. **56**,3187 (1997).
6. *Cavity Quantum Electrodynamics*, edited by H. J. Kimble, Academic Press, San Diego, 1994.
7. R. J. Thompson, G. Rempe, and H. J. Kimble, Phys. Rev. Lett. **68**, 1132 (1992).
8. H. Mabuchi, Q. A. Turchette, M. S. Chapman, and H. J. Kimble, Opt. Lett. **21**, 1393 (1996).
9. C. J. Hood, M. S. Chapman, T. W. Lynn, and H. J. Kimble, Phys. Rev. Lett. **80**, 4157 (1998).
10. H. Mabuchi, J. Ye, H. J. Kimble, Appl. Phys. B **68**, 1095 (1999).
11. P. Münstermann, T. Fischer, P. W. H. Pinkse, and G. Rempe, Opt. Comm. **159**, 63 (1999).

NARROW-LINE COOLING OF CALCIUM

U. STERR, T. BINNEWIES, G. WILPERS, F. RIEHLE, J. HELMCKE
Physikalisch-Technische Bundesanstalt, Bundesallee 100, 38116 Braunschweig, Germany
E-mail: uwe.sterr@ptb.de

A cloud of ^{40}Ca atoms has been cooled and trapped to a temperature as low as 6 μK by operating a magneto-optical trap on the spin-forbidden intercombination transition. The scattering rate was enhanced by a factor of 15 by quenching the long-lived excited state with an additional laser, while still preserving a high selectivity in velocity. With this method more than 10 % of pre-cooled atoms from a standard magneto-optical trap have been transferred to the ultracold trap.

1 Introduction

Neutral alkaline earths are of special interest for atom-interferometry and optical frequency standards because of their narrow intercombination lines and their single ground state. Alkaline earths can be laser-cooled readily on their strong resonance lines in the singlet system. However, the temperature is limited to the Doppler-limit due to the lack of a ground-state splitting in contrast to the sub-Doppler temperatures in the microkelvin range achievable with atoms that posses multiple Zeeman or hyperfine levels in the ground-state. To further decrease the temperature, second-stage cooling on the intercombination transition has been successfully applied to prepare an ultracold ensemble of strontium atoms [1]. For the lighter earth-alkalines calcium and magnesium, the scattering rate on this transition is much lower and the resulting scattering force is in the order (for Ca) or even less (for Mg) than the gravitational force.

We have increased the low scattering rate associated with the narrow line width with a repumping or 'quenching' laser which de-excites the atoms via an intermediate level. This additional laser excites atoms from the $4s4p\ ^3P_1$ to the $4s4d\ ^1D_2$ state, which decays back to the ground state within a few microseconds. The alternative quench transition $4s4p\ ^3P_1$ - $4s5s\ ^1S_0$ was used by Curtis et al. [2] for a 1D cooling scheme.

2 Experimental Setup

We realized this novel cooling scheme in an experiment, where we started with a standard magnetooptical trap (MOT) on the cooling transition $^1S_0$ -

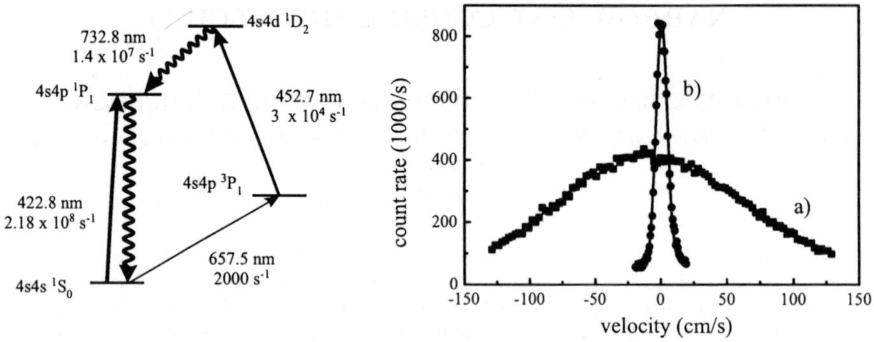

Figure 1. Energy diagram of ^{40}Ca with wavelengths and A coefficients relevant to the cooling scheme. The graph on the right shows the velocity distribution before (a) and after 25 ms of narrow-line cooling (b).

$^1P_1$ made of three retroreflected laser beams with $\lambda = 423$ nm. With a total power of 30 mW approximately 10^7 atoms were cooled and trapped directly from a thermal beam. The narrow-line MOT at 657 nm was also realized by three retroreflected circularly polarized beams (power approx. 7 mW per beam with a diameter of 5 mm), that were parallel to the 423 nm MOT beams. The quench laser beam ($\lambda = 453$ nm, diameter ≈ 3 mm) was fed through the setup three times and finally retroreflected, similarly to a MOT configuration, to make efficient use of the available power ($P \approx 20$ mW).

As the recoil shift due to one absorbed 657 nm photon is larger than the linewidth we have broadened the laser spectrum by harmonic frequency modulation (peak-to-peak amplitude $\delta f_{pp} = 1.4$ MHz, modulation frequency $f_{mod} = 15$ kHz). The typical detuning of the high frequency end was $\Delta \approx 2\pi \times 280$ kHz.

After pre-cooling the atoms in the broad-line MOT to $v_{\rm rms} \approx 0.7$ m/s, the 423 nm laser was switched off and the magnetic field gradient was lowered within 100 μs from 60×10^{-4} T/cm to 0.3×10^{-4} T/cm and the 657 nm cooling laser and the quench laser were switched on. The small gradient accounts for the much smaller acceleration that can be reached by cooling on the narrow line, even with quenching.

The velocity distribution of the atomic ensemble was obtained by measuring the remaining Doppler broadening of the intercombination transition (fig. 1). Comparing the areas of the two curves in fig. 1 we calculate a transfer efficiency of $\eta_N = 12$ %.

The temperature as a function of the cooling time (fig. 2) shows in the

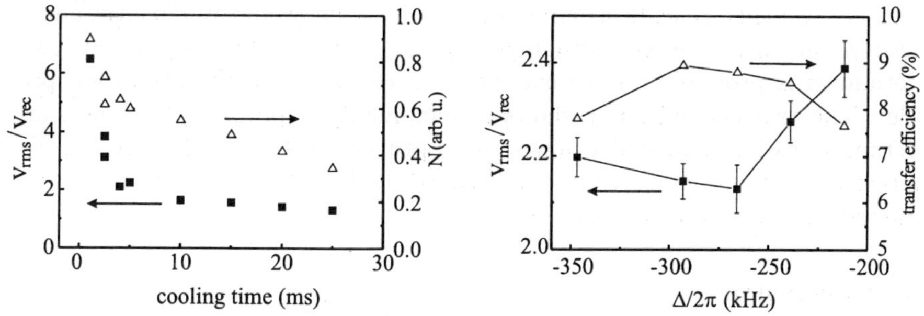

Figure 2. Velocity (squares) in units of the combined effective recoil velocity and transfer efficiency (triangles) versus cooling time (left) and as a function of the detuning (right).

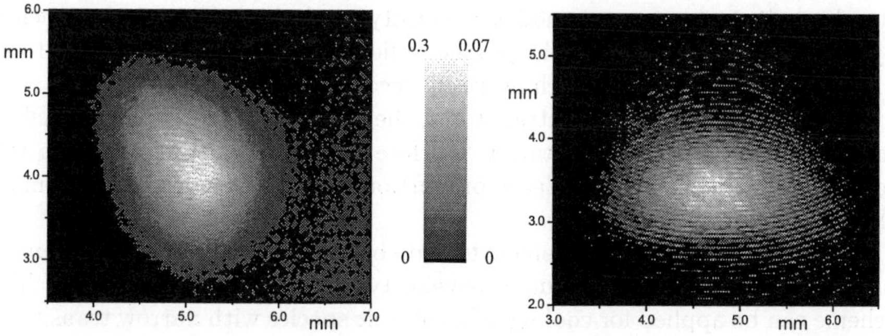

Figure 3. Absorption images of the precooled cloud (left) and the ultracold cloud (right), the scale indicates the optical density.

beginning an exponential decay with a time constant of 1.4 ms and for longer cooling times a further slow reduction. A temperature of (5.6 ± 0.4) μK after 100 ms of cooling has been observed, which corresponds to an rms velocity of 3.4 cm/s. This is close to the effective recoil calculated as the quadratic sum of the three photon momenta involved in the quenching process ($v_{\text{rec}} \approx$ 3.5 cm/s). The time dependence of the total number of atoms shows first a fast loss due to atoms that could not be captured by the narrow-line MOT and then an almost linear decrease in atom number with cooling time. From absorption images of the ultracold atomic cloud after a cooling time of 15 to 20 ms (fig. 3) we obtain rms radii of 0.8 mm and 0.4 mm in the horizontal and vertical directions, respectively.

The velocity reaches a minimum around a detuning of 270 kHz and is nearly constant for larger detunings (see fig. 2, right). The increase at lower detunings is probably due to excitation in the wings of the excitation profile. The transfer efficiency shows a maximum at approximately the same detuning, as for smaller detunings the capturing probability is reduced. For larger detunings the loss increases because the trap becomes larger than the intersection region of the laser beams.

3 Conclusion

This quench cooling scheme has the distinct advantage that the scattering rate can be tailored for different purposes by adjusting the power of the quench laser, e.g. using a high scattering rate for efficient capturing and then lowering the rate to reach the lowest temperatures. For use in an atomic frequency standard ultracold atoms at moderate density are desired. This can be reached by completely switching off the magnetic field and finally cooling the atoms in an optical molasses to reach even sub-recoil temperatures.

In optical clocks with neutral atoms, the residual first oder Doppler effect introduces the biggest uncertainty [3]. Here the use of ultracold atoms will readily dimish this effect at least proportional to the velocity, i.e. by more than one order of magnitude.

In conclusion, we have demonstrated cooling of a ^{40}Ca atomic ensemble in three dimensions to a residual rms-velocity close to one effective recoil. The scheme can be applied for cooling other atomic species with narrow transitions that can be artificially broadened to overcome the limitations due to the associated small forces.

Acknowledgments

We thank J. Keupp and S. Baluchev for technical assistance and T. E. Mehlstäubler, E. M. Rasel, and W. Ertmer for helpful discussions. The work was supported by the Deutsche Forschungsgemeinschaft under SFB 407 and by the European Commission through the Human Potential Programme (Cold Atoms and Ultra-precise Atomic Clocks; CAUAC).

References

1. T. Ido *et al*, *Phys. Rev.* A **61**, 061403 (2000).
2. A. Curtis *et al*, *Phys. Rev.* A **64**, 031403(R) (2001).
3. F. Riehle *et al*, *IEEE Trans. Instrum. Meas.* IM **48**, 613 (1999).

MTHPC FLUORESCENCE AS A PH-INSENSITIVE TUMOR MARKER IN A COMBINED PHOTODYNAMIC DIAGNOSIS AND PHOTODYNAMIC THERAPY TREATMENT OF MALIGNANT BRAIN TUMORS

MONIKA RITSCH-MARTE, ANDREAS ZIMMERMANN,

Institute for Medical Physics, University of Innsbruck, 6020 Innsbruck, Austria
E-mail: monika.ritsch-marte@uibk.ac.at

AND HERWIG KOSTRON

Department of Neurosurgery, University of Innsbruck, 6020 Innsbruck, Austria
E-mail: herwig.kostron@uibk.ac.at

The feasibility and the effectiveness of a combination of photodynamic therapy (PDT) and photodynamic diagnosis (PDD) mediated by the same photosensitizer (mTHPC) has been assessed in the treatment of 22 patients with malignant brain tumors. Since the fluorescence excitation efficiency is of great importance for PDD and since there is a difference in the interstitial pH between normal and tumor tissue, we have also performed *in vitro* measurements on the pH-dependence of the fluorescence emission intensity of mTHPC.

1 Introduction

For malignant brain tumors there exists a correlation between patient survival and the thoroughness of tumor eradication [1]. Diffuse growth into normal brain parenchyma, as occurring e.g. in *Glioblastoma multiforme*, makes the differentiation between normal and tumor tissue difficult. Especially in specific areas in the brain, the removal of pathological material well into adjacent healthy tissue is not an option because of possible induction of a neurological deficit. Thus intraoperative (real-time !) spectral differentiation between normal and malignant brain tissue is an important issue.

Photosensitizers are organic substances which are designed to 1) accumulate in the tumor and 2) lead to a phototoxic reaction in the tumor tissue upon irradiation with a suitable light source. The various photosensitizers are characterized by their tumor-selective concentration and/or retention and by their fluorescence emission. The ultimate goal of photodynamic therapy and diagnosis is the detection and the eradication of tumor cells using one and the same agent, acting simultaneously as a fluorophore for tumor cell labeling and as a photosensitizer for tumor cell destruction.

We have made this logical step to combine photodynamic diagnosis with subsequent photodynamic therapy based on the porphyrin meta-tetrahydroxyphenylchlorin (mTHPC) in the treatment of gliomas. Besides the specific photophysical properties (high phototoxicity at low activation energies, high concentration ratios), mTHPC also shows a strong fluorescence. On the basis of its high affinity to neoplastic tissue the fluorescence of mTHPC by means of blue light excitation can be exploited for intraoperative visualization of hardly recognizable tumor tissue, thereby essentially maximizing the extent of resection.

The second generation photosensitizer mTHPC has already been used as an exogenous photoactive agent for photodynamic therapy in a wide field of cancer treatments (see [2,3,4] and References therein). But to our knowledge this is the first time that a combined mTHPC-mediated PDT&PDD treatment has been carried out in the brain.

2 Combined photodynamic therapy and photodynamic diagnosis treatment

The spectral properties of the photosensitizer mTHPC, with a main fluorescence emission peak at 652nm which can be excited e.g. by a filtered xenon lamp around 417nm (Fig.1), are well suited for simultaneous fluorescence imaging. It is possible to use compact imaging systems, which were devised for intraoperative PDD with other photosensitizers and which have recently become commercially available, with only minor adaptions.

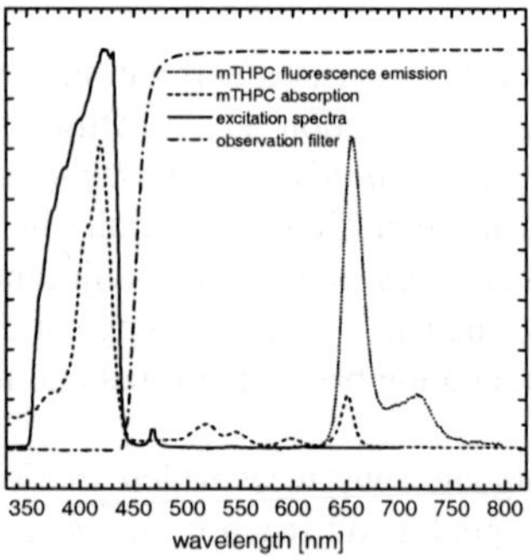

Figure 1. Spectral characterization of mTHPC as a photosensitizer: Absorption maxima at 417nm (fluorescence excitation for PDD) and 652nm (absorption peak for PDT), mTHPC fluorescence emission with a peak wavelength of 652nm and 718nm; also shown: spectrum of the blue excitation light and transmission of the observation filter.

Several commercially available fluorescence diagnostic systems were investigated for their applicability for clinical practice. We have adapted and optimized a diagnostic system which includes a surgical microscope, an excitation light source (filtered to 370-440nm), a video camera detection system, and a spectrometer for clear identification of the mTHPC fluorescence emission at 652nm.

For selected pixels of interest from ‘suspicious’ areas in the recorded fluorescence image, a more precise spectral characterization was obtained by a spectrometer, providing additional information for the surgeon’s difficult decision on the necessity of further resection. Especially in regions of faint fluorescence it proved essential to maximize the spectral information by optimizing and matching the spectral properties of all components, such as excitation source, video camera and color filters.

3 pH-dependence of the photophysical properties of the mTHPC

Since photochemical properties of some photosensitizers are known to depend on pH [5,6], and since tumor tissue usually has a lower interstitial pH compared to normal tissue, we have measured the influence of pH on the mTHPC fluorescence (Fig. 2). Selecting e.g. the two excitation wavelengths 405nm and 436nm, which showed a significant pH-influence in the absorption curve, allows the derivation of a calibration curve by dividing the fluorescence intensities corresponding to these two excitation wavelengths ("fluorescence ratio" in Fig. 2). Outside the physiological range (for pH < 6), a strong dependence on pH was found; thus the calibration curve can be used for the optical determination of the pH value (for ‘pH-imaging’) in the range of pH 3.5 – 6.

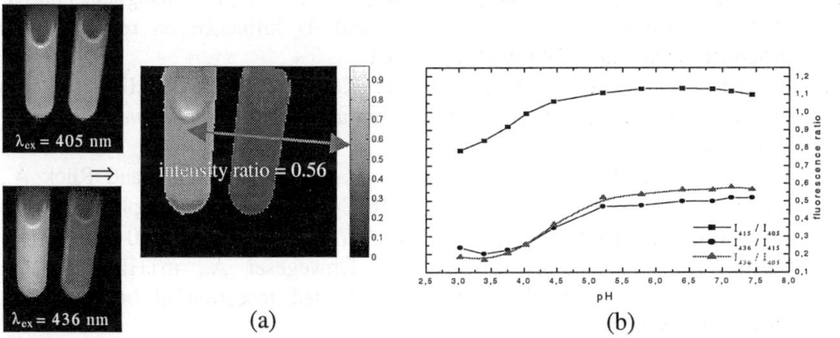

Figure 2. pH-dependence of mTHPC-fluorescence: (a) image of ratio I_{405}/I_{436} of fluorescence intensities for 405nm and 436nm excitation, respectively; (b) calibration curves.

However, in the range of interest for the PDT-PDD treatment (pH = 6.5 – 7.5) the calibration curve is flat. From these *in vitro* results we therefore conclude that there is no *directly* pH-induced decrease in mTHPC-fluorescence. In brain tissue *in vivo*, however, there might be other indirect influences, as for instance pH-dependent cellular uptake of mTHPC, which are not accounted for in our *in vitro* situation, but which might have an impact on diagnostic and therapeutic issues [6].

4 Conclusions

Until now a mTHPC-mediated combined intraoperative phototherapeutic and photodiagnostic treatment supplementing the conventional tumor resection has been carried out on 22 patients. Based on the data acquired from 138 tissue samples during the last few years we have been able to demonstrate that our approach of a combined treatment, which could be characterized as *"to see and to treat"*, is a feasible concept giving rise to a good specificity and sensitivity: Correlation with the histological analysis yielded a sensitivity of 87.9%, a specificity of 95.7%, and an accuracy of 90.6%.

5 Acknowledgements

This work was supported by the Austrian Science Fund (P13458-MED).

References

1. Albert F., Forstinger M., Sartor K., Adams H. and Kunze S., Early postoperative magnetic resonance imaging after resection of malignant glioma: objective evaluation of residual tumor and its influence on regrowth and prognosis. *Neurosurg.* **34** (1994) pp. 45-61.
2. Zimmermann A., Ritsch-Marte M. and Kostron H. A., mTHPC-mediated photodynamic diagnosis of malignant brain tumors, *Photochem. Photobiol.* (2001), in press.
3. Kostron H. A., Obwegeser A., Jakober R., A. Zimmermann and Rück A., Experimental and clinical results of mTHPC-(Foscan-)mediated photodynamic therapy for malignant brain tumors, *Proc. SPIE* **3247** (1998) pp. 40–45.
4. Kostron H. A., Zimmermann A. and Obwegeser A., mTHPC-mediated photodynamic detection for fluorescence-guided resection of brain tumors, *Proc. SPIE* **3262** (1998) pp. 259–264.
5. Fuchs C., Riesenberg R., Siegert J. and Baumgartner R., ph-dependent formation of 5-aminolaevulinic acid-induced protporphyrin IX in fibrosarcoma cells, *Photochem. Photobiol. B* **40** (1997) pp. 49–54.
6. Ma L. W., Bjorklund E. and Moan J., Photochemotherapy of tumors with mTHPC is pH-dependent, *Cancer Letters* **138** (1999) pp. 197–201.